Lepton Scattering, Hadrons and QCD

NATIONAL INSTITUTE FOR
Theoretical Physics

SPECIAL RESEARCH
CENTRE FOR THE
SUBATOMIC
STRUCTURE OF MATTER

Proceedings of the Workshop on

Lepton Scattering, Hadrons and QCD

Adelaide, Australia 26 March – 6 April 2001

Editors

W. Melnitchouk
Adelaide University, Australia
Jefferson Laboratory, USA

A. W. Schreiber
A. W. Thomas
Adelaide University, Australia

P. C. Tandy
Kent State University, USA

World Scientific
New Jersey • London • Singapore • Hong Kong

Published by

World Scientific Publishing Co. Pte. Ltd.
P O Box 128, Farrer Road, Singapore 912805
USA office: Suite 1B, 1060 Main Street, River Edge, NJ 07661
UK office: 57 Shelton Street, Covent Garden, London WC2H 9HE

British Library Cataloguing-in-Publication Data
A catalogue record for this book is available from the British Library.

Front Cover: The cover shows the city of Adelaide at night, looking southeast across the River Torrens; this view was enjoyed by participants who attended the workshop banquet. The photo is used by courtesy of Tourism SA.

LEPTON SCATTERING, HADRONS AND QCD

ISBN 981-02-4749-4

Printed in Singapore by Uto-Print

Foreword

A group of about 65 physicists, from 9 countries, gathered in Adelaide in March 2001 to examine recent developments in the exploration of hadronic structure through lepton scattering and to advance the description of hadron physics directly from lattice QCD and non-perturbative QCD models.

This workshop was jointly organized and funded by the National Institute for Theoretical Physics (NITP) and Adelaide University's Special Research Centre for the Subatomic Structure of Matter (CSSM). Of the participants, 36 came from overseas and represented some 26 institutions in North America, Europe, and Japan. As with previous workshops hosted by the CSSM there were formal sessions of invited talks each morning and only one or two talks after coffee in the afternoons. Informal discussions and working sessions constituted the remainder of the time. As usual, this format was successful in stimulating considerable discussion and collaborative research.

The program covered the following topics: reviews of deep inelastic lepton scattering data; presentations of the most recent experimental results from DESY, HERA, SLAC, JLab, NIKHEF, and RIKEN; unpolarized and polarized structure functions, generalized parton distributions; nuclear effects; quark-hadron duality; electromagnetic form factors; structure functions and hadron properties from lattice QCD; models based on the Dyson-Schwinger equations. The prospects for small-x physics at RHIC were also discussed. The speakers and topics were chosen in an effort to strengthen the links between these activities, and topics that attracted especial interest included the implications of chiral symmetry for lattice data, the first presentation of new G_E^p/G_M^p data, leading protons/neutrons at DESY and quark-hadron duality.

On behalf of all the participants we would like to express our thanks to the group of people whose efforts enabled this workshop to run smoothly and ensured a stimulating atmosphere. Sharon Johnson and Sara Boffa of the CSSM office handled all of the logistics and the organizational matters with much generosity and skill; Ramona Adorjan provided very valuable professional assistance in computational matters.

All contributions in this volume have been independently refereed.

W. Melnitchouk
A. W. Schreiber
P. C. Tandy
A. W. Thomas
Adelaide, July 2001

Contents

Foreword v

1. PARTONIC STRUCTURE OF HADRONS

Nucleon Form Factors and Structure Functions from Lattice QCD
M. Göckeler, R. Horsley, D. Pleiter, P. E. L. Rakow and G. Schierholz 3

Chiral Extrapolation of Lattice Structure Function Calculations
W. Detmold 13

Charge Symmetry in Parton Distributions
J. T. Londergan 21

Distortions in the Negative Energy Dirac sea: Violation of the Gottfried Sum Rule and $\Delta\bar{u}$ in the Proton
K. Tsushima, A. W. Thomas and G. V. Dunne 29

Parton Distributions for the Pion in a Chiral Quark Model
E. Ruiz Arriola 37

Generalized Parton Distributions and Distribution of Partons in the Transverse Plane
M. Burkardt 45

Exclusive Processes at HERMES
N. Bianchi 53

Soft Pion Production Associated with Deeply Virtual Compton Scattering
L. Mosse 61

2. SPIN STRUCTURE OF HADRONS

An Experimental Review of the Nucleon Spin Structure Functions
G. G. Petratos 71

Polarized Structure Functions in QCD
J. Kodaira 81

Single Spin Asymmetries and Quark Fragmentation
M. Anselmino, D. Boer, U. D'Alesio and F. Murgia 89

Determining ΔG with Polarized Photo- and Hadroproduction of Heavy Quarks
I. Bojak 96

Flavour Symmetry Breaking in the Polarized Nucleon Sea
F.-G. Cao and A. I. Signal 102

3. PERTURBATIVE — NONPERTURBATIVE QCD TRANSITION

Lepton Scattering and Quark-Hadron Duality Studies at JLab
R. Ent 113

Quark-Hadron Duality in Inclusive Electron-Hadron Scattering
W. Melnitchouk 122

Signals of Local Duality from a Perturbative QCD Analysis of Inclusive *ep* Scattering
S. Liuti 130

Gluons, Quarks and the Transition from Nonperturbative to Perturbative QCD
A. G. Williams 138

Estimating Low Energy Model Parameters from Deep Inelastic Scattering
L. P. Hoyt and A. I. Signal 146

4. FORM FACTORS

Measurement of G_{E_p}/G_{M_p} to $Q^2 = 5.6$ GeV^2 via Recoil Polarization at Jefferson Lab
O. Gayou 157

Physical Hadron Properties from Lattice Data at Large Quark Masses
A. W. Thomas 165

Meson Cloud Considerations in the Strange Magnetic Moment of the Nucleon from Lattice QCD
D. B. Leinweber and A. W. Thomas 174

Electromagnetic Interactions in Light Front Dynamics
J.-F. Mathiot 183

Soft QCD Modeling of Meson Electromagnetic Form Factors
P. C. Tandy 192

Nucleon Form Factors in the Covariant Diquark-Quark Model
R. Alkofer and M. Oettel 201

5. HADRON EXCITATIONS, CONFINEMENT AND CHIRAL SYMMETRY BREAKING

Experimental Studies of the Hadron Spectrum
J. Napolitano 211

The Character of Goldstone Bosons
M. B. Hecht, C. D. Roberts and S. M. Schmidt 219

Confinement from Coulomb Gauge QCD and Exotic Phenomenology
A. P. Szczepaniak 228

Regulator Free Dyson-Schwinger Equation Studies of Non-Perturbative Field Theory
A. Kızılersü, A. W. Schreiber, T. Sizer and A. G. Williams 236

Domain-Like Structures in the QCD Vacuum, Confinement and Chiral Symmetry Breaking
A. C. Kalloniatis and S. N. Nedelko 244

Glueball Properties in Anisotropic SU(3) Lattice QCD with an Improved Action
N. Ishii, H. Suganuma and H. Matsufuru 252

Deconfining by Winding
R. Hofmann 260

Charmonium, Glueballs and Exotic Hybrids in a Relativistic Many-Body Approach
S. R. Cotanch 270

6. SMALL-x PHYSICS AND NUCLEAR MEDIUM EFFECTS

Small x Physics and the Initial Conditions in Heavy Ion Collisions
A. Krasnitz and R. Venugopalan 281

Leading Nucleon Production at HERA
G. Levman 291

Nuclear Medium Effects at HERMES
P. di Nezza 299

Non-Singlet Structure Function of the ^{3}He $-^3$ H System and Divergence of the Gottfried Integral
V. Guzey, A. W. Thomas, K. Tsushima, K. Saito and M. Strikman 307

Quark-Diquark Structure of the Nucleon: Structure Functions, Static Properties and Nuclear Matter Equation of State
W. Bentz, H. Mineo, A. W. Thomas and K. Yazaki 314

Physics Motivation for a Polarized Electron-Ion Collider
J. M. Cameron and J. T. Londergan 322

List of participants 331

1. Partonic Structure of Hadrons

NUCLEON FORM FACTORS AND STRUCTURE FUNCTIONS FROM LATTICE QCD

M. GÖCKELER[1], R. HORSLEY[2], D. PLEITER[2], P. E. L. RAKOW[1], G. SCHIERHOLZ[2,3]

[1]*Institut für Theoretische Physik, Universität Regensburg, D-93040 Regensburg, Germany*

[2]*John von Neumann-Institut für Computing NIC, Deutsches Elektronen-Synchrotron DESY, D-15735 Zeuthen, Germany*

[3]*Deutsches Elektronen-Synchrotron DESY, D-22603 Hamburg, Germany*

In this talk we highlight recent lattice calculations of the nucleon form factors and structure functions.

1 Introduction

The lattice formulation of *QCD* is, at present, the only known way of obtaining low energy properties of the theory in a direct way, without any model assumptions. Quantities within the grasp of lattice *QCD* involving light quarks include the hadron mass spectrum, quark masses, the Λ parameter, the chiral condensate, the sigma term, decay constants, the axial and tensor charge of the nucleon, form factors and twist two and four polarised and unpolarised moments of structure functions for the nucleon, pion and rho, to name only a few of the possibilities. Our group (the *QCDSF* collaboration) has been actively involved for the last few years in attempting to determine some of these quantities, all characterised by their non-perturbative nature. Rather than giving an exhaustive progress report of recent developments in the field[a] we shall, due to lack of space, focus our attention on two topics: nucleon form factors (including the axial charge) and moments of structure functions and how the lattice method can lead to their determination.

The lattice approach involves first euclideanising the *QCD* action and then discretising space-time (with lattice spacing a). The path integral then becomes a very high dimensional partition function, which is amenable to Monte Carlo methods of statistical physics. This allows correlation functions (which can be related to *QCD* matrix elements) to be determined. Progress in the field is slow. First our 'box' must be large enough to fit our correla-

[a]The annual Lattice Conference is a good source of information, the last being held in Bangalore, India[1].

tion functions into. Then, often, a chiral extrapolation must be made from a quark mass region around the strange quark mass to the light up and down quarks. Finally the continuum limit must be taken, i.e. $a \to 0$. This is all very time consuming as, in the statistical mechanics picture, we are approaching a second order phase transition, with all its attendant problems. Additionally, simply to save computer time, the fermion determinant in the action is often discarded. This 'quenched' or 'valence' procedure is an uncontrolled approximation. Recently, however, simulations with two mass-degenerate sea quarks have begun appearing, allowing a first look at the possible effects of quenching. Finally, in addition to the above problems, to be able to compare with phenomenological or experimental results, matrix elements must be renormalised. In total this is an ambitious programme.

2 Generalities

2.1 The processes

Lepton–nucleon elastic scattering, $lN \to lN$, in which a photon is exchanged between the lepton (usually an electron) and the nucleon (usually a proton), has been studied for many years. Indeed there has been a resurgence of interest in these processes as part of the Jefferson Laboratory (Jlab) 'hadron' physics programme. The scattering matrix element can be decomposed into a known electromagnetic piece and an unknown QCD matrix element with decomposition:

$$\langle N(\vec{p}\,',\vec{s}\,')|J^\mu(\vec{q})|N(\vec{p},\vec{s})\rangle = \overline{u}(\vec{p}\,',\vec{s}\,')\left[\gamma^\mu F_1(Q^2) + i\sigma^{\mu\nu}\frac{q_\nu}{2m_N}F_2(Q^2)\right]u(\vec{p},\vec{s}), \tag{1}$$

where $q = p' - p$ is the momentum transfer and $Q^2 = -q^2 > 0$. The values at $Q^2 = 0$ are $F_1^p(0) = 1$ due to the conservation of the vector current and $F_2^p(0) = \mu^p - 1$, the anomalous magnetic moment, in units of $e/2m_N$ or magnetons. ($F_1^n(0) = 0$, $F_2^n(0) = \mu^n$.) Experimentally, it is more convenient to define the Sachs form factors

$$G_e(Q^2) = F_1(Q^2) - \frac{Q^2}{(2m_N)^2}F_2(Q^2),$$
$$G_m(Q^2) = F_1(Q^2) + F_2(Q^2).$$

Earlier experimental results give phenomenological (dipole) fits of

$$G_e^p(Q^2) = \frac{G_m^p(Q^2)}{\mu^p} = \frac{G_m^n(Q^2)}{\mu^n} = \left(1 + \frac{Q^2}{m_V^2}\right)^{-2}, \qquad G_e^n(Q^2) = 0, \tag{2}$$

with $m_V \approx 0.82\,\mathrm{GeV}$, $\mu^p \approx 2.79$ and $\mu^n \approx -1.91$.

Similarly neutrino–nucleon scattering, for example $\nu_\mu n \to \mu^- p$ mediated by a W^+ exchange, leads to an unknown axial current hadronic matrix element between neutron and proton states, which, with the use of current algebra and isospin invariance, may be re-written between p states alone and has Lorentz decomposition

$$\langle p(\vec{p}\,', \vec{s}\,')|A_\mu^{u-d}(\vec{q})|p(\vec{p}, \vec{s})\rangle = \overline{u}(\vec{p}\,', \vec{s}\,') \left[\gamma_\mu \gamma_5 g_A(Q^2) + i\gamma_5 \frac{q_\mu}{2m_N} h_A(Q^2)\right] u(\vec{p}, \vec{s}) \tag{3}$$

where $A_\mu^{u-d} = \overline{u}\gamma_\mu\gamma_5 u - \overline{d}\gamma_\mu\gamma_5 d$. Experimental results give phenomenological fits

$$g_A(Q^2) = g_A(0)\left(1 + \frac{Q^2}{m_A^2}\right)^{-2}, \tag{4}$$

with $m_A \approx 1.00\,\mathrm{GeV}$. From the β decay we know[2] $g_A \equiv g_A(0) = 1.2670(35)$.

At higher momentum transfer, the nucleon is broken up by the photon (or $W^\pm$) probe. We enter the regime of *deep inelastic scattering* experiments $eN \to eX$ (or $\nu_\mu N \to \mu^- X$). The operator product expansion, OPE, leads to relations between moments of the structure functions and certain nucleon matrix elements. For example,

$$\int_0^1 dx x^{n-2} F_2(x, Q^2) = \frac{1}{3} E_n^{\overline{MS}}(\frac{\mu^2}{Q^2}, g^{\overline{MS}}) v_n^{\overline{MS}}(g^{\overline{MS}}) + O(\frac{1}{Q^2}). \tag{5}$$

Here x is the Bjorken variable and $v_n^{\overline{MS}} \propto \langle N|\mathcal{O}_n|N\rangle^{\overline{MS}}$, where $\mathcal{O}_n \sim \overline{q}\gamma_{\mu_1} D_{\mu_2} \ldots D_{\mu_n} q$ are operators bilinear in the quarks, each containing $n-1$ covariant derivatives. For eN scattering only even moments contribute, while for νN all moments are allowed.

All the matrix elements briefly discussed above can be determined non-perturbatively using lattice *QCD*.

2.2 Lattice Technicalities

The general method of determining matrix elements is by forming ratios of three-point to two-point correlation functions:

$$R_{\alpha\beta}(t, \tau; \vec{p}) = \frac{\langle N_\alpha(t; \vec{p})\mathcal{O}(\tau; \vec{0})\overline{N}_\beta(0; \vec{p})\rangle}{\langle N(t; \vec{p})\overline{N}(0; \vec{p})\rangle}, \tag{6}$$

where N is the three quark baryon operator and $\mathcal{O}$ is a bilinear quark operator. Provided that $t \gg \tau \gg 0$, then R can be shown to be directly proportional to the matrix element $\langle N|\mathcal{O}|N\rangle$. ($R$ may be easily generalised to allow for momentum transfer.) If we re-write the numerator of eq. (6) using quark propagators, we see that we have two classes of diagrams: one in which the inserted operator is attached to quark propagators to the nucleon, and a second class where the quark propagator from the operator is disjoint from the baryon. In this second class the interactions between the inserted operator and the baryon take place only via gluon exchange. Numerically, due to large ultra-violet fluctuations, no useful signal is seen. So either we must presently assume that their effects are small, or calculate 'non-singlet' (i.e. $u-d$) matrix elements where these second class terms cancel.

The matrix elements must be renormalised. There are several possibilities. We may simply consider a ratio of matrix elements between different hadronic states, so that the renormalisation constant cancels. However, this is not very useful. The axial and vector renormalisation constants (the currents being partially conserved) may be determined by demanding that their (continuum) Ward Identities are obeyed, which produces non-perturbative renormalisation constants. (Incidently they are also renormalisation scheme independent.) Perturbation theory can be applied, but due to technical problems only one loop results are known. Even with this restriction, perturbation theory can be improved leading to 'tadpole improved' or *TI* perturbation theory. However it still suffers from unknown systematic errors. The ALPHA Collaboration has developed techniques[3] allowing some renormalisation constants to be determined by stepping up from a low energy scale to a high energy scale where the Z can be matched to (known) perturbation theory. Finally one can attempt to 'mimic' the perturbation theory method by calculating the operator numerically between quark states in the Landau gauge. Also note that it is desirable to make the errors in the determination to be of $O(a^2)$ rather than $O(a)$, because then the approach to the continuum is faster[4]. This introduces further irrelevant operators, with 'improvement' coefficients which also must be determined.

At present, for the axial and vector currents most renormalisation constants and improvement coefficients are known non-perturbatively. For v_n we rely on *TI* perturbation theory.

3 Results

3.1 The Axial Charge

Using the quenched approximation we have made simulations at three lattice spacings: $a = 0.093, 0.068$ and 0.051 fm[b]. For each lattice size simulations are

[b]Corresponding to $a^{-1} = 2.12, 2.90$ and $3.85\,\mathrm{GeV}$, with scale[5] $r_0 = 0.5\,\mathrm{fm}$.

made at four or more quark masses, which are then linearly extrapolated to the chiral limit. (A picture for v_n will be shown later in section 3.3.) We also have one unquenched result at $a \approx 0.11$ fm. In Fig. 1 we show all these numbers[6]. The results all lie somewhat lower than the experimental value, and at present

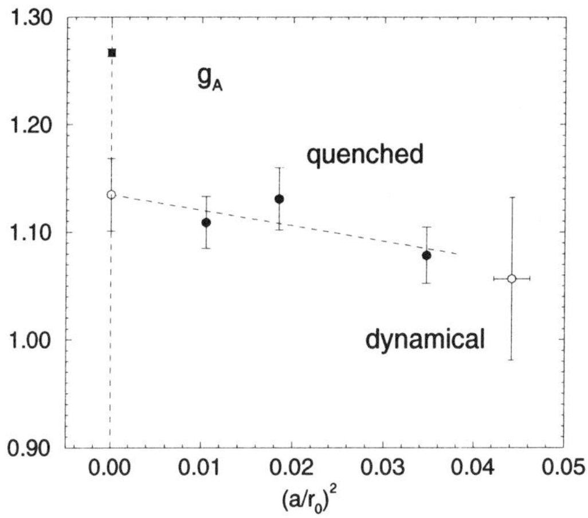

Figure 1. The continuum extrapolation of g_A.

there is no clear difference between quenched and unquenched fermions. (Note that for the quenched results, this is a fully non-perturbative calculation; for the unquenched result, there is still a little uncertainty.) While it is possible to extrapolate the quenched data to the continuum limit (although it would be very desirable to have more points because the noise is uncomfortably large at present), we see that we are only at the beginning of obtaining results with dynamical fermions.

3.2 Form Factors

In Fig. 2 we show proton electric and magnetic form factors[7], for our quenched lattice $a = 0.068$fm. Other lattice values and the unquenched results are very similar. Due to the momentum transfer the results are rather noisy. However the present trend is clear: the dipole fit gives values of m_V too large in comparison with the phenomenological value. This is confirmed in Fig. 3. For the nucleon we find $r_{\rm rms} \equiv \sqrt{12}/m_V \approx 0.60$ fm in the quenched approximation and $r_{\rm rms} \approx 0.70$ fm for the dynamical case. This is to be compared with the phenomenological value $r_{\rm rms} = 0.83$ fm. Perhaps the 'pion

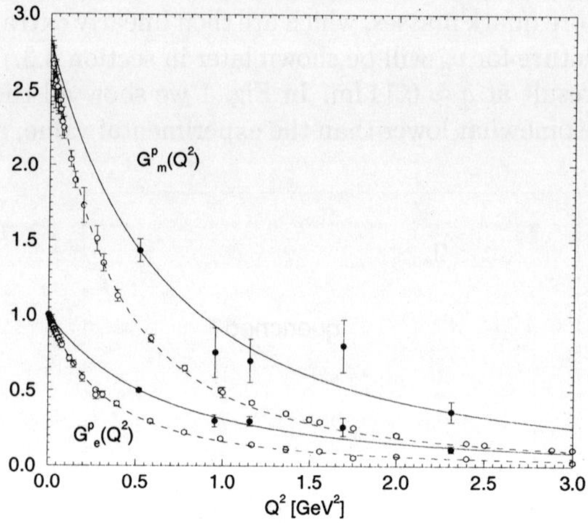

Figure 2. G_e^p and G_m^p compared to experimental data for quenched fermions at $a = 0.068$fm.

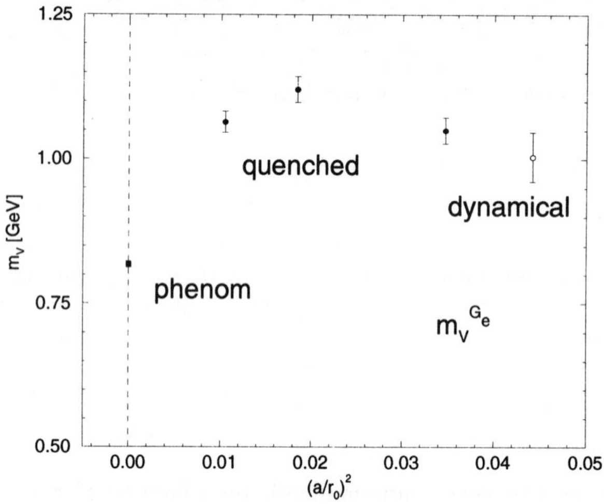

Figure 3. $m_V^{G_e}$ compared to its phenomenological value.

cloud' has not developed at our dynamical quark masses, which makes the nucleon appear to be smaller than it really is. μ^p is roughly consistent with the phenomenological value (but with large errors), while $m_V^{G_m}$ is again too large.

The recent Jlab results[8] for G_e^p/G_m^p indicate a functional difference between the electric and magnetic form factors. This is shown in Fig. 4. If

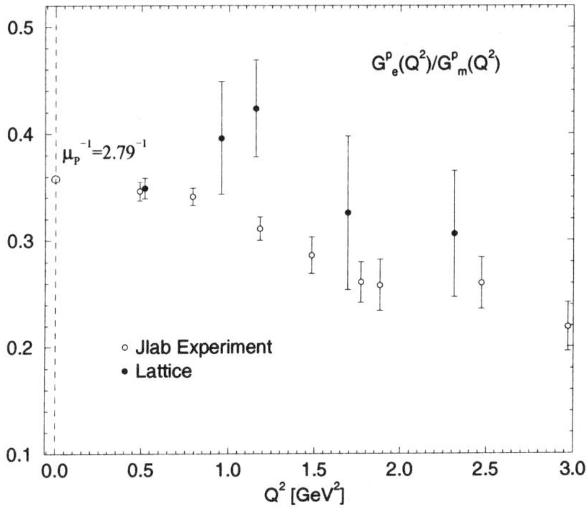

Figure 4. G_e^p/G_m^p compared to experimental data for quenched fermions at $a = 0.068$fm.

eq. (2) holds then the ratio of form factors should be constant. Unfortunately at present the lattice data is too noisy to draw any conclusion.

Finally in Fig. 5 we show the axial form factor. Again the previous comments as for G_e^p, G_m^p hold verbatim.

3.3 Moments of the nucleon structure function

Some years ago, we published our first results[9] for v_n, $n = 1, 2, 3$. The replotted data for $v_n^{(u)\overline{MS}} - v_n^{(d)\overline{MS}}$ is shown again in Fig. 6. for unimproved Wilson fermions. A comparison is made with *MRS* phenomenological results[10]. While higher moments tend to agree (although large error bars might hide any discrepancy), it seems that there is a discrepancy for the lowest moment. This may be due to renormalisation and discretisation effects, higher-twist contributions, the chiral extrapolation[11], or perhaps the phenomenological distribution functions, being *global* fits, do not fit this moment very well. Naively this result is also perhaps expected as v_2 is part of the energy-momentum sum rule, $\sum v_2^{\text{valence}} + v_2^{\text{sea}} = 1$, and due to quenching the sea contribution is reduced. For higher moments this is less of an effect, as the gluon contribution is more important for smaller x.

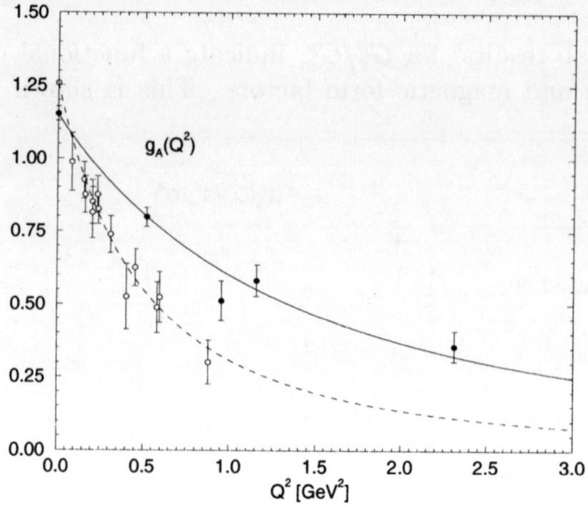

Figure 5. g_A^p compared to experimental data for unquenched fermions at $a = 0.068$fm.

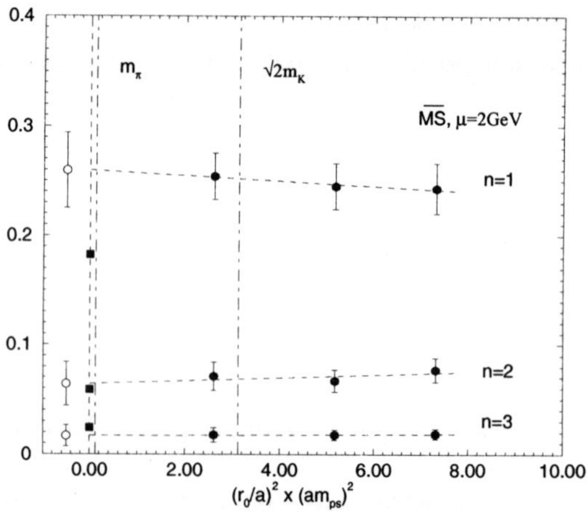

Figure 6. The chiral extrapolation for $v_n^{(u)\overline{MS}} - v_n^{(d)\overline{MS}}$ for quenched fermions at $a =$ 0.093fm (circles), compared to the MRS fit function[10] (filled squares). The dot-dashed lines show the positions of m_π and a (hypothetical) strange quark pseudoscalar particle.

First let us look at the experimental data. The cleanest *direct* determination of $v_n^{(u)\overline{MS}} - v_n^{(d)\overline{MS}}$ is given from the $F_2^p - F_2^n$ *NMC* data[12], as shown in

Fig. 7, because the one experiment uses both p *and* n targets. Note that there

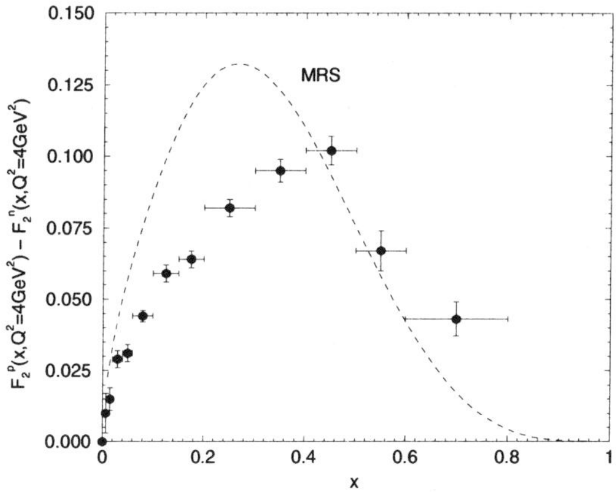

Figure 7. The experimental binned data for $F_2^p - F_2^n$ at $Q^2 = 4\mathrm{GeV}^2$ from the *NMC* Collaboration[12] compared with the *MRS* parton fit function [10] .

is a paucity of data at large $x \sim 1$, which is the region where the moments are most sensitive to. (We make a simple linear extrapolation from the largest data point to $x = 1$.) While the *MRS* fit does not quite follow the data, the area under the curve $\propto v_2^{(u)\overline{MS}} - v_2^{(d)\overline{MS}}$ is about the same, so there is indeed no problem.

Secondly we have investigated the continuum extrapolation for the lowest moment, in Fig. 8 using improved fermions. Surprisingly, perhaps, there seems to be little a^2 trend in the data (c.f. this with g_A, Fig. 1). Thus the previous discrepancy still holds.

Acknowledgements

The unquenched configurations were generated in collaboration with *UKQCD* Collaboration. The propagators were computed using the *T3E* at ZIB and the *APE100* at NIC (Zeuthen).

References

1. Proceedings of the XVIIIth International Symposium on Lattice Field Theory, Bangalore, India, editors T. Bhattacharya, R. Gupta and A.

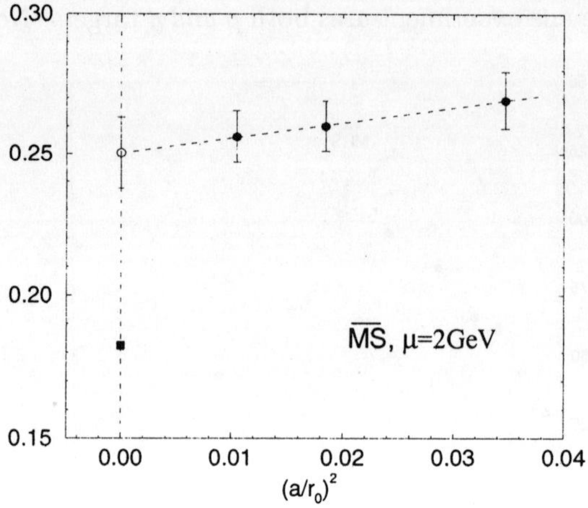

Figure 8. The continuum extrapolation for $v_2^{(u)\overline{MS}} - v_2^{(d)\overline{MS}}$.

Patel, *Nucl. Phys. Proc. Suppl.* **94** (2001).

2. D. E. Groom et al, *Eur. Phys. Jour.* C **15**, 1 (2000).
3. R. Sommer, Schladming Lectures (1997), hep-ph/9711243.
4. M. Lüscher, Les Houches Lectures (1997), hep-ph/9711205.
5. M. Guagnelli, R. Sommer and H. Wittig, *Nucl. Phys.* B **535**, 389 (1998), hep-lat/9806005.
6. R. Horsley, Lattice 2000, *Nucl. Phys. Proc. Suppl.* **94**, 307 (2001), hep-lat/0010059.
7. S. Capitani, M. Göckeler, R. Horsley, B. Klaus, H. Oelrich, H. Perlt, D. Petters, D. Pleiter, P. E. L. Rakow, G. Schierholz, A. Schiller and P. Stephenson, *Nucl. Phys. Proc. Suppl.* **73**, 294 (1999), hep-lat/9809172.
8. M. K. Jones et al., *Phys. Rev. Lett.* **84**, 1398 (2000), nucl-ex/9910005.
9. M. Göckeler, R. Horsley, E.-M. Ilgenfritz, H. Perlt, P. Rakow, G. Schierholz and A. Schiller, *Phys. Rev.* D **53**, 2317 (1996), hep-lat/9508004.
10. A. Martin, R. G. Roberts and W. J. Stirling, *Phys. Lett.* B **354**, 155 (1995), hep-ph/9502336.
11. W. Detmold, W. Melnitchouk, J. W. Negele, D. B. Renner and A. W. Thomas, hep-lat/0103006.
12. A. Arneodo et al., *Phys. Rev.* D **50**, R1 (1994).

CHIRAL EXTRAPOLATION OF LATTICE STRUCTURE FUNCTION CALCULATIONS

W. DETMOLD

Special Research Centre for the Subatomic Structure of Matter, and Department of Physics and Mathematical Physics, Adelaide University, 5005, Australia

An improved method for the chiral extrapolation of lattice QCD calculations of moments of structure functions from heavy quark masses to the physical region is presented. This extrapolation is based on the general constraints of the chiral symmetry of QCD and is able provide an excellent fit to the lattice data and the experimental values for each moment. Using the extrapolated moments, the Björken x dependence of the $u - d$ structure function is also investigated.

1 Introduction

Parton distribution functions (PDFs) are light cone correlation functions which, in the infinite momentum frame, can be interpreted as probability distributions for finding the specific partons within the nucleon. Consequently, their measurement (and in particular, surprising discoveries such as the proton spin crisis[1], the SU(2) flavor symmetry violation of the proton sea[2], and the nuclear EMC effect[3]) have had an important impact on our understanding of the structure of hadrons. Future experiments will explore further aspects of the nonperturbative origin of parton distributions and test theoretical predictions, thereby deepening that understanding.

Parton distribution functions are not accessible in perturbation theory, but have been investigated in numerous QCD-motivated quark models, and some important model independent results[4] exist. Ultimately one would like to explore these phenomena directly from QCD. Currently, lattice field theory is the only quantitative tool available which is based on QCD from first principles. However, naive extrapolation of existing lattice simulations of structure functions to physical quark masses typically lie some 50% above the experimental data. Here I will present the results of a recent analysis[5] of lattice QCD data on the first three nontrivial moments of the nonsinglet parton distribution $u - d$ based on a consistent chiral extrapolation of the moments to the physical region, which for the first time resolves this discrepancy.

2 Lattice calculations of structure function moments

Lattice QCD is formulated in Euclidean space, so direct calculations of parton distribution functions (PDFs), which are light-cone correlation functions, are not possible. However, using the operator product expansion one can calculate the *moments* of the PDFs, defined as:

$$\langle x^n \rangle_q = \int_0^1 dx\, x^n \; \left(q(x,Q^2) + (-1)^{n+1}\bar{q}(x,Q^2)\right) \; , \tag{1}$$

where the quark distribution $q(x, Q^2)$ is a function of the Björken scaling variable x and the momentum scale Q^2. The operator product expansion relates the moments $\langle x^n \rangle_q$ to forward nucleon matrix elements of certain local twist-2 operators (see Ref. 5 for further details) which can be accessed in lattice simulations.

Early calculations of moments of structure functions within lattice QCD were performed by Martinelli and Sachrajda[6]. However, the more recent data used in this analysis are taken from the QCDSF[7] and MIT[8] groups and shown in Fig. 1 for the $n = 1$, 2 and 3 moments of the $u - d$ difference in the $\overline{\rm MS}$ scheme. These calculations have been performed for both full and quenched[a] QCD using a variety of quark actions and for quark masses, m_q, ranging from 30 to 190 MeV. Full details are given in the appropriate references and summarised in Ref. 5. Due to technical details (also discussed in Ref. 5), only calculations for the lowest three nontrivial moments of the non-singlet quark distributions can be reliably compared with phenomenology. So far only the unpolarised, non-singlet PDF $u(x) - d(x)$ has been considered, but a similar analysis for the corresponding polarised distribution is currently underway.

To compare the lattice results with the experimentally measured moments, an extrapolation of the data from the lowest quark mass used ($\sim$ 50 MeV) to the physical value ($\sim$ 5–6 MeV) is required. Naively this is done assuming that the moments depend linearly on the quark mass. However, as shown in Fig. 1 (long dashed lines), a linear extrapolation of the world lattice data for the $u - d$ moments typically overestimates the experimental values by 50%. This suggests that important physics is still being omitted from the lattice calculations and their extrapolations.

[a] In this work, I will present an extrapolation formula whose structure stems from the properties of full QCD and use it to analyse both full and quenched lattice data. While the issue of whether this is appropriate will be relevant at small quark masses, where the effects of pion loops will dominate the m_q dependence, at the presently large quark masses where lattice calculations are performed the effects of pion loops are suppressed and one does not expect large differences. The unquenched lattice data shown in Fig. 1 indeed confirm this.

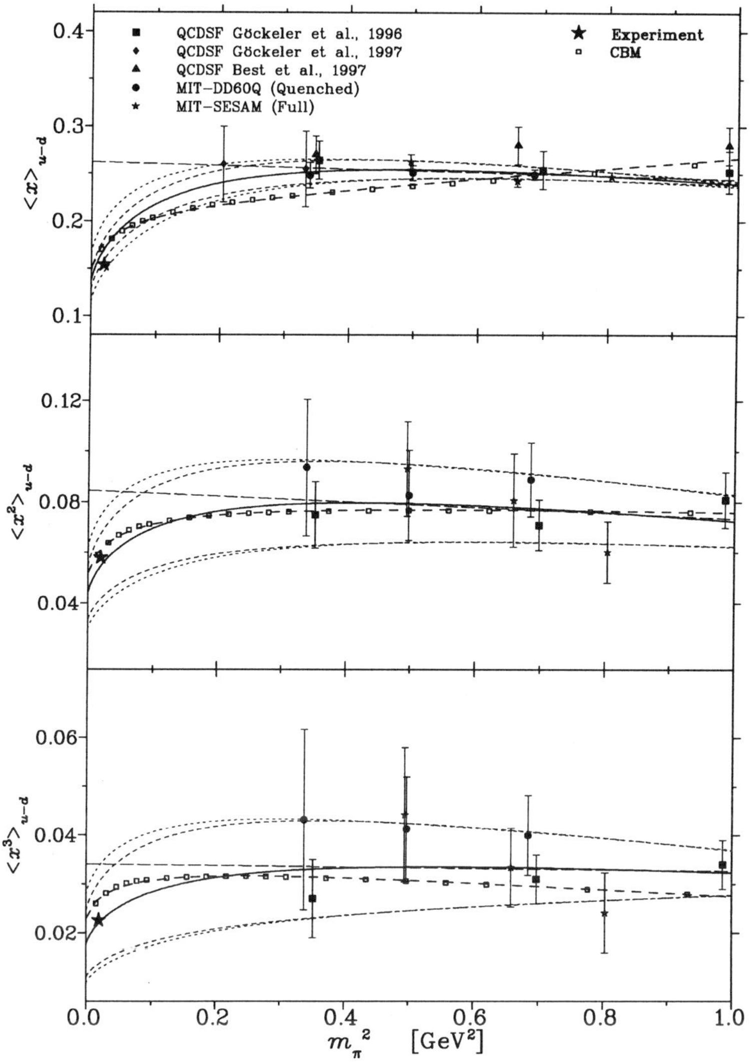

Figure 1. Moments of the $u-d$ quark distribution. Data is taken from Refs. 7,8 and labelled by year of publication. MIT-SESAM data are from full (unquenched) simulations while the other data are from quenched simulations. The straight (long-dashed) lines are linear fits to the data, while the curves have the correct LNA behavior in the chiral limit. For each moment, the best fit to the lattice data using Eq. (3) is shown by the solid curve (with $\mu = 550$ MeV), while the inner envelope about this represents the statistical errors in the data. The effect of the uncertainty in the parameter μ is illustrated by the outer lower (upper) short-dashed curves, which correspond to $\mu = 450$ (650) MeV. The small squares are the CBM results, and the dashed curve through them best fits using Eq. (3). The star represents the phenomenological values taken from NLO fits[9] in the $\overline{\text{MS}}$ scheme.

3 Nonanalytic behaviour in the pion cloud of the nucleon

One knows on very general grounds that a linear extrapolation in $m_q \sim m_\pi^2$ must fail as it omits crucial nonanalytic structure associated with chiral symmetry breaking. Even at the lowest quark mass accessed on the lattice, the pion mass is over 300 MeV. Earlier studies of chiral extrapolations of lattice data for hadron masses[10], magnetic moments[11] and charge radii[12] have shown that for quark masses above 50–60 MeV, hadron properties behave very much as one would expect in a constituent quark model, with relatively slow, smooth behavior as a function of the quark mass. However, for $m_q < 50$ MeV one typically finds rapid, nonlinear variation due to the nonanalytic behaviour of Goldstone boson loops[13].

In general, contributions to the physical properties of hadrons from intermediate states involving the surrounding meson cloud gives rise to unique terms which are nonanalytic in the quark mass. These stem from the infra-red behavior of the chiral loops and are *model independent*. The leading nonanalytic (LNA) term for the u and d distributions in the physical nucleon arises from the single pion loop dressing of the bare nucleon and has been shown[4] to behave as:

$$\langle x^n \rangle_q^{\mathrm{LNA}} \sim m_\pi^2 \log m_\pi \; . \tag{2}$$

Intermediate states containing a Δ isobar also produce terms at this order, but are small due to the N–Δ mass difference. Other contributions give rise to further nonanalytic, but sub-leading behaviour[14].

4 Improved chiral extrapolation of structure function moments

4.1 Extrapolation Formula

Experience with the chiral behavior of masses and magnetic moments shows that the LNA terms alone are not sufficient to describe lattice data for $m_\pi >$ 200 MeV [10,11]. So in order to fit the lattice data at larger m_π while preserving the correct chiral behavior of moments as $m_\pi \to 0$, a low order, analytic expansion in m_π^2 is also included in our extrapolation and the moments of $u - d$ are fitted with the form:

$$\langle x^n \rangle_{u-d} = a_n + b_n\, m_\pi^2 + a_n\, c_{\mathrm{LNA}}\, m_\pi^2 \ln\left(\frac{m_\pi^2}{m_\pi^2 + \mu^2}\right) , \tag{3}$$

where the coefficient $c_{\mathrm{LNA}} = -(3g_A^2 + 1)/(4\pi f_\pi)^2$ in chiral perturbation theory[15] and $c_{\mathrm{LNA}} = -4g_A^2/(4\pi f_\pi)^2$ in a Fock state expansion of the nucleon in the infinite momentum frame[4]. The parameters a_n, b_n and μ are *a priori* undetermined. The mass μ determines the scale above which pion loops no longer yield rapid variation and corresponds to the upper limit of the momentum integration if one applies a sharp cut-off in the pion loop integral. Multi-meson loops and other contributions cannot give rise to the LNA behaviour in Eq. (2) and thus near the chiral limit Eq. (3) is the most general form for moments of the PDFs at $\mathcal{O}(m_\pi^2)$ which is consistent with chiral symmetry.

4.2 A Model Calculation

While the current lattice data are at values of m_π too high to display any deviation from constituent quark behavior, it is not obvious why a lowest order form should be able to fit data up to $m_\pi \sim 1$ GeV. This question of whether Eq. (3) can indeed provide a decent parameterization of the m_π dependence of PDF moments, was addressed using the cloudy bag model[16] (CBM) which is consistent with the chiral properties of QCD, and is known to reproduce the LNA behavior of chiral perturbation theory[12,13]. Studies of the N and Δ masses[10] and the nucleon magnetic moments[11] also indicate that the CBM can describe the lattice data over a wide range of quark mass (at least in a qualitative manner). The details[17] of structure function calculations in the CBM are well known, and since the model is not our main focus here, we simply present the results for the first three moments in Fig. 1. The small squares are the results of the model calculation while the dashed curves through them are χ^2 fits to these using the form Eq. (3). Clearly, this form provides an excellent fit to the CBM data, which are also in qualitative agreement with the calculated lattice moments. These results give us confidence that a fit to the lattice data based on Eq. (3) should be reasonable.

4.3 Lattice Extrapolation

Having motivated the functional form of the extrapolation formula, we now apply Eq. (3) to the lattice data. Unfortunately, data are not yet available at quark masses low enough such that a reliable determination of the mass parameter μ could be made. Consequently, for the central curves in each panel of Fig. 1 the value that is most consistent with all experimental moments was chosen, $\mu = 550$ MeV. With μ thus fixed, the results of the best χ^2 fit (for parameters a_n and b_n) to the lattice data for each moment are given by the central solid lines in Fig. 1. To estimate the error in the extrapolated value (for a fixed μ), we also fit to the extrema of the error bars on the data as is shown in Fig. 1 by the inner envelopes around these curves.

The value of μ required here is similar to the scale at which the behavior found in other observables, such as magnetic moments and masses, switches from smooth and constituent quark-like (slowly varying with respect to the current quark mass) to rapidly varying and dominated by Goldstone boson loops. The similarity of these scales for the various observables simply reflects the common scale at which the Compton wavelength of the pion becomes comparable to the size of the bare hadron. This value of μ is also similar to the value required by the χ^2 fits to the CBM. However, it is stressed that μ is not yet determined by the lattice data and it is indeed possible to consistently fit both the lattice data and the experimental values with μ ranging from 400 MeV to 700 MeV. This dependence on μ is illustrated in Fig. 1 by the difference between the inner and outer envelopes on the fits. The former are the best fits to the lower (upper) limits of the error bars, while the latter use $\mu = 450$ (650) MeV instead of the central value of $\mu = 550$ MeV. Data at smaller quark masses are therefore crucial to constrain this parameter and perform an accurate extrapolation.

4.4 Parton Distribution Functions

It is interesting to see to what extent the lowest moments can be used to reconstruct the PDF itself. In order to compare with standard parameterisations[18], the form

$$xq(x) = \alpha x^b(1-x)^c(1+\epsilon\sqrt{x}+\gamma x),$$

is assumed for the distribution. Correspondingly the nth moment will be given by

$$\langle x^n\rangle_{u-d} = \alpha\left[B(1+c,1+b+n)+\epsilon B(1+c,\tfrac{3}{2}+b+n)+\gamma B(1+c,2+b+n)\right],$$

where $B(x,y)$ is the β–function. Leaving the parameters ϵ and γ fixed to their MRS values (see Ref. 14 for further details), the extrapolated lattice moments are used to fit the parameters b and c (α is fixed by requiring $\langle x^0\rangle_{u-d} = 1$). This procedure is performed at the physical quark mass using both the naive, linear and our improved chiral extrapolations with vastly differing results as shown in Fig. 2.

5 Conclusions

In summary, the quark mass dependence of moments of the $u-d$ distribution function have been investigated, paying particular attention to the behavior in the chiral limit. As was found for hadron masses, magnetic moments and

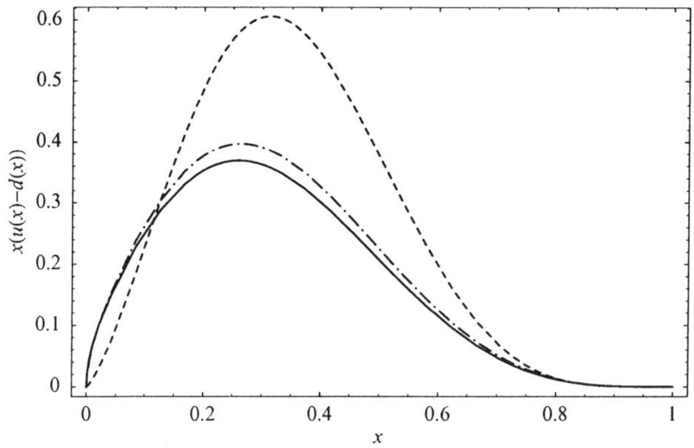

Figure 2. The PDF $x(u(x) - d(x))$ is shown at the physical quark mass as extracted using a linear extrapolation of the lattice data (dashed curve) and our improved chiral extrapolation (solid curve). For comparison the MRS[18] parameterisation is shown (dashed–dotted curve).

charge radii, for large quark masses the moments of structure functions exhibit only slowly varying, linear dependence on m_q, but when the mass decreases towards the chiral limit, this dependence becomes increasingly nonlinear as the effects of chiral loops become dominant. The transition occurs when the pion Compton wavelength becomes larger than the pion source.

A low order formula for the mass dependence of moments has been proposed, which embodies this nonanalytic behavior expected from the chiral properties of QCD, and has been used to extrapolate the available lattice data to the physical region. The applicability of a low order expansion for the lattice data is also supported by phenomenological chiral quark model studies. Compared with linear extrapolations, which badly overestimate the experimental values, the improved extrapolation presented here shows that within the current errors there is no evidence of a discrepancy between the lattice data and experiment once the correct dependence on quark mass near the chiral limit is incorporated. However, in order to properly constrain the extrapolation purely from data, statistically compelling lattice calculations are needed at quark masses ~ 20 MeV.

Acknowledgments

The work on which this presentation is based was performed in collaboration with W. Melnitchouk, J. W. Negele, D. B. Renner and A. W. Thomas and was supported by the Australian Research Council. I am grateful for useful discussions with many workshop participants, in particular with G. Schierholz.

References

1. J. Ashman *et al.*, *Phys. Lett.* B **206**, 364 (1988)
2. P. Amaudruz *et al.*, *Phys. Rev. Lett.* **66**, 2712 (1991)
   E.A. Hawker *et al.*, *Phys. Rev. Lett.* **80**, 3715 (1998)
3. J.J. Aubert *et al.*, *Phys. Lett.* B **123**, 275 (1983)
4. A.W. Thomas, W. Melnitchouk and F.M. Steffens, *Phys. Rev. Lett.* **85**, 2892 (2000)
5. W. Detmold *et al.*, hep-lat/0103006, submitted to *Phys. Rev. Lett.*
6. G. Martinelli and C.T. Sachrajda, *Phys. Lett.* B **196**, 184 (1987); *Nucl. Phys.* B **306**, 865 (1988)
7. M. Göckeler *et al.*, *Phys. Rev.* D **53**, 2317 (1996) M. Göckeler *et al.*, *Nucl. Phys. Proc. Suppl.* **53**, 81 (1997) C. Best *et al.*, hep-ph/9706502.
8. D. Dolgov *et al.*, hep-lat/0011010, and to be published; D. Dolgov, Ph.D. thesis, MIT, Sep. 2000.
9. H.L. Lai *et al.*, *Eur. Phys. J. C* **12**, 375 (2000) A.D. Martin *et al.*, *Eur. Phys. J. C* **14**, 133 (2000) M. Gluck, E. Reya and A. Vogt, *Eur. Phys. J. C* **5**, 461 (1998)
10. D.B. Leinweber, A.W. Thomas, K. Tsushima and S.V. Wright, *Phys. Rev.* D **61**, 074502 (2000)
11. D.B. Leinweber *et al.*, *Phys. Rev.* D **60**, 034014 (1999) D.B. Leinweber and A.W. Thomas, *ibid.* D **62**, 074505 (2000); E.J. Hackett-Jones *et al.*, *Phys. Lett.* B **489**, 143 (2000)
12. M.A.B. Bég and A. Zepeda, *Phys. Rev.* D **6**, 2912 (1972) J. Gasser, M.E. Sainio and A. Svarc, *Nucl. Phys.* B **307**, 779 (1988) D.B. Leinweber and T.D. Cohen, *Phys. Rev.* D **47**, 2147 (1993) E.J. Hackett-Jones, D.B. Leinweber and A.W. Thomas, *Phys. Lett.* B **494**, 89 (2000)
13. S. Weinberg, *Physica (Amsterdam)* **96 A**, 327 (1979) J. Gasser and H. Leutwyler, *Ann. Phys.* **158**, 142 (1984)
14. W. Detmold *et al.*, in preparation.
15. D. Arndt and M.J. Savage, nucl-th/0105045; J.-W. Chen and X. Ji, hep-ph/0105197.
16. S. Théberge, G.A. Miller and A.W. Thomas, *Phys. Rev.* D **22**, 2838 (1980) A.W. Thomas, *Adv. Nucl. Phys.* **13**, 1 (1984)
17. A.W. Schreiber, A.I. Signal and A.W. Thomas, *Phys. Rev.* D **44**, 2653 (1991) ; C. Boros and A.W. Thomas, *Phys. Rev.* D **60**, 074017 (1999)
18. A.D. Martin, *et al.*, *Phys. Rev.* D **50**, 6734 (1994)

CHARGE SYMMETRY IN PARTON DISTRIBUTIONS

J.T. LONDERGAN
Dept. of Physics and Nuclear Theory Center, Indiana University
Bloomington, IN 47405, USA;
E-mail: tlonderg@indiana.edu

Theoretical calculations suggest charge symmetry violation [CSV] for parton valence distributions of a few percent. Precision measurements of structure functions in muon and neutrino DIS set rather stringent experimental limits on CSV. Previous data appeared to show substantial CSV effects in the nucleon sea; this effect has largely disappeared upon re-analysis of CCFR data. We suggest further experiments which could test parton CSV.

Precision tests of approximate quark symmetries provide a "window" to non-perturbative physics. There is now solid evidence for flavor symmetry violation in the light quark sea (u,d,s). The s quark distribution is found to be substantially smaller than the non-strange sea. Experiments by the NMC [1], E866 [2] and HERMES [3] collaborations have now confirmed that $\bar{d}(x) > \bar{u}(x)$. Semi-quantitative explanations for this effect have been obtained in the "meson-cloud" picture, and also from chiral soliton models.

Charge Symmetry Violation in Parton Distributions

Although the aforementioned experiments appear to show large flavor symmetry effects, *all these results depend on the assumption of parton charge symmetry.* Ma[4] showed that the $\bar{d}/\bar{u}$ results could be reproduced, even in the case of exact flavor symmetry, by assuming (amazingly large) violation of parton charge symmetry. In this talk we review the following questions: *1) What are theoretical predictions of parton CSV? 2) What are the present experimental limits on parton charge symmetry? 3) What experiments could improve the current limits on parton CSV?*

A Model for Parton Charge Symmetry Violation

Charge symmetry for valence parton distributions has been investigated by several groups [5,6]. We use the Adelaide model, with which valence parton distributions can be constructed from quark model matrix elements through the relation,

$$q(x,\mu^2) = M \sum_X |\langle X|\psi_+(0)|N\rangle|^2 \, \delta(M(1-x) - p_X^+) \tag{1}$$

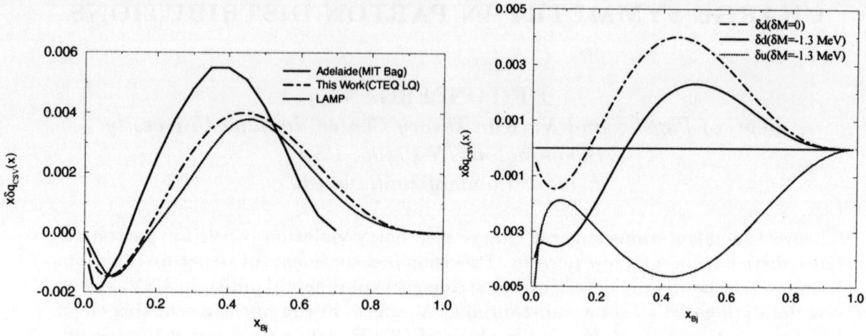

Figure 1: (L) Dot-dashed curve: analytic approximation of Eq. 2, compared with numerical calculations using Eq. 1. (R) CSV valence quark distributions $x\delta u(x)$ (neg) and $x\delta d(x)$ (pos), from Ref. [6].

In Eq. 1, $\psi_+ = (1+\alpha_3)\psi/2$, X represents a complete set of eigenstates of the Hamiltonian H, and the delta function enforces four-momentum conservation.

CSV effects in parton distributions arise from differences in up and down quark masses and electromagnetic effects in Eq. 1. For moderate x values, CSV effects can be estimated using an analytic approximation for the change in the valence quark distribution $\delta q_v(x)$ which arises from a change δm in the diquark mass [6],

$$\delta q_v(x) \approx \frac{2m\,\delta m(1-x)}{M^2(1-x)^2+m^2}\frac{dq_v(x)}{dx} \tag{2}$$

Using Eq. 2, CSV effects can be estimated directly from phenomenological parton distributions without using quark models. The results obtained are in good agreement with direct quark model calculations of CSV effects through Eq. 1, as is shown on the left-hand side of Fig. 1. Fig. 1 shows calculations of parton charge symmetry violation which include both diquark mass contributions and nucleon mass differences; parton CSV terms are defined by

$$\delta d_{\rm v}(x) = d^p_{\rm v}(x) - u^n_{\rm v}(x); \qquad \delta u_{\rm v}(x) = u^p_{\rm v}(x) - d^n_{\rm v}(x) \tag{3}$$

With this notation, all quark distributions henceforth refer to parton distributions in the proton.

The parton CSV effects are predicted to be approximately equal and opposite, $\delta u_v(x) \approx -\delta d_v(x)$, as is shown on the right-hand side of Fig. 1. Since at large x, $d_v(x) << u_v(x)$, the fractional CSV term will be much larger for the "minority quark" term $d_v(x)$ than for the "majority" quark distribution

$u_v(x)$. The quantity $\delta d_v(x)/d_v(x)$ is predicted to be of order 3-6% at large x. Although this fractional CSV effect is rather large, identifying it requires experiments which specifically probe the very small minority quark distribution at large x.

Experimental Status of Parton Charge Symmetry

The most sensitive experimental test of parton charge symmetry to date is the "charge ratio", sometimes called the "$5/18^{ths}$ rule." This is a simple relation between the F_2 structure functions for charged lepton DIS and neutrino charged current reactions on an isoscalar target N_0:

$$\begin{aligned} R_c(x) &= \frac{F_2^{\gamma N_0}(x)}{\frac{5}{18}F_2^{\nu N_0}(x) - xS(x)/3} \approx 1 + \frac{\bar{s}(x) - s(x)}{\bar{Q}(x)} \\ &+ \frac{4(\delta u_v(x) - \delta d_v(x)) + \delta\bar{u}(x) - \delta\bar{d}(x)}{5\bar{Q}(x)} \\ \bar{Q}(x) &= \sum_{j=u,d,s} Q_j(x) - \frac{3\,S(x)}{5} \end{aligned} \tag{4}$$

In Eq. 4, parton distributions in capital letters denote the sum of quark plus antiquark probabilities for that flavor. The relation $R_c(x) = 1$ should hold for all x and Q^2, with no QCD corrections. The only things which break this relation are parton CSV terms, or contributions from $s(x) \neq \bar{s}(x)$. The dependence on strange quarks can be removed by using the average of ν and $\bar{\nu}$ structure functions, $< F_2^{W\,N_0} >$. This gives

$$< R_c(x) > \equiv \frac{F_2^{\gamma N_0}(x)}{\frac{5}{18} < F_2^{W\,N_0}(x) > -xS(x)/3} \approx 1 - \frac{3(\delta D(x) - \delta U(x))}{10\bar{Q}(x)} \ . \tag{5}$$

There now exist precise μ-D structure functions from the NMC group [1], and ν-Fe structure functions from the CCFR collaboration [7]. The statistical errors are very good, and the groups have made a concerted effort to calculate all corrections to this formula. The left-hand side of Fig. 2 plots $R_c(x)$, the ratio of NMC μ-D structure functions to CCFR (corrected for strange quark probabilities), following the initial analysis of the CCFR data. The solid circles show $R_c(x)$, when the heavy target corrections are calculated specifically for neutrinos [8]. For intermediate values $x > 0.1$ the agreement between structure functions is very good, and we can set upper limits of a few percent on parton CSV. However, in the region $x < 0.1$, R_c deviated from unity by as much as

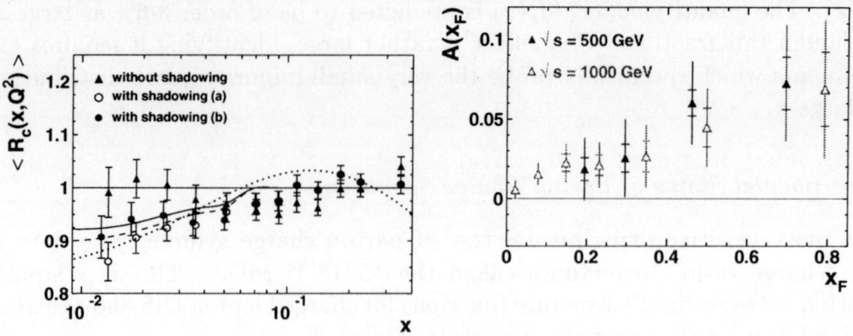

Figure 2: (L) Charge ratio $R_c(x)$ of Eq. 4, obtained using NMC μ-D structure functions and CCFR ν-iron, using old CCFR analysis. Solid circles: shadowing corrections calculated explicitly for ν; open circles: ν target corrections taken from lepton DIS shadowing, from Ref. [8]. (R) Forward-backward asymmetry for W production at two RHIC energies.

10%. Boros *et al.* [9] concluded that the data appeared to reveal a very large sea quark CSV effect (roughly 25%) at small x.

The apparent violation of parton charge symmetry was so large that it was difficult to reconcile with the very small CSV effects seen in nuclear reactions. The CCFR group has recently re-analysed their neutrino data[10], including two effects which significantly change the low-x structure functions. First, neutrino DIS measures a linear combination of F_2 and F_3 structure functions. The original CCFR analysis used a theoretical estimate for F_3. With improved data, the two structure functions could be separated because of their different y dependence. A correction is also necessary for the charm quark mass, since much of the data is taken at energies near charm threshold. The previous analysis used the "slow-rescaling" approximation[11], which replaces $x \to \xi = x(1 + m_c^2/Q^2)$ in the "charm-producing" sector. In their re-analysis, the CCFR group carried out NLO calculations for the cross sections. Using this method, it was unnecessary to adopt the slow-rescaling approximation. Boros *et al.* [12] pointed out the significant effects this method would have on the low-x structure functions. The two effects adopted by CCFR tend to remove nearly all of the low-x discrepancy, as is shown in Fig. 3, which plots the charge ratio of Eq. 4 vs. Q^2 for various x values, using the CCFR ν results and various muon structure functions. The solid curves are NLO theoretical calculations (assuming charge symmetry) using different methods for including heavy flavor effects. The apparent discrepancy at low x has largely disappeared. Note, however,

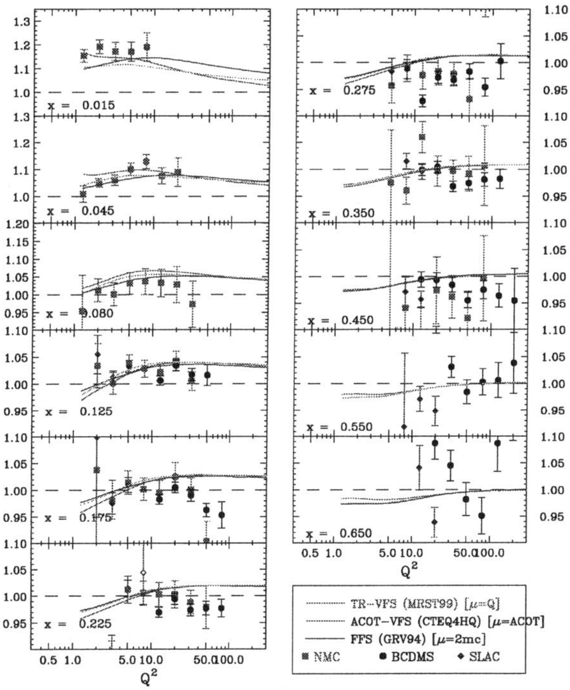

Figure 3: Charge ratio of Eq. 4, from recent CCFR analysis, Ref. [10]. CCFR ν data is compared with muon data from NMC and other experiments. Curves are NLO calculations using various methods for including heavy flavor effects.

that the F_3 structure function inferred from the data is substantially larger than theoretical predictions using phenomenological parton distributions.

Possible Experimental Tests of Parton Charge Symmetry

Now that very large CSV effects appear unlikely, future tests of parton charge symmetry will involve extremely precise measurements of structure functions. The first topic mentioned here, $\nu - D$ charged-current DIS, would be relatively clean and straightforward to analyse, provided that precise cross section measurements could be obtained. In other processes, separation of the CSV signal generally requires accurate evaluation of other competing processes.

$\nu - D$ Structure Functions

The charge ratio requires a comparison of F_2 structure functions for lepton-nucleus DIS with those from charge-changing weak interactions. The $\mu - D$ measurements from the NMC group are ideal for this purpose. To date, the $\nu - Fe$ measurements from CCFR are the most reliable neutrino structure functions yet measured, since the iron target provides sufficient counts for precision measurements and for kinematic separation of the different structure functions. Neutrino structure functions from deuterium, as measured by the BCDMS collaboration [13], are not sufficiently precise to provide firm limits on parton CSV. Following the main injector upgrade at FNAL, one now has the prospect of neutrino beams orders of magnitude more copious than were previously available. This offers the possibility of precision measurements of $\nu - D$ structure functions, without the significant target corrections (*e.g.*, corrections for non-isoscalar contributions in iron, and large small-x shadowing corrections) which make detailed comparisons quite difficult.

At present the mini-BooNE experiment is trying to confirm neutrino oscillation signals seen in the LSND experiment. Future upgrades of this experiment could measure neutrino scattering cross sections. If a deuterium target were included, it would be possible to obtain charged-current $\nu - D$ DIS which could be directly compared to the NMC muon results without requiring the significant corrections from the iron data. This would provide the most precise limits to date on parton CSV.

W Production Asymmetry at a Hadron Collider

At sufficiently high energies, the charge-changing structure functions on an isoscalar target are equal except for contributions from valence quark CSV, and possible strange or charmed quark terms. It is thus possible to test parton charge symmetry by comparing processes such as $p + D \to W^+ + X$ and $p + D \to W^- + X$. It would be possible to measure these at hadron colliders such as RHIC or LHC, with the availability of deuteron beams. If we define $\sigma^\pm = (d\sigma/dx_F)^{W^\pm}$, then the forward-backward asymmetry is given by

$$A(x_F) = \frac{\sigma^+(x_F) + \sigma^-(x_F) - \sigma^+(-x_F) - \sigma^-(-x_F)}{\sigma^+(x_F) + \sigma^-(x_F) + \sigma^+(-x_F) + \sigma^-(-x_F)} \ ,$$
$$A(x_F) \propto (U(x_1) - D(x_1))\delta\bar{d}(x_2) + (\bar{d}(x_2) - \bar{u}(x_2))\delta d(x_1) \ . \tag{6}$$

In deriving the second line of Eq. 6, we have dropped terms proportional to $\sin^2 \Theta_c$, and we have assumed $\delta d \approx -\delta u$ as suggested from theoretical considerations. On the right side of Fig. 2 we show a calculation of the W

asymmetry $A(x_F)$ of Eq. 6 by Boros *et al.* [9]. Note that the CSV terms used here are those extracted from the original CCFR analysis; as these terms will be much smaller following the re-analysis of this data, the expected W asymmetry will be significantly reduced from that shown in Fig. 2.

Reactions Involving Pions and Isoscalar Nuclei

One possible test of parton charge symmetry is a comparison of Drell-Yan processes for charged pions on isoscalar targets in the "valence-dominated" region, where both x_π and x are large. Here the DY process occurs predominantly through a valence $\bar{d}$ ($\bar{u}$) in a $\pi^+(\pi^-)$ annihilating a down (up) valence quark in the nucleon. CSV effects will contribute to differences between DY cross sections for π^+ and π^- beams on an isoscalar target. This has been investigated by Londergan et al.[14]. In order to measure charge symmetry-violating amplitudes, it is necessary to extract the CSV "signal" from valence-sea interference terms. With sufficiently intense pion beams and excellent systematics, this process could be used to probe valence quark CSV in the nucleon.

Another potentially sensitive probe of parton CSV effects is semi-inclusive meson electroproduction, in a process like $e + N_0 \to \pi^{+(-)} + X$. Comparison of π^+ and π^- rates is sensitive to CSV effects [15]. This could in principle be investigated at the HERMES experiment at HERA, which has deuteron targets, and which has been studying semi-inclusive meson production. Extraction of the CSV term requires precise knowledge of several factors, and in particular both the "favored" and "unfavored" pion fragmentation functions must be very accurately measured.

Conclusions

Theoretical calculations predict both the magnitude and sign of valence quark charge symmetry violation. CSV contributions in the nucleon sea are much less well known, although calculations of these effects are ongoing. The most precise test of parton charge symmetry is comparison of the F_2 structure functions for both charged-lepton DIS and neutrino charge-changing reactions. With recent data, and new analyses, it is now possible to place strong upper limits of a few percent on CSV effects in parton distributions in the region $0.1 \leq x \leq 0.4$. Hints of extremely large CSV effects in the nucleon sea for $x < 0.1$ have now disappeared upon re-analysis of the CCFR data. There are prospects for further precision CSV tests, although the most promising course would be to use high-flux neutrino beams at FNAL for ν DIS measurements on deuterium.

Acknowledgments

This research was performed in collaboration with A.W. Thomas and other collaborators, notably C. Boros and C. Benesh. The author thanks the Special Research Centre for the Subatomic Structure of Matter for its hospitality. This research was supported in part by the NSF under contract NSF-PHY-0070368, and by the Australian Research Council.

References

1. P. Amaudruz et al. (NMC Collaboration,) *Phys. Rev. Lett.* **66**, 2712 (1991).
2. E.A. Hawker et al. (E866 Collaboration), *Phys. Rev. Lett.* **80**, 3715 (1998).
3. K. Ackerstaff et al. (HERMES Collaboration), *Phys. Rev. Lett.* **81**, 5519 (1998).
4. B.-Q. Ma, *Phys. Lett.* B **274**, 111 (1992).
5. E. Sather, *Phys. Lett.* B **274**, 433 (1992); E. Rodionov, A.W. Thomas and J.T. Londergan, *Int. J. Mod. Phys. Lett.* A **9**, 1799 (1994); C.J. Benesh and T. Goldman, *Phys. Rev.* C **55**, 441 (1997).
6. C.J. Benesh and J.T. Londergan, *Phys. Rev.* D **58**, 1218 (1998).
7. W.G. Seligman et al. (CCFR Collaboration), *Phys. Rev. Lett.* **79**, 1213 (1997).
8. C. Boros, J.T. Londergan and A.W. Thomas, *Phys. Rev.* D **58**, 114030 (1998).
9. C. Boros, J.T. Londergan and A.W. Thomas, *Phys. Rev. Lett.* **81**, 4075 (1998); *Phys. Rev.* D **59**, 074021 (1999).
10. U.K. Yang et al. (NuTeV/CCFR Collaboration), *Phys. Rev. Lett.*, to be published (2001).
11. H. Georgi and D. Politzer, *Phys. Rev.* D **14**, 1829 (1976).
12. C. Boros, F.M. Steffens, J.T. Londergan, and A.W. Thomas, *Phys. Lett.* B **468**, 161 (1999).
13. A.C. Benvenuti et al. (BCDMS Collaboration), *Phys. Lett.* B **237**, 592 (1990).
14. J.T. Londergan, G.T. Garvey, G.Q. Liu, E.N. Rodionov and A.W. Thomas, *Phys. Lett.* B **340**, 115 (1994).
15. J.T. Londergan, Alex Pang and A.W. Thomas, *Phys. Rev.* D **54**, 3154 (1996).

DISTORTIONS IN THE NEGATIVE ENERGY DIRAC SEA: VIOLATION OF THE GOTTFRIED SUM RULE AND $\Delta\bar{u}$ IN THE PROTON

K. TSUSHIMA[1], A. W. THOMAS[2]
Special Research Centre for the Subatomic Structure of Matter and Department of Physics and Mathematical Physics, Adelaide University, Adelaide SA 5005, Australia
E-mail: [1]ktsushim@physics.adelaide.edu.au, [2]athomas@physics.adelaide.edu.au

G.V. DUNNE
Department of Physics, University of Connecticut, Storrs, CT 06269-3046, USA
E-mail: dunne@phys.uconn.edu

We report on recent work concerning the effect which the change in vacuum structure (negative energy Dirac sea), in the presence of a confining scalar field, has on the nucleon structure functions and parton distributions. Using the Dirac equation in 1+1 dimensions, we show that distortions in the Dirac sea are responsible for part of the violation of the Gottfried sum rule – i.e., part of the flavor asymmetry in the proton sea. Our basic argument is that, even if isospin is an exact symmetry, the presence of a confining potential changes the vacuum structure, and inevitably leads to a violation of SU(2) flavour symmetry in a hadron with a different number of valence u and d quarks. The same mechanism also leads to a prediction for $\Delta\bar{u}$ and $\Delta\bar{d}$.

1 Introduction

The study of the violation of the Gottfried sum rule[1] has attracted considerable interest since its experimental confirmation by the New Muon Collaboration (NMC)[2]. At present, there are several possible explanations for it, including mesonic (mainly pionic[3]) contributions and the Pauli exclusion principle at the quark level[4] – for recent reviews see Refs.[5,6,7,8].

Here we report our recent work[9], on this problem, aimed at putting the phenomenon into a more general context. Our approach is based on a totally nonperturbative effect, namely the change in the vacuum structure (including distortion of the negative energy Dirac sea) in the presence of a confining potential. Such changes in vacuum structure have been studied in great detail in other contexts (including the phenomenon of fractional charge[10,11,12]), since the initial work of Jackiw and Rebbi[10].

In the case of the nucleon structure functions, self-consistent studies of the vacuum structure have been made within a chiral quark soliton model[13,14,15,16].

However, it is not easy to see how the change in vacuum structure is directly reflected in the calculated results.

In this report we will explicitly show that the existence of a (quark confining) potential distorts the negative energy Dirac sea, and leads to a violation of the valence quark number normalization. Thus, it inevitably leads to the violation of the Gottfried sum rule. We emphasize that our argument holds even if charge symmetry and flavor SU(2) symmetry are assumed to be valid. We also predict $\Delta\bar{u}$ in the proton based on the same mechanism.

2 The Dirac equation in 1+1 dimensions

Consider massless ($m = 0$) fermions in $1+1$ dimensions, in the presence of a static scalar potential $V(z)$. The fermion spectrum is determined from the Dirac Hamiltonian,

$$H\,\varphi(z) = \left[-i\alpha\frac{d}{dz} + \beta V(z)\right]\varphi(z),$$
$$= \begin{pmatrix} 0 & -\frac{d}{dz} + V(z) \\ \frac{d}{dz} + V(z) & 0 \end{pmatrix}\varphi(z) = E\varphi(z), \quad (1)$$

where we have chosen the Dirac matrix representation with $\alpha = \sigma_2$ and $\beta = \sigma_1$. We choose $V(z)$ to be an odd function, with a kink-like profile such that $V(\infty) = V_0 = -V(-\infty)$. Then the spectrum is necessarily symmetric about $E = 0$, and there are continuum thresholds at $E = \pm V_0$. There is a bound state at $E = 0$, but it is irrelevant for our discussion. Instead, we consider bound states in the gap $|E|(\neq 0) < |V_0|$, which occur in pairs symmetrically placed about $E = 0$.

We choose the profile for the scalar potential: $V(z) = V_0 \tanh z$, where $V_0 > 0$. Then the $E \neq 0$ bound state wave functions with energy $E_n^{(\pm)} = \pm\sqrt{n(2V_0 - n)}$, are given by

$$\chi_{b,n}^{(\pm)}(z) = N_n \begin{pmatrix} \sqrt{\frac{n}{2V_0 - n}}\, P_{V_0}^{V_0 - n}(\tanh z) \\ \pm P_{V_0 - 1}^{V_0 - n}(\tanh z) \end{pmatrix}, \quad (2)$$

for $n = 1, 2, \ldots [V_0]$, where $[V_0]$ means the greatest integer less than V_0. In Eq. (2), $P_\nu^\mu(z)$ is the associated Legendre function. If V_0 is an integer then there are real (but non-normalizable) threshold states obtained by setting $n = V_0$ in Eq. (2). The continuum state wave functions are given by

$$\varphi_p^{(\pm)}(z) = N_p \begin{pmatrix} \sqrt{\frac{V_0 - ip}{V_0 + ip}}\, P_{V_0}^{ip}(\tanh z) \\ \pm P_{V_0 - 1}^{ip}(\tanh z) \end{pmatrix}, \quad (3)$$

with energy $E_p^{(\pm)} = \pm\sqrt{V_0^2 + p^2}$. For simplicity, we choose $V_0 = \frac{3}{2}$, where there is a zero energy state, and two other bound states with energies $E_1^{(\pm)} = \pm\sqrt{2}$:

$$\chi_{b,1}^{(\pm)}(z) = \begin{pmatrix} \sqrt{2}\tanh z \,\mathrm{sech}^{1/2} z \\ \pm\mathrm{sech}^{1/2} z \end{pmatrix}. \tag{4}$$

These solutions are to be compared to the free Dirac equation solutions of positive and negative energies ($E_p = \pm|p|$):

$$u_p(z) = \frac{1}{\sqrt{2}} \begin{pmatrix} -ip/|p| \\ 1 \end{pmatrix} e^{ipz}, \qquad v_p(z) = \frac{1}{\sqrt{2}} \begin{pmatrix} ip/|p| \\ -1 \end{pmatrix} e^{-ipz}. \tag{5}$$

Then, the field operator can be expanded in terms of either $\{u_p, v_p, -\infty < p < \infty\}$ or $\{\chi_{b,n}, \varphi_p^{(+)}, \varphi_p^{(-)}, -\infty < p < \infty\}$, each of which is a complete, orthonormal set [12]:

$$\psi(z) = \int_{-\infty}^{\infty} \frac{dp}{2\pi} \left[b_p u_p(z) + d_p^\dagger v_p(z)\right] \tag{6}$$

$$= e_n \chi_{b,n} + \int_{-\infty}^{\infty} \frac{dp}{2\pi} \left[a_p \varphi_p^{(+)}(z) + c_p^\dagger \varphi_p^{(-)}(z)\right]. \tag{7}$$

The creation and annihilation operators for the set with $V_0 \neq 0$ satisfy,

$$\{e_n^\dagger, e_m\} = \delta_{nm}, \ \{a_p^\dagger, a_q\} = 2\pi\delta(p-q), \ \{c_p^\dagger, c_q\} = 2\pi\delta(p-q), \tag{8}$$

and all other anticommutators vanish. The operator sets, e.g., $\{e_n, a_p, c_p^\dagger\}$ and $\{b_p, d_p^\dagger\}$ can be related using the orthogonality (Bogoliubov transformation):

$$e_n = \langle \chi_{b,n} | u_p \rangle b_p + \langle \chi_{b,n} | v_p \rangle d_p^\dagger, \tag{9}$$

$$a_k = \langle \varphi_k^{(+)} | u_p \rangle b_p + \langle \varphi_k^{(+)} | v_p \rangle d_p^\dagger, \tag{10}$$

$$c_k^\dagger = \langle \varphi_k^{(-)} | u_p \rangle b_p + \langle \varphi_k^{(-)} | v_p \rangle d_p^\dagger, \tag{11}$$

where the bra-ket notation implies, $\int_{-\infty}^{\infty} \frac{dp}{2\pi}$.

Let us focus on the bound state, $|\chi_{b,1}\rangle$, where the wave function is normalized to unity and its expression is given by Eq. (4). Using the relation of Eq. (9) and the property of orthonormality, the normalization of the bound state can be written by:

$$1 = \langle \chi_{b,1} | \chi_{b,1} \rangle = \int_{-\infty}^{\infty} \frac{dp}{2\pi} \left(|\langle \chi_{b,1} | u_p \rangle|^2 + |\langle \chi_{b,1} | v_p \rangle|^2 \right). \tag{12}$$

In Eq. (12), the term $|\langle\chi_{b,1}|v_p\rangle|^2$ ($|\langle\chi_{b,1}|u_p\rangle|^2$), can be interpreted as the probability density distributions with which the bound state $|\chi_{b,1}\rangle$ would be filled by the negative (positive) energy free states, if the potential were turned off suddenly (but not adiabatically).

We show in Fig. 1 the quantity $|\langle\chi_{b,1}|u_p\rangle|^2 + |\langle\chi_{b,1}|v_p\rangle|^2$, (square of the wave function overlaps) as a function of energy for the free plane wave solutions, $E_p(free)$. Surprisingly, one can immediately notice that the positive energy

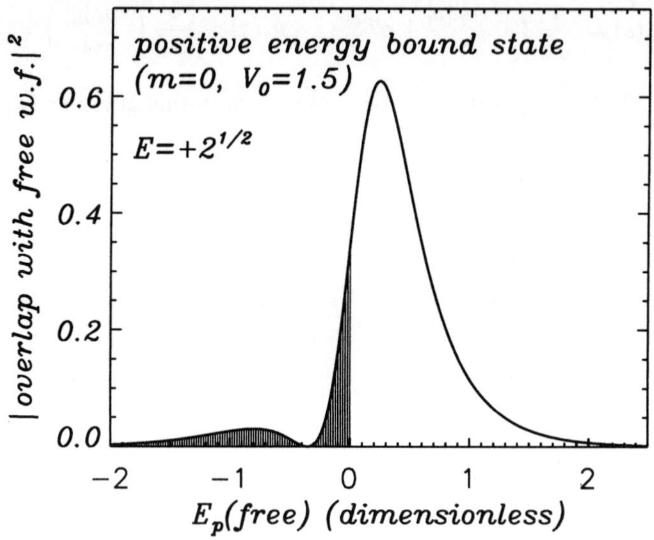

Figure 1: Bound state number density distributions as a function of free plane wave energy, $E_p(free)$, when projected onto free plane wave solutions.

bound state wave function, $\chi_{b,1}$, has a non-zero overlap with the negative energy free plane wave solutions (the shaded area in Fig. 1). If we integrate over $E_p(free)$, it gives unity by construction, i.e. by probability conservation, as it should be. We should also mention that the negative (positive) energy continuum state wave functions of Eq. (3) also have non-zero overlap with the positive (negative) energy free plane wave solutions.

The conclusion of this section is that the positive energy bound state, appearing in the Fock space built in the presence of the potential, has a non-zero overlap with the negative energy free plane wave solutions. It is with this point that we want to make a connection with the deep-inelastic scattering (DIS) and the parton model.

3 Connection with the Parton Model

In this section we consider the interpretation of the results obtained in the previous section in connection with the parton model. It may be useful to recall the assumptions which leads to the standard parton model according to Ref. [17]: *(1) The current couples to quarks. Then the contributions to the forward Compton amplitude can be classified by the flow and interactions of quark lines. (2) At large value of Q^2 the currents, but not the states, may be treated as in free field theory. Thus, final state interactions and vertex corrections are ignored.*

Let us now identify the positive energy bound state wave function of Eq. (4) as that of the valence quark in a simple, model proton. We recall that in Fig. 1 the quark number density distribution is projected onto free plane wave solutions of energy $E_p(free)$. As mentioned above, these free plane wave solutions are now to be identified with the partons, which appear in free field theory. As expected, the valence quark wave functions in the proton have non-zero overlap with the negative energy free plane waves, and this means that the distributions are non-zero in the region where Bjorken x is negative. As a result the integral of the valence distribution over the physical region $x > 0$ will not be one. The resolution of this problem lies in the fact that the Dirac sea in the presence of a scalar potential is also different from that of free space. In particular, the Dirac sea in the presence of the soliton in our 1D example (or the confining potential for the proton) consists of having all negative energy states occupied. If re-expressed in terms of free plane waves, however, one finds an equal number of positive energy free states occupied and negative energy states (holes) empty. That is, when the occupied negative energy states are expressed in terms of free states one finds an intrinsic sea of $q\bar{q}$ pairs. In the absence of valence quarks this sea carries no net flavour (or in 3D, spin), relative to the free space vacuum.

However, once we add a single valence quark (say a u quark), the negative energy tail will fill some of the holes in the (free space) Dirac sea. An equal number of occupied positive energy states from the u quark Dirac sea will then be interpreted as the missing valence normalization (because u_v is **defined** to be $u - \bar{u}$). The net result is that we have one valence u quark, together with a reduction in the number of $u\bar{u}$ pairs in the sea of the proton. That is, one will have $\bar{d} > \bar{u}$ by an amount equal to the area under the valence curve to the left of $E_p = 0$ in Fig. 1. If instead we had one valence u and one valence d the intrinsic sea would be flavour symmetric again, while for two valence u's and one valence d, as in the proton, we have $\bar{d} > \bar{u}$ by the same amount as for the single valence u (the area under the valence curve to the left of $E_p = 0$ in Fig.

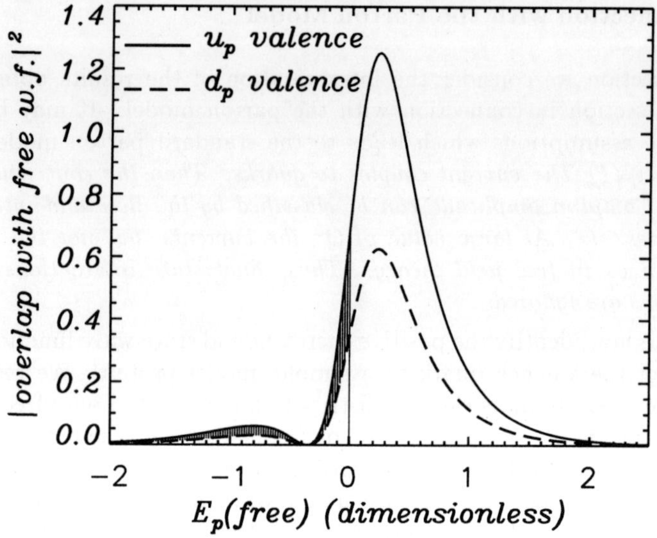

Figure 2: u and d valence quark number density distributions in the "proton" as a function of free energy, $E_P(free)$, projected onto free plane waves.

1).

The situation with respect to the Gottfried sum rule is therefore illustrated in Fig. 2. Because there are more valence u's, the u sea is depleted more than the sea of d's. Mathematically, the expression for the Gottfried sum rule is then:

$$
\begin{aligned}
S_G &= \int_0^1 \frac{dx}{x}\left[F_2^{\mu p}(x) - F_2^{\mu n}(x)\right], \\
&= \frac{1}{3}\int_0^1 dx[u_v(x) - d_v(x)] - \frac{2}{3}\int_0^1 dx[\bar{d}(x) - \bar{u}(x)], \\
&= \frac{1}{3} - \frac{2}{3}\int_{-\infty}^{\infty} \frac{dp}{2\pi}|\langle v_p|\chi_{b,1}\rangle|^2 .
\end{aligned}
\tag{13}
$$

(Note that we have assumed charge symmetry to set $d^n = u^p \equiv u$, etc.)

4 Conclusion

By considering the simple 1D problem outlined here, we aimed to make the physics origins of one source of flavour asymmetry in the proton explicit and easy to understand. While one cannot access spin in the 1D case, the simplicity of the argument for flavour makes the corresponding argument for $\Delta\bar{u}$ (and $\Delta\bar{d}$) clear. Since in a spin up proton described by an SU(6) wave function the valence u quark is predominantly spin up, one will be missing spin up u quarks in the sea of the proton. Thus $\Delta\bar{u}$ will be positive and of a similar magnitude to that of $\bar{d} - \bar{u}$ from the same source [4,18]. Since the d quarks tend to have spin down in the proton, one also expects $\Delta\bar{d}$ to be negative and considerably smaller in magnitude than $\Delta\bar{u}$. Since any contribution to the observed flavour asymmetry from the nucleon's pion cloud is unpolarized [3,19], the experimental determination of $\Delta\bar{u}$ and $\Delta\bar{d}$ would be extremely valuable in reaching a quantitative understanding of the roles of these two very different physical mechanisms in generating the non-perturbative structure of the proton sea.

Acknowledgments

We would like to thank W. Melnitchouk and A.W. Schreiber for many helpful discussions. K.T. would like to acknowledge the hospitality at KFA Jülich, where this report was completed. This work was supported by the Australian Research Council and Adelaide University. GD is supported by the US DOE, and thanks the CSSM at Adelaide for hospitality during a visit in which this work was begun.

References

1. K. Gottfried, *Phys. Rev. Lett.* **18**, 1174 (1967).
2. P. Amaudruz *et al.*, *Phys. Rev. Lett.* **66**, 2712 (1991); M. Arneodo *et al.*, *Phys. Rev.* D **50**, R1 (1994).
3. A. W. Thomas, *Phys. Lett.* B **126**, 97 (1983).
4. A. I. Signal and A. W. Thomas, *Phys. Rev.* D **40**, 2832 (1989); A.W. Schreiber, PhD thesis, University of Adelaide, (1990); A. W. Schreiber, A. I. Signal and A. W. Thomas, *Phys. Rev.* D **44**, 2653 (1991).
5. S. Kumano, *Phys. Rep.* **303**, 183 (1998).
6. J. Speth and A.W. Thomas, *Adv. Nucl. Phys. Vol.* **24**, 83 (1998).
7. J.T. Londergan, A.W. Thomas, *Prog. Part. Nucl. Phys.* **41**, 49 (1998).
8. R. Vogt, *Prog. Part. Nucl. Phys.* **45**, S105 (2000).
9. G.V. Dunne, A.W. Thomas, K. Tsushima, in preparation.
10. R. Jackiw and C. Rebbi, *Phys. Rev.* D **13**, 3398 (1976); see also, R. Jackiw, hep-th/9903255, *Dirac Prize Lecture*, Trieste, Italy - March 1999.
11. J. Goldstone and F. Wilczek, *Phys. Rev. Lett.* **47**, 986 (1981); J. Goldstone and R. L. Jaffe, *Phys. Rev. Lett.* **51**, 1518 (1983); For a review, see: A. Niemi and G. Semenoff, *Phys. Rep.* **135**, 99 (1986).
12. R. MacKenzie and F. Wilczek, *Phys. Rev.* D **30**, 2194 (1984).
13. M. Wakamatsu, *Phys. Rev.* D **44**, R2631 (1991); *Phys. Lett.* B **269**, 394 (1991); *Phys. Rev.* D **46**, 3762 (1992).
14. D. Diakonov *et al.*, *Nucl. Phys.* B **480**, 314 (1996).
15. P.V. Pobylitsa *et al.*, *Phys. Rev.* D **59**, 034024 (1999); K. Goeke *et al.*, *Nucl. Phys.* A **680**, 307 (2000); B. Dressler *et al.*, *Prog. Part. Nucl. Phys.* **44**, 293 (2000).
16. H. Weigel, E. Ruiz Arriola and L. Gamberg, *Nucl. Phys.* B **560**, 383 (1999).
17. R. L. Jaffe, in *Relativistic Dynamics and Quark-Nuclear Physics*, pp 537 - 618, M.B. Johnson and A. Picklesimer (Eds), (Wiley, New York, 1986).
18. F. Cao and A. I. Signal, hep-ph/0103113.
19. A. W. Thomas, W. Melnitchouk and F. M. Steffens, *Phys. Rev. Lett.* **85**, 2892 (2000) [hep-ph/0005043].

PARTON DISTRIBUTIONS FOR THE PION IN A CHIRAL QUARK MODEL[a]

E. RUIZ ARRIOLA

Departamento de Fisica Moderna, Universidad de Granada,
18071 GRANADA, Spain
E-mail: earriola@ugr.es

Parton distributions for the pion are studied in a chiral quark model characterized by a quark propagator for which a spectral representation is assumed. Electromagnetic and chiral symmetry constraints are imposed through the relevant Ward-Takahashi identities for flavoured vertex functions. Finiteness of the theory, requires the spectral function to be non-positive definite. Straightforward calculation yields the result that the pion structure function becomes one in the chiral limit, regardless of the details of the spectral function. LO and NLO evolution provide a satisfactory description of phenomenological parameterizations of the valence distribution functions but fails to describe gluon and sea distributions.

1 Introduction

Deep inelastic scattering (DIS) is regarded as one of the traditional ways to unveil the quark substructure of hadrons [1]. The hadronic tensor for electroproduction on a (unpolarized) hadron is defined as

$$W_{\mu\nu}(p,q) = \frac{1}{4\pi} \int d^4\xi e^{iq\cdot\xi} \langle p \left| \left[J_\mu^{\rm em}(\xi), J_\nu^{\rm em}(0) \right] \right| p \rangle \tag{1}$$

where p is the hadron momentum, $J_\mu^{\rm em}(\xi) = \bar{q}(\xi)\hat{Q}\gamma_\mu q(\xi)$ the electromagnetic current with $\hat{Q} = {\rm diag}(e_u, e_d, e_s)$ the quark charge matrix and q the photon momentum transfer. Introducing the relativistically invariant Euclidean momentum transfer $Q^2 = -q^2$ and the Bjorken variable $x = Q^2/2p \cdot q$, in the Bjorken limit ($Q^2 \to \infty$ with x fixed), Eq.(1) can be expressed in terms of the quark and antiquark distribution functions, $q_i(x)$ and $\bar{q}_i(x)$

$$W_{\mu\nu} \xrightarrow{\rm Bj} F(x) \left[-g_{\mu\nu} + \frac{q_\mu q_\nu}{q^2} - \frac{1}{q^2} \left(p_\mu - \frac{q_\mu}{2x} \right) \left(p_\nu - \frac{q_\nu}{2x} \right) \right] \tag{2}$$

with $F(x) = \sum_i e_i^2 \left[q_i(x) + \bar{q}_i(x) \right]/2$ as a consequence of Lorentz gauge invariance, scaling and the underlying spin 1/2 nature of quarks. Logarithmic scaling violations due to perturbative QCD radiative corrections [2], relate in

[a] Talk given at the Workshop on " Lepton Scattering, Hadrons and QCD " March 26 – Friday April 6, 2001. ; Adelaide (Australia)

a linear fashion the structure functions at a given reference scale, Q_0^2, to the scale of interest, Q^2,

$$F(x, Q^2) = U(Q^2, Q_0^2) F(x, Q_0^2). \tag{3}$$

Here, $U(Q^2, Q_0^2)$ is a linear matrix operator, fulfilling renormalization group properties $U(Q_1^2, Q_2^2)U(Q_2^2, Q_3^2) = U(Q_1^2, Q_3^2)$ and $U(Q^2, Q^2) = 1$ and which can be evaluated for high Q^2. Thus, one is left with the calculation of structure functions at a certain scale Q_0 (the so called initial condition, $F(x, Q_0^2)$) and the assumption that the scale is high enough to make perturbative matching meaningful.

In spite of the wealth of experimental and theoretical information [3,4,5,6,7], first principle calculations of the initial condition, $F(x, Q_0^2)$, remain, so far, elusive. Thus, theoretical quark models of hadrons and quark-hadron duality assumptions are used. This obviously raises embarrassing questions on confinement, because the intermediate states in the commutator of Eq.(1) should be physical hadron states and, instead, one uses the virtual Compton amplitude [9]

$$T_{\mu\nu}(p, q) = i \int d^4\xi e^{iq\cdot\xi} \langle p \left| T\left\{ J_\mu^{\rm em}(\xi) J_\nu^{\rm em}(0) \right\} \right| p \rangle \tag{4}$$

with $W_{\mu\nu} = 1/(2\pi)\mathrm{Im}T_{\mu\nu}$. Introducing light-cone (LC) coordinates $x = (x^+, x^-, \vec{x}_\perp)$ with $x^\pm = x^0 \pm x^3$ one gets the following expression for the initial condition

$$F(x, Q_0^2) = -\frac{i}{4\pi} \int \frac{dk^- d^2\vec{k}_\perp}{(2\pi)^3} \mathrm{tr}\left[\hat{Q}^2 \gamma^+ \chi(p, k) \right] \Big|_{k^+ = p^+ = mx} \tag{5}$$

where the forward quark-target scattering amplitude is defined

$$\chi(p, k) = -i \int d^4\xi\, e^{i\xi\cdot k} \langle p | T\{q(\xi)\bar{q}(0)\} | p \rangle \tag{6}$$

$\chi(p, k)$ corresponds to the unamputated vertex. Eq.(5) holds under the assumption of scaling and finiteness in the Bjorken limit.

Among the hadrons, pseudoscalar mesons and, in particular the pion, appear as the simplest and best understood states from a theoretical viewpoint, since most of its properties can be explained in terms of spontaneous chiral symmetry breaking. Its bound state properties are not expected to depend crucially on confinement, at least for small excitation energies. Furthermore, the pion structure functions have been deduced from phenomenological QCD studies [5,7] and there also exist lattice calculations of the first few moments [10]. Thus, we do not expect to achieve a better theoretical understanding of a DIS process than in the case of the pion. Hence, structure functions of the pion have been computed several times in different quark-loop models [11,12,13,14,15,16,17,18,19]. The issue of implementing gauge invariance and hence proper normalization of $F(x,Q_0^2)$ is problematic and therefore requires some special care.

2 Quark propagator and Ward-Takahashi identities

I present here another model [20] for pion structure based in a spectral representation of the quark propagator and chiral Ward-Takahashi identities (WTI) for the flavoured vertex functions, deduced from conservation of the vector current (CVC) and partial conservation of the axial current (PCAC)

$$J_V^{\mu,a}(x) = \bar{q}(x)\gamma^\mu \frac{\tau_a}{2} q(x) \qquad \partial_\mu J_V^{\mu,a}(x) = 0 \tag{7}$$

$$J_A^{\mu,a}(x) = \bar{q}(x)\gamma^\mu \gamma_5 \frac{\tau_a}{2} q(x) \qquad \partial_\mu J_A^{\mu,a}(x) = \bar{M}_0 \bar{q}(x) i\tau_a \gamma_5 q(x) \tag{8}$$

with $\bar{M}_0$ the average up and down current quark masses.

We assume a spectral representation for the quark propagator

$$S(p) = \int dw \frac{\rho(w)}{\not{p} - w} = -i \int d^4x e^{ip\cdot x} \langle 0 \left| T\left\{ q(x)\bar{q}(0) \right\} \right| 0 \rangle \tag{9}$$

The requirement of a physical Hilbert space would imply the positivity of the spectral function $\rho(w)$, but we are not assuming this here. Predictions based on Eq. (9) require a specific form of $\rho(w)$ as input, but it turns out [20] that in the chiral limit, $\bar{M}_0 \to 0$ in Eq. (8), many properties of interest are independent of $\rho(w)$. Thus we will focus on those observables and consider the chiral limit. The vector and axial unamputated vertex functions satisfy the WTI

$$(p'-p)_\mu S(p')\Gamma_V^{\mu,a}(p',p)S(p) = \frac{\tau_a}{2} S(p) - S(p')\frac{\tau_a}{2} \tag{10}$$

$$(p'-p)_\mu S(p')\Gamma_A^{\mu,a}(p',p)S(p) = S(p')\frac{\tau_a}{2}\gamma_5 + \gamma_5 \frac{\tau_a}{2} S(p) \tag{11}$$

The gauge technique, in the form introduced in QED [21] consists of writing *a solution* for the vector and axial vertices in the form

$$S(p')\Gamma_V^{\mu,a}(p',p)S(p) = \int dw \rho(w) \frac{1}{\not{p}' - w} \gamma^\mu \frac{\tau_a}{2} \frac{1}{\not{p} - w} \tag{12}$$

$$S(p')\Gamma_A^{\mu,a}(p',p)S(p) = \int dw \rho(w) \frac{1}{\not{p}' - w} \left(\gamma^\mu - \frac{2wq^\mu}{q^2} \right) \gamma_5 \frac{\tau_a}{2} \frac{1}{\not{p} - w} \tag{13}$$

with $q = p' - p$. An interesting consequence of the axial Ward identity (13) is the presence of a massless pseudoscalar pole, which is identified with the pion.

3 Properties of the spectral function

Some properties of the spectral function $\rho(w)$ can be derived by studying specific processes and comparing with known QCD results. For instance, vacuum polarization can be computed closing the vector-quark-quark vertex in Eq. (12)

$$i\int d^4x e^{-ip\cdot x}\langle 0|T\left\{J_\mu^{\rm em}(x)J_\nu^{\rm em}(0)\right\}|0\rangle = \left(-g_{\mu\nu}+\frac{p_\mu p_\nu}{p^2}\right)\Pi^{\rm em}(p^2) \quad (14)$$

The $e^+e^-\to$ hadrons cross-section is proportional to the imaginary part of the vacuum charge polarization operator. Asymptotically, one has

$$\sigma(e^+e^-\to \text{hadrons}) \to \frac{4\pi\alpha^2}{3s}\left(\sum_i e_i^2\right)\int dw\rho(w) \quad (15)$$

Thus, the proper QCD asymptotic result is obtained if

$$\int dw\rho(w) = 1 \quad (16)$$

This condition is equivalent to impose that $\lim_{p\to\infty} \not{p}S(p) = 1$.

The weak pion decay constant can be computed from the axial-axial correlation function, since

$$\langle 0\left|J_A^{\mu a}(x)\right|\pi_b(p)\rangle = if_\pi p_\mu \delta_{a,b}e^{ip\cdot x} \quad (17)$$

and then, inserting a complete set of eigenstates between the currents,

$$\int d^4x e^{-ip\cdot x}\langle 0|T\left\{J_A^{\mu a}(x)J_A^{\nu b}(0)\right\}|0\rangle = if_\pi^2\delta_{ab}\frac{p^\mu p^\nu}{p^2}+\dots \quad (18)$$

where the dots indicate regular pieces in the limit $p^2\to 0$. Again, closing the axial-quark-quark vertex in Eq. (13) to make a quark loop, and going to the pion pole, $p^2\to 0$, one gets

$$f_\pi^2 = \frac{4N_c}{(4\pi)^2}\int dw\rho(w)\frac{1}{i}\int\frac{d^4p}{(2\pi)^4}\frac{w^2}{(p^2-w^2)^2} \quad (19)$$

The momentum integral is logarithmically divergent. Thus, if

$$\int dw\, w^2\rho(w) = 0 \qquad \text{then} \qquad f_\pi^2 = \frac{4N_c}{(4\pi)^2}\int dw\, w^2(-\log w^2)\rho(w) \quad (20)$$

As we see, $\rho(w)$ cannot be a positive definite function, regardless of its support. f_π^2 is positive if $\rho(w)$ is positive around $w=0$. Notice also that the first condition of Eq. (20) implies that the value of f_π does not depend on any particular scale used to make the logarithm dimensionless. If $\rho(w)$ had only one zero, the scale of f_π would be set by it.

A direct application of the previous conditions (16) and (20) can be found in the calculation of the pion electromagnetic form factor $F_\pi^{\rm em}(q^2)$, for which complete evaluation requires an explicitly knowledge of the spectral function $\rho(w)$. Besides a proper normalization, $F_\pi^{\rm em}(0) = 1$, the mean squared radius becomes

$$\langle r_\pi^2 \rangle = -\frac{1}{6}\frac{dF}{dq^2}\bigg|_{q^2=0} = \frac{N_c}{4\pi^2 f_\pi^2}\int dw \rho(w) \tag{21}$$

which is the value obtained using the unregularized quark loop if the normalization condition (16) is used. Thus, the pion has a finite size in this framework.

The amplitude for the (anomalous) neutral pion decay $\pi^0(p) \to \gamma(p_1,\mu) + \gamma(p_2,\nu)$ can also be computed from the triangle diagram yielding for on-shell massless pions $p_1^2 = p_2^2 = 0$, the result

$$M_{\mu\nu}(p_1,p_2) = \frac{N_c}{12\pi^2 f_\pi}\epsilon_{\mu\nu\alpha\beta}p_2^\alpha p_1^\beta \int dw \rho(w) \tag{22}$$

which, again, coincides with the standard one if Eq.(16) is used.

4 Pion Structure Function

An advantage of our method is that calculations can be directly done in Minkowski space and hence we may directly profit from LC coordinates. For the evaluation of the structure function of the pion the relevant vertex is constructed from two axial currents and is defined

$$\begin{aligned}&(2\pi)^4\delta(p'+q'-p-q)\ \chi_{AA}^{\mu,a;\nu,b}(p',q';p,q) \qquad (23)\\ &= i\int d^4x d^4x' d^4y' d^4y\ \langle 0|T\left\{J_A^{\mu,a}(x)J_A^{\nu,b}(x')q(y)\bar{q}(y')\right\}|0\rangle e^{i[q'\cdot x'+p'\cdot y'-q\cdot x-p\cdot y]}\end{aligned}$$

and fulfills the WTI

$$\begin{aligned}iq^\mu \chi_{AA}^{\mu,a;\nu,b}(p',q';p,q) &= \epsilon_{bac}S(p')\Gamma_V^{\nu,c}(p',\ p\ )S(p) \qquad (24)\\ &+\frac{\tau_a}{2}\gamma_5 S(p'-q)\Gamma_A^{\nu b}(p'-q,p)S(p) + S(p')\Gamma_A^{\nu b}(p',p+q)S(p+q)\frac{\tau_a}{2}\gamma_5\end{aligned}$$

Up to transverse pieces one gets the solution

$$\begin{aligned}\chi_{AA}^{\mu,a;\nu,b}(p',q';p,q) &= \int dw\rho(w)\frac{1}{\not{p}'-w}\Big\{\frac{wq'^\nu q^\mu}{q'^2q^2}\delta_{ab} \qquad (25)\\ &+\left(\gamma_\nu+\frac{2wq'^2}{q'^2}\right)\gamma_5\frac{\tau_b}{2}\frac{1}{\not{p}+\not{q}-w}\left(\gamma_\mu-\frac{2wq^\mu}{q^2}\right)\gamma_5\frac{\tau_a}{2}+\text{crossed}\Big\}\frac{1}{\not{p}'-w}\end{aligned}$$

At the pion poles, $q^2, q'^2 \to 0$, we get the $\pi q \to \pi q$ unamputated amplitude in the forward direction ($q = q'$)

$$\chi_{AA}^{\mu,a;\nu,b}(p,q;p,q) \to \frac{q^\nu q^\mu}{q^4} f_\pi^2 \chi^{ba}(p,q) \tag{26}$$

and direct use of Eq.(5) together with the conditions (16) and (20) yields the initial condition for the quark distributions in π^+,

$$u_\pi(x, Q_0^2) = \bar{d}_\pi(1-x, Q_0^2) = \theta(x)\theta(1-x) \tag{27}$$

This result has been derived in the chiral limit, is independent of the spectral function $\rho(w)$ and has proper support and normalization. By construction, Eq.(27) is consistent with chiral symmetry, although is not necessarily a consequence of it. The result has previously been obtained by several means within the NJL model either using Pauli-Villars regularization [13,17] on the virtual Compton amplitude (4) or imposing a transverse cut-off [16] upon the quark-target amplitude (5). Within the present approach it can be shown that the forward Compton scattering amplitude (4) in the Bjorken limit yields the finite result presented here [20] based on Eq.(5). To really appreciate this point, let us mention that Eq.(27) disagrees with various NJL calculations, which use different regularizations. If Eq.(4) is used with a four-dimensional cut-off [11] or Eq.(5) is used with Lepage-Brodsky regularization [16], different shapes for the quark distributions are obtained. The null-plane [12], NJL model [11] and spectator model [15] calculations also produce different results. In all cases, the use of momentum dependent form factors or non-gauge invariant regularizations make the connection between Eq.(4) and Eq.(5) doubtful and, furthermore, spoil normalization. The results based on a quark loop with momentum dependent quark masses [18,19] seem to produce a non-constant distribution.

5 QCD evolution

Eq.(27) can also be understood [23] in terms of phase space arguments and point couplings (i.e., constant matrix elements) which for N constituents gives

$$xV(x, Q_0^2) = N(N-1)x(1-x)^{N-2} \qquad \langle xV \rangle \equiv \int_0^1 dx V(x, Q_0^2) x = 1 \tag{28}$$

Of course, it is tempting to look at the nucleon case $N = 3$, although we obviously expect mass corrections to be more important both in the momentum dependence of the $Nqqq$ matrix element and in the phase space, and, in addition we do not have as much theoretical support as in the pion case. As usual,

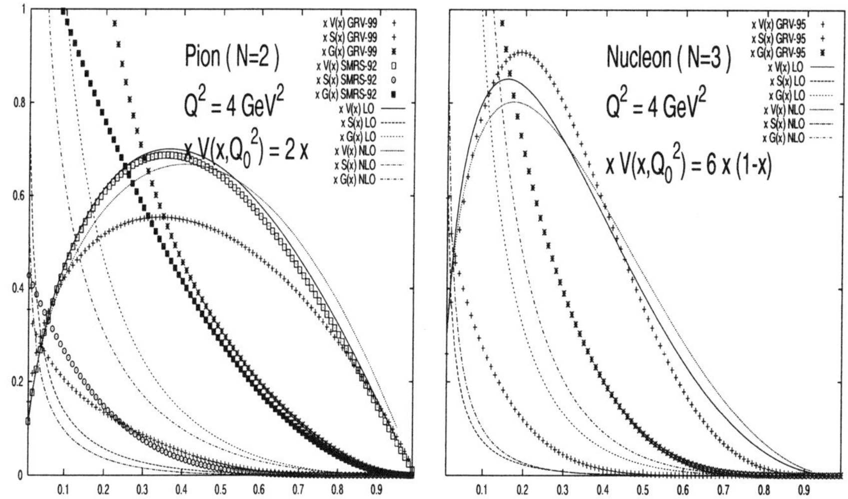

Figure 1: Valence, gluon and sea distributions in the pion (left) and the nucleon (right) at $Q^4 = 4\text{GeV}^2$ compared with phenomenological analysis for the pion SMRS92 [7], GRV99 [8] and for the nucleon GRV95 [3]. We take $\langle xV \rangle_\pi = 0.47$ and $\langle xV \rangle_N = 0.40$ at $Q^2 = 4\text{GeV}^2$. The sea and gluon distributions fulfill $S(x, Q_0^2) = G(x, Q_0^2) = 0$.

we evolve Eq. (28) at LO and NLO exactly implementing the group properties mentioned after Eq.(3) and employed to look for low energy unitarity limits [22]. Assuming vanishing initial sea and gluon distributions, the result for the dynamically generated valence, sea and gluon distributions for a hadron with $N = 2$ and $N = 3$ constituents compared with known parameterizations for the pion [7,8] and the nucleon [3] respectively can be seen in Fig 1. The agreement with the phenomenological valence distributions is good. Actually, since normalization of the valence distributions is preserved by evolution and Q_0^2 is determined by constraining the valence momentum fraction at $Q^2 = 4\text{GeV}^2$ to the phenomenological ones, $\langle xV \rangle_\pi = 0.47$ and $\langle xV \rangle_N = 0.40$ the bulk of the distribution is reproduced. Sea and gluon distributions do not obey such strong constraints and provide more information regarding the quality of a model. As we see, they are not as good. Nevertheless, a more definite statement might be made if uncertainties in the distributions were considered on the QCD evolution side [24]. The estimate of systematic errors in the model, i.e. corrections to the initial condition, remains yet a difficult and open problem.

Acknowledgments

I warmly thank the organizers for the invitation. I thank R. Delbourgo for discussions at CSSM where this work was started. Support from DGES (Spain) Project PB98-1367 and by the Junta de Andalucía is acknowledged.

References

1. R. G. Roberts, *The Structure of the Proton*, Cambridge Monographs on Mathematical Physics, 1990.
2. G. Altarelli and G. Parisi, *Nucl. Phys.* B **126**, 298 (1977).
3. M. Glück, E. Reya and A. Vogt, *Z. Phys.* C **67**, 433 (1995).
4. A. D. Martin, R.G. Roberts, W.J. Stirling and R.S. Thorne, *Eur. Phys. J.* C **4**, 463 (1998).
5. M. Glück, E. Reya and A. Vogt, *Eur. Phys. J.* C **5**, 461 (1998).
6. H. L. Lai *et al.*, *Eur. Phys. J.* C **12**, 375 (2000).
7. P.J. Sutton et al., *Phys. Rev.* D **45**, 2349 (1992).
8. M. Gluck, E. Reya, I. Schienbein, *Eur. Phys. J.* C **10**, 313 (1999).
9. R. L. Jaffe, *Relativistic Dynamics and Quark Nuclear Physics*, proceedings of the Los Alamos School, 1985, edited by M. B. Johnson and A. Picklesimer (Wiley, New York, 1986).
10. C. Best *et al. Phys. Rev.* D **56**, 2743 (1997).
11. T. Shigetani, K. Suzuki and H. Toki, *Phys. Lett.* B **308**, 383 (1993).
12. T. Frederico and G. A. Miller, *Phys. Rev.* D **50**, 210 (1994).
13. R. M. Davidson and E. Ruiz Arriola, *Phys. Lett.* B **359**, 273 (1995).
14. C.M. Shakin, Wei-Dong Sun, *Phys. Rev.* C **51**, 2171 (1995).
15. R. Jakob, P.J. Mulders, J. Rodrigues, *Nucl. Phys.* A **626**, 937 (1997).
16. W. Bentz et al., *Nucl. Phys.* A **651**, 143 (1999).
17. H. Weigel, E. Ruiz Arriola and L. Gamberg, *Nucl. Phys.* B **560**, 383 (1999).
18. A.E. Dorokhov and L. Tomio, *Phys. Rev.* D **62**, 014016 (2000).
19. M.B. Hecht, C.D. Roberts and S.M. Schmidt, *Phys. Rev.* C **63**, 025213 (2001).
20. E. Ruiz Arriola, to be submitted.
21. P. West and R. Delbourgo, *J. Phys.* A **10**, 1049 (1977); see also R. Delbourgo, *Nuovo Cimento* A **49**, 484 (1979) and *Aust. J. Phys.* **52**, 681 (1999). and references therein.
22. E. Ruiz Arriola, *Nucl. Phys.* A **641**, 461 (1998).
23. E. Ruiz Arriola, Talk given at Mini-workshop on Hadrons as Solitons, Bled, Slovenia, 1999. *hep-ph/9910382*
24. K. Nagano, in DIS2001, http://dis2001.bo.infn.it/wg/sfwg.html.

GENERALIZED PARTON DISTRIBUTIONS AND DISTRIBUTION OF PARTONS IN THE TRANSVERSE PLANE

M. BURKARDT

Department of Physics, New Mexico State University, Las Cruces, NM 88011, USA
E-mail: burkardt@nmsu.edu

I discuss the physical interpretation of GPDs in the limit $\xi = 0$, where the t dependence contains information about the distribution of partons in the transverse plane. GPDs thus allow a simultaneous determination of the longitudinal momentum and transverse position of partons in the infinite momentum frame, which also sheds some light on the physical interpretation of Ji's angular momentum sum rule.

1 Motivation

Recently, there has been a lot of interest in Generalized parton distributions (GPDs) because on the one hand GPDs can be probed in real and deeply virtual Compton scattering (DVCS) and on the other hand they can be related to the total angular momentum (including the orbital part!) carried by the quarks [1]

$$\boxed{\text{DVCS}} \quad \Leftrightarrow \quad \boxed{\text{GPDs}} \quad \Leftrightarrow \quad \vec{J}_q. \tag{1}$$

However, even though GPDs clearly are interesting observables, it is not *a priory* clear whether they have a simple physical interpretation [2]. In these notes, we will investigate the physics of GPDs in the limit where the momentum transfer on the target is purely transverse and we will show that the t-dependence of GPDs in this limit can be identified with the Fourier transform of parton distributions in the transverse plane w.r.t. the impact parameter.

2 Generalized parton distributions

GPDs are defined very similar to regular parton distributions

$$\begin{aligned} F_q(x,\xi,t) &= \int \frac{dx^-}{4\pi} e^{ip^+x^-x} \left\langle P' \left| \bar{q}\left(-\frac{x^-}{2}\right) \gamma^+ q\left(\frac{x^-}{2}\right) \right| P \right\rangle \\ &= H_q(x,\xi,t)\frac{1}{2p^+}\bar{u}(P')\gamma^+ u(P) + E_q(x,\xi,t)\frac{1}{2p^+}\bar{u}(P')\frac{i\sigma^{+\nu}\Delta_\nu}{2M}u(P) \end{aligned} \tag{2}$$

where light front (LF) coordinates are defined as $x^\pm = \frac{1}{\sqrt{2}}\left(x^0 \pm x^3\right)$, $\mathbf{x}_\perp = (x^1, x^2)$, and $\Delta = P' - P$.

Unlike the familiar (forward) parton distribution functions (PDFs)

$$q(x) = \int \frac{dx^-}{4\pi} e^{ip^+x^-x} \left\langle P \left| \bar{q}\left(-\frac{x^-}{2}\right) \gamma^+ q\left(\frac{x^-}{2}\right) \right| P \right\rangle, \tag{3}$$

which have the physical meaning of a momentum density in the infinite momentum frame (IMF), GPDs do not have the meaning of a density but that of a probability amplitude: GPDs describe the amplitude for finding a quark with momentum fraction $x-\xi$ (in the IMF) in the nucleon and replacing it in the nucleon with a momentum transfer Δ^μ.

Several limiting cases are familiar: for $P' = P$ one recovers the usual PDFs

$$H(x,0,0) = q(x) \qquad \tilde{H}(x,0,0) = \Delta q(x), \tag{4}$$

while integration over x yields the familiar form factors, e.g.

$$\int dx H(x,\xi,t) = F_1(t) \qquad \int dx E(x,\xi,t) = F_2(t). \tag{5}$$

Therefore another simple physical interpretation is that **GPDs 'measure' the contribution of quarks with longitudinal momentum fraction x to the corresponding form factor**, i.e. a GPD is the form factor for the operator that 'filters out' quarks with longitudinal momentum fraction x in the proton (in conventional PDFs, the quark is inserted back into the nucleon without momentum transfer).

Given the latter analogy, and given the fact that one can associate the Fourier transform of form factors with charge distributions in position space, it is very tempting to expect that GPDs also contain some information about the distribution of partons in position space.

3 The physical interpretation of $H(x,0,t)$

In this section we will focus on the helicity non-flip case (unpolarized) [a]. Furthermore, we consider a situation when the momentum transfer between the initial and final state in Eq. (2) is purely transverse. In this limit, $F_q(x,0,t) = H(x,0,t)$ can be written as overlap integrals between LF wave functions (Fock space amplitudes)[5] $\Psi_N(x,\mathbf{k}_\perp)$

$$H(x,-\mathbf{\Delta}_\perp^2) = \sum_N \sum_j \int [dx]_N \int \left[d^2\mathbf{k}_\perp\right]_N \delta(x-x_j) \Psi_N^*(x_i,\mathbf{k}'_{\perp,i}) \Psi_N(x_i,\mathbf{k}_{\perp,i}) \tag{6}$$

[a] Note that our results can be easily generalized to the polarized counterpart $\tilde{H}(x,0,t)$.

where $\mathbf{k}'_{\perp,i} = \mathbf{k}_{\perp,i} - x_i \mathbf{\Delta}_\perp$ for $i \neq j$ and $\mathbf{k}'_{\perp,j} = \mathbf{k}_{\perp,j} + (1-x_j)\mathbf{\Delta}_\perp$. Eq. (6) is very similar to the expression for the form factor in the Drell-Yan frame, except that the x of the 'active' quark is not integrated over, and it is exact if one knows the Ψ_N for *all* Fock components. The highly compact notation in Eq. (6) is illustrated by considering as an example the contributions from the two and three particle Fock components to $H(x, \mathbf{q}_\perp)$ for the pion

$$\begin{aligned} H(x, -\mathbf{\Delta}_\perp^2) &= \int d^2\mathbf{k}_\perp \psi^*_{\mathbf{\Delta}_\perp}(x, \mathbf{k}_\perp + \mathbf{\Delta}_\perp)\psi_{\mathbf{0}_\perp}(x, \mathbf{k}_\perp) \\ &+ \int d^2\mathbf{k}_\perp d^2\mathbf{l}_\perp dy\, \psi^*_{\mathbf{\Delta}_\perp}(x, \mathbf{k}_\perp + \mathbf{\Delta}_\perp, y, \mathbf{k}_\perp)\psi_{\mathbf{0}_\perp}(x, \mathbf{k}_\perp, y, \mathbf{l}_\perp) + \dots, \end{aligned} \tag{7}$$

where '...' indicates contributions from 4 and more particle Fock components.

3.1 Nonrelativistic form factors and charge density in position space

Eq. (7) resembles expressions for the nonrelativistic (NR) form factor. For example for NR two & three body systems, the form factor reads respectively

$$F(\vec{q}) = \int d^3\vec{k}\psi^*_{\vec{q}}(\vec{k} + \vec{q})\psi_{\vec{0}}(\vec{k}) \tag{8}$$

$$F(\vec{q}) = \int d^3\vec{k}_1 d^3\vec{k}_2\, \psi^*_{\vec{q}}(\vec{k}_1 + \vec{q}, \vec{k}_2)\, \psi_{\vec{0}}(\vec{k}_1, \vec{k}_2). \tag{9}$$

Like GPDs, $F(\vec{q})$ is also off-diagonal in the momentum since the form factor is the non-forward ($\vec{P}' \neq \vec{P}$) matrix element of the vector current. In order to relate $F(\vec{q})$ to position space densities, one first uses that NR boosts are purely kinematic, i.e.

$$\vec{k}'_i = \vec{k}_i + x_i\vec{q}, \tag{10}$$

with $x_i = \frac{m_i}{M}$, to express the boosted wave functions in terms of wave functions in the rest frame

$$\begin{aligned} \psi_{\vec{q}}(\vec{k}) &= \psi_{\vec{0}}(\vec{k} - x\vec{q}) \\ \psi_{\vec{q}}(\vec{k}_1, \vec{k}_2) &= \psi_{\vec{0}}(\vec{k}_1 - x_1\vec{q}, \vec{k}_2 - x_2\vec{q}). \end{aligned} \tag{11}$$

One can thus rewrite $F(\vec{q})$ as an autocorrelation of the $\vec{P} = \vec{0}$ wave function

$$F(\vec{q}) = \int d^3\vec{k}\, \psi^*_{\vec{0}}(\vec{k} + (1-x)\vec{q})\, \psi_{\vec{0}}(\vec{k}) \tag{12}$$

$$F(\vec{q}) = \int d^3\vec{k}_1 d^3\vec{k}_2\, \psi^*_{\vec{0}}(\vec{k}_1 + (1-x_1)\vec{q}, \vec{k}_2 - x_2\vec{q})\, \psi_{\vec{0}}(\vec{k}_1, \vec{k}_2),$$

for two and three particle systems respectively. Finally, upon Fourier transforming to position space, one can turn the autocorrelation function in Eq. (12) into a density. More precisely, one can express $F(\vec{q})$ in terms of the charge density $\rho(\vec{r})$ at distance $\vec{r}$ from the center of mass $\vec{R}_{CM} \equiv \sum_i \frac{m_i}{M}\vec{r}_i$.

$$F(\vec{q}) = \int d^3 r e^{-i\vec{q}\cdot\vec{r}} \rho(\vec{r}). \tag{13}$$

3.2 Relevance for GPDs

Purely transverse boosts in the LF frame also form a Galilean subgroup

$$x_i \longrightarrow x_i \qquad \mathbf{k}_{\perp,i} \longrightarrow \mathbf{k}_{\perp,i} + x_i \mathbf{q}_\perp, \tag{14}$$

where the momentum fractions x_i play the role of the mass fractions $\frac{m_i}{M}$ in the NR case. Therefore, LF Fock space amplitudes transform under purely $\perp$ boosts very similar to the way NR wave functions transform. For example, for the two and three body Fock components one finds

$$\begin{aligned} \psi_{\mathbf{q}_\perp}(x, \mathbf{k}_\perp) &= \psi_{\mathbf{0}_\perp}(x, \mathbf{k}_\perp - x\mathbf{q}_\perp) \\ \psi_{\mathbf{q}_\perp}(x, \mathbf{k}_\perp, y, \mathbf{l}_\perp) &= \psi_{\mathbf{0}_\perp}(x, \mathbf{k}_\perp - x\mathbf{q}_\perp, y, \mathbf{l}_\perp - y\mathbf{q}_\perp). \end{aligned} \tag{15}$$

Because of the very similar boost properties and the very similar expressions for GPDs one the one hand (7) and NR form factors on the other hand (8), one can proceed with GPDs in analogy with NR form factors to express them in terms of a density. First one expresses $H(x, \mathbf{q}_\perp)$ in terms of autocorrelations of Fock space amplitudes in the $\mathbf{p}_\perp = \mathbf{0}_\perp$ frame

$$\begin{aligned} H(x, -\mathbf{\Delta}_\perp^2) &= \int d^2\mathbf{k}_\perp \psi^*_{\mathbf{0}_\perp}(x, \mathbf{k}_\perp + (1-x)\mathbf{\Delta}_\perp)\psi_{\mathbf{0}_\perp}(x, \mathbf{k}_\perp) \\ &+ \int d^2\mathbf{k}_\perp d^2\mathbf{l}_\perp dy \psi^*_{\mathbf{0}_\perp}(x, \mathbf{k}_\perp + (1-x)\mathbf{\Delta}_\perp, y, \mathbf{k}_\perp - x\mathbf{\Delta}_\perp)\psi_{\mathbf{0}_\perp}(x, \mathbf{k}_\perp, y, \mathbf{l}_\perp) + ..., \end{aligned} \tag{16}$$

and then one Fourier transforms the $\perp$ coordinates to position space. In complete analogy with NR form factors this allows one to express $H(x,t)$ as the Fourier transform of parton distributions in the $\perp$ plane

$$H(x, -\mathbf{\Delta}_\perp^2) = \int d^2\mathbf{b}_\perp e^{-i\mathbf{\Delta}_\perp \cdot \mathbf{b}_\perp} f(x, \mathbf{b}_\perp), \tag{17}$$

where $f(x, \mathbf{b}_\perp)$ **is the probability density to find a quark with momentum fraction** x **at** $\perp$ **distance** $\mathbf{b}_\perp$ **from the** $\perp$ **center of momentum** $\mathbf{R}_\perp^{CM} \equiv \sum_i x_i \mathbf{r}_{i,\perp}$.

Note that $f(x, \mathbf{b}_\perp)$ is gauge invariant: to prove this, one uses the formal operator definition of $f(x, \mathbf{b}_\perp)$, which in LF gauge is given by [3]

$$f(x, \mathbf{b}_\perp) = \int \frac{dx^-}{4\pi} e^{ip^+x^-x} \left\langle \psi_{loc} \left| \bar{q}\left(-\frac{x^-}{2}, \mathbf{b}_\perp\right) \gamma^+ q \left(\frac{x^-}{2}, \mathbf{b}_\perp\right) \right| \psi_{loc} \right\rangle \quad (18)$$

Here $|\psi_{loc}\rangle \equiv \int d^2p_\perp \psi(\mathbf{p}_\perp) |p\rangle$ is a wave packet of plane wave proton states which is very localized in the transverse direction, but still has a sharp longitudinal momentum. This can be accomplished by taking $\psi(\mathbf{p}_\perp) = const.$, which guarantees that the $\perp$ center of momentum (the variable conjugate to the total $\perp$ momentum!) is at the origin. To make Eq. (18) manifestly gauge invariant one inserts a straight line gauge string from $(-\frac{x^-}{2}, \mathbf{b}_\perp)$ to$(\frac{x^-}{2}, \mathbf{b}_\perp)$.

This example also illustrates why the $\perp$ center of momentum is of significance here: first of all, the intrinsic dynamics of the hadron is independent of the overall $\perp$ momentum in the IMF ($\perp$ boosts are purely kinematical), which allows one to localize the wave packet in the $\perp$ direction through a wave packet. Since the variable conjugate to the overall $\perp$ momentum is $\mathbf{R}_\perp^{CM}$, localizing the wave packet is equivalent to working in the frame where $\mathbf{R}_\perp^{CM} = 0$. [b]

4 Discussion

By means of a Fourier transform, $H(x, -\mathbf{\Delta}_\perp^2)$ tells us how partons are distributed in the transverse plane as a function of the distance from the $\perp$ center of momentum. First of all this provides us, at least for $\xi = 0$, with a very simple interpretation for GPDs in terms of a density. A similar interpretation exists for $\tilde{H}(x, -\mathbf{\Delta}_\perp^2)$. in terms of polarized impact parameter dependent quark distributions in a longitudinally polarized target. These fundamental results by themselves are already important observations because they illustrate the kind of physics that one can learn about hadrons by studying GPDs. However, above and beyond this physics insight, our results also have a number of practical application because they allow one to use geometric insights about hadron structure to model GPDs as illustrated by the following examples:

1. First of all, the mere fact that the $\mathbf{b}_\perp$ distribution is measured w.r.t. $\mathbf{R}_\perp^{CM}$ necessarily implies that the width of the $\mathbf{b}_\perp$ distribution goes to zero as $x \to 1$, since the active quark becomes the $\perp$ center of mass in that limit. In momentum space this implies that $H(x, t)$ should become t-independent as $x \to 1$. Many models for GPDs are consistent with this kind of behavior, but our analysis indicates that this is a general result.

[b] Of course the same reasoning also applies to the familiar case of the NR form factor.

2. Eq. (18) illustrates that the impact parameter dependent parton distributions can be expressed in the form

$$f(x,\mathbf{b}_\perp) \sim \left\langle \psi_{loc} \left| b^\dagger(xp^+,\mathbf{b}_\perp) b(xp^+,\mathbf{b}_\perp) \right| \psi_{loc} \right\rangle = \left| b(xp^+,\mathbf{b}_\perp) |\psi_{loc}\rangle \right|^2, \tag{19}$$

where $b(xp^+,\mathbf{b}_\perp)$ creates quarks of long. momentum xp^+ at $\perp$ position $\mathbf{b}_\perp$. This implies that $f(x,\mathbf{b}_\perp)$ has a probabilistic interpretation, and therefore it has to satisfy positivity constraints, i.e.

$$f_q(x,\mathbf{b}_\perp) \equiv \int d^2\mathbf{\Delta}_\perp H_q(x,0,-\mathbf{\Delta}_\perp^2) e^{i\mathbf{\Delta}_\perp\cdot\mathbf{b}_\perp} \tag{20}$$

should be positive for all x and all $\mathbf{b}_\perp$ (for antiquarks the same holds modulo an obvious sign). This positivity condition provides us with a new constraint on all models for GPDs.

3. Since very little is known about the actual t-dependence of GPDs, it is often convenient to use simple parameterizations. A commonly used form (motivated by LF constituent models) is $H(x,0,-\mathbf{\Delta}_\perp^2) = q(x)e^{-a\mathbf{\Delta}_\perp^2\frac{1-x}{x}}$. However, such a $\mathbf{\Delta}_\perp$-dependence yields an unacceptably rapid growth for the $\perp$ hadron size $\langle \mathbf{b}_\perp^2 \rangle \sim \frac{1}{x}$ in the small x region. Space time descriptions of parton structure in the small x region[6] suggest a $\perp$ size of hadrons that grows like $\alpha \ln \frac{1}{x}$. Translated into GPDs this implies that the small x behavior should be parameterized by a functional form like

$$H(x,0,-\mathbf{\Delta}_\perp^2) = q(x)e^{-\alpha\mathbf{\Delta}_\perp^2 \ln\frac{1}{x}} \quad \text{or} \quad q(x)e^{-\alpha\mathbf{\Delta}_\perp^2(1-x)\ln\frac{1}{x}} \tag{21}$$

(the 2nd ansatz is also consistent with the Drell-Yan-West relation).

4. At a more qualitative level, one expects quarks at large x to come from the more localized valence 'core' of the hadron, while the small x region should also receive contributions from the much wider meson 'cloud' and therefore one would in general expect a gradual increase of the t-dependence of $H(x,0,t)$ as one goes from larger to smaller values of x

5 The Physics of $E(x,0,t)$

So far we have only considered GPDs in an unpolarized (or longitudinally) polarized nucleon. For polarized nucleon states (we use an IMF helicity basis here [7]) one finds the following useful relations when $\Delta^+ = 0$

$$\int \frac{dx^-}{4\pi} e^{ip^+x^-x} \left\langle P+\Delta,\uparrow \left| \bar{q}(0)\gamma^+ q(x^-) \right| P,\uparrow \right\rangle = H(x,0,-\mathbf{\Delta}_\perp^2) \tag{22}$$

$$\int \frac{dx^-}{4\pi} e^{ip^+x^-x} \left\langle P+\Delta,\uparrow \left| \bar{q}(0)\gamma^+ q(x^-) \right| P,\downarrow \right\rangle = -\frac{\Delta_x - i\Delta_y}{2M} E(x,0,-\mathbf{\Delta}_\perp^2),$$

i.e. if we take for example a nucleon polarized in the x direction (in the IMF)

$$F_q(x,0,-\mathbf{\Delta}_\perp) = H(x,0,-\mathbf{\Delta}_\perp^2) + i\frac{\Delta_y}{M}E(x,0,-\mathbf{\Delta}_\perp^2). \tag{23}$$

Eq. (23) allows us to draw the following conclusions for the physics of $E(x,0,t)$

1. We know already that $H(x,0,-\mathbf{\Delta}_\perp^2)$ describes the distribution of unpolarized partons in the $\perp$ plane for a nucleon that is either unpolarized or polarized in the z direction. From Eq. (23) we can immediately read off that $\frac{\Delta_\perp}{M}E(x,0,-\mathbf{\Delta}_\perp^2)$ **describes how the distribution of partons in the $\perp$ plane depends on the polarization of the nucleon**. For example, for a nucleon polarized in the x direction the z-momentum distribution of partons on the +y side will differ from the one on the $-y$ side — which makes sense if one considers the classical picture of a sphere that spins around the x-direction while moving in the z-direction. The z-momenta due to spinning add to the translatory momenta for 'partons' on the $+y$ side and subtract on the $-y$ side and hence it is not surprising to find a difference between the parton distributions on the $\pm y$ sides.

2. Since the probabilistic interpretation should not depend on the polarization, the interpretation given in (1) provides us with another positivity constraint relating the Fourier transform of E to the one of H:

$$\left|\frac{\nabla_{b_\perp}}{M}\int d^2\mathbf{b}_\perp e^{i\mathbf{b}_\perp\cdot\mathbf{\Delta}_\perp}E(x,0,-\mathbf{\Delta}_\perp^2)\right| < \int d^2\mathbf{b}_\perp e^{i\mathbf{b}_\perp\cdot\mathbf{\Delta}_\perp}H(x,0,-\mathbf{\Delta}_\perp^2). \tag{24}$$

3. Finally, we are also able to illustrate the physics of Ji's angular momentum sum rule $\langle J_q\rangle = \frac{1}{2}\int dx x\left[H_q(x,0,0)+E_q(x,0,0)\right]$. For this purpose, we first note that GPDs for $\xi = 0$ allow the simultaneous determination of the momentum of partons in the z direction and their position in the $\perp$ direction (which is very much reminiscent of angular momentum $L_x = xp_z - zp_x$) and it should thus not surprise that GPDs allow a determination of the total angular momentum of the quarks. In order to illustrate this connection, let us consider a nucleon polarized in the x direction.[c] Because we are interested in the angular momentum in the rest frame of the target, we again need to consider again nucleons polarized in the x-direction, but now in the nucleon rest frame. This differs

[c] I was only able to find a simple physical interpretation for the connection between GPDs and the $\perp$ components of $\vec{J}_q$. However, by rotational invariance, one can apply the result also to J_q^z, and, by boost invariance, J_q^z should also be he same for a nucleon moving in the z-direction.

from the IMF by a Melosh rotation and one finds, up to terms linear in $\Delta_\perp$

$$F_q(x,0,\mathbf{\Delta}_\perp) = H(x,0,-\mathbf{\Delta}_\perp^2) + i\frac{\Delta_y}{2M}\left[H(x,0,-\mathbf{\Delta}_\perp^2) + E(x,0,-\mathbf{\Delta}_\perp^2)\right]. \quad (25)$$

For the x component of the angular momentum carried by the quarks one thus finds [d]

$$\begin{aligned}\langle J_q^x\rangle &= \langle yT^{0z}\rangle - \langle zT^{0y}\rangle = 2\langle yT^{0z}\rangle = 2\frac{M}{2}\int dx \int d^2\mathbf{b}_\perp\, b_y x f(x,\mathbf{b}_\perp) \quad (26)\\ &= Mi\,\frac{d}{d\Delta_y}\int dx\, x F_q(x,0,\mathbf{\Delta}_\perp)\Big|_{\Delta=0} = \frac{1}{2}\int dx\, x\left[H(x,0,0)+E(x,0,0)\right]\end{aligned}$$

which is the angular momentum relation derived in Ref. [1]. What is new is that through the geometric interpretation of GPDs, we are now able to understand the physics of this sum rule by relating GPDs to the expectation value of $\langle yp_z\rangle$ for a nucleon polarized in the x direction.

Acknowledgments

This work was supported by a grant from DOE (FG03-95ER40965) and through Jefferson Lab by contract DE-AC05-84ER40150 under which the Southeastern Universities Research Association (SURA) operates the Thomas Jefferson National Accelerator Facility.

1. X. Ji, *Phys. Rev. Lett.* **78**, 610 (1997).
2. X. Ji, in *Future Directions in Quark Nuclear Physics*, eds. A.W. Thomas and K. Tsushima (World Scientific, Singapore, 1999), p. 159.
3. M. Burkardt, *Phys. Rev.* D **62**, 071503 (2000).
4. M. Burkardt, hep-ph/0007036; hep-ph/0008051; hep-ph/0010082.
5. M. Diehl *et al.*, *Eur. Phys. Jour.* C**8**, 409 (1999); *Nucl. Phys.* B **596**, 33 (2001); B.C. Tiburzi and G.A. Miller, hep-ph/0104198.
6. V.N. Gribov, hep-ph/0006158.
7. S.J. Brodsky et al., *Nucl. Phys.* B **593**, 311 (2001).

[d] In order to calculate $\langle J_x\rangle$, it first seems that one needs both $\langle yT^{0z}\rangle$ and $\langle zT^{0y}\rangle$, but for stationary localized states the two expectation values are (up to a sign) the same and it is thus sufficient to calculate $\langle yT^{0z}\rangle$. Furthermore, in the rest frame one can replace the usual momentum density in $\langle yT^{0z}\rangle$ by the LF momentum density T^{++}, whose matrix elements can be expressed by the GPDs.

EXCLUSIVE PROCESSES AT HERMES

NICOLA BIANCHI
FOR THE HERMES COLLABORATION
INFN - Laboratori Nazionali di Frascati, Frascati I-00044, Italy
E-mail: bianchi@hermes.desy.de

Preliminary results for exclusive electroproduction of vector mesons, pseudoscalar mesons and real photons from HERMES are presented. Different observables like cross sections and single spin azimuthal asymmetries are studies. Exclusive processes are connected with the recent formalism of the Generalized Parton Distributions.

1 Introduction

Inclusive and semi-inclusive lepton-nucleon scattering experiments played a mayor role in our understanding of the internal structure of the nucleon. Recent progress in this field is related to exclusive reactions and the *Skewed Parton Distributions* (SPD)[1] which take into account the dynamical correlations between partons of different momenta. The SPD-framework can be used to evaluate a wide range of observables, such as the electromagnetic form factors, inclusive parton distributions and exclusive meson-production cross sections. In particular the SPD have attracted increased interest with the suggestion of the possibility to have access the total angular momentum of the partons.

Four leading-twist distribution functions enter in the SPD description and the quantum numbers of final states select different SPD:

- Vector mesons : $\rho^0_L \rightarrow$ unpolarized SPD
- Pseudoscalar mesons : $\pi^0 \rightarrow$ polarized SPD
- Pseudoscalar mesons : $\pi^+ \rightarrow$ polarized SPD + meson exchange
- DVCS : $\gamma \rightarrow$ polarized and unpolarized SPD

Therefore it is clear that only measurements of all the above listed reactions will provide a complete picture of the SPD framework.

HERMES was mainly designed for SIDIS to study ordinary polarized parton distributions[2] and only recently HERMES started to investigate its capability to study exclusive reactions. In this talk, recent and mainly preliminary HERMES results on exclusive processes are presented.

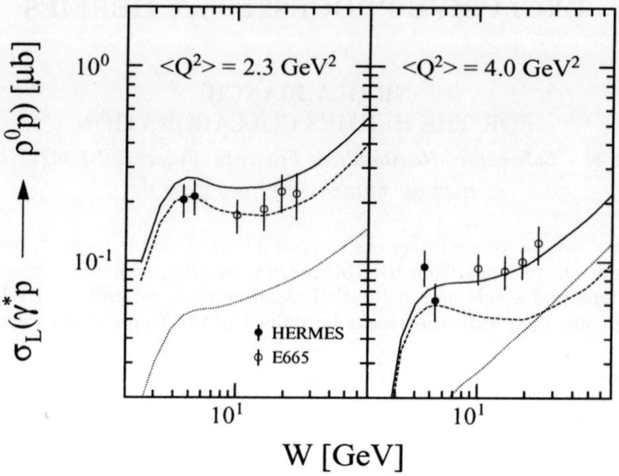

Figure 1. The longitudinal component of the virtual-photoproduction cross section. The solid lines represent the results of a calculation based on SPD. The dashed (dotted) curves represent the quark (two-gluon) exchange contributions within this calculation.

2 Vector mesons

Exclusive ρ meson electroproduction has been measured by HERMES by detecting two hadrons of opposite charges into the spectrometer[3]. Since the factorization proof for the process holds only for the longitudinal component, this was extracted from the decomposition of the decay-angle distribution which is related to the polarization density matrix. In fig. 1 is shown the longitudinal component of the cross section for the exclusive ρ electroproduction[4]. The HERMES result are in agreement with a SPD prediction[5] which includes the quark-exchange and the two-gluon exchange mechanism. In the HERMES energy range the former mechanism is predicted to dominate.

Preliminary results for the exclusive electroproduction of higher-mass vector mesons are presented in fig. 2. As it is seen the photoproduction extrapolation of HERMES data for both ω and Φ mesons is in agreement with the available data, while the electroproduction results allow to fill the missing region between low energy and the high energy data. In particular the result for the Φ-meson is in agreement with the energy dependence measured by NMC, H1 and ZEUS at higher energies.

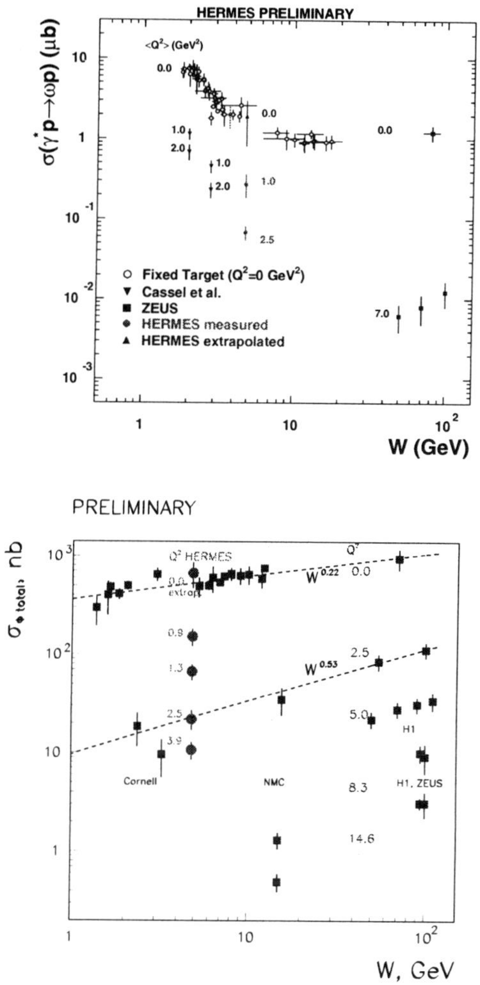

Figure 2. Real and virtual exclusive photoproduction cross section for the ω (top) and Φ (bottom). HERMES results are compared with world data.

3 Pseudoscalar mesons

The SIDIS electroproduction of charged and neutral pions has been extensively studied by HERMES. [6] In particular the isospin invariance of the fragmen-

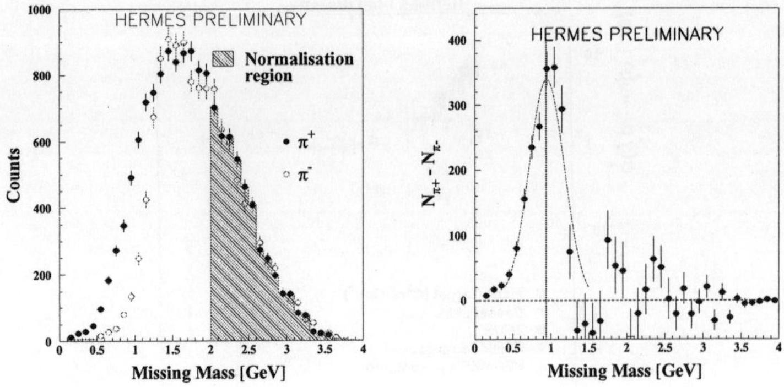

Figure 3. Measured π^+ and π^- yield as a function of missing mass (left). Right figure shows the difference of π^+ and normalized π^- yield versus missing mass resulting in a clear peak around the nucleon mass indicated by the line.

tation process has been shown to be valid up to $z \sim 0.75$. At higher values of z, which corresponds to small values of M_X in the process $e^+p \to e^+\pi X$, a clear isospin breaking is observed and the charged pion production is larger than the neutral one. At large z, or at low M_X, the exclusive production of pion is dominant over the fragmentation process.

The exclusive π^+ channel was identified by using the π^- SIDIS distribution as a reliable estimate of the background for the π^+ channel once is correctly normalized outside the exclusive region. In fig. 3 the background subtraction procedure and the remaining exclusive π^+ peak is shown. Within the current experimental setup a missing mass resolution of 0.2 GeV was obtained from the data for values of M_X around the nucleon mass in agreement with Monte Carlo simulation based on a GPD event generator. The evaluation of the exclusive π^+ cross section is under study.

A powerful tool in the study of the structure of the nucleon is provided by the measurement of the single spin azimuthal distributions. The kinematics of such a process is illustrated in fig. 4, where the scaling variables are defined. The relevant azimuthal angle Φ is the angle between the lepton scattering plane and the hadron production plane. By measuring the azimuthal distribution of pions when only the target is longitudinally polarized, moments $A_{UL}^{\sin\Phi}$ of the cross section weighted by $\sin\Phi$ were extracted:

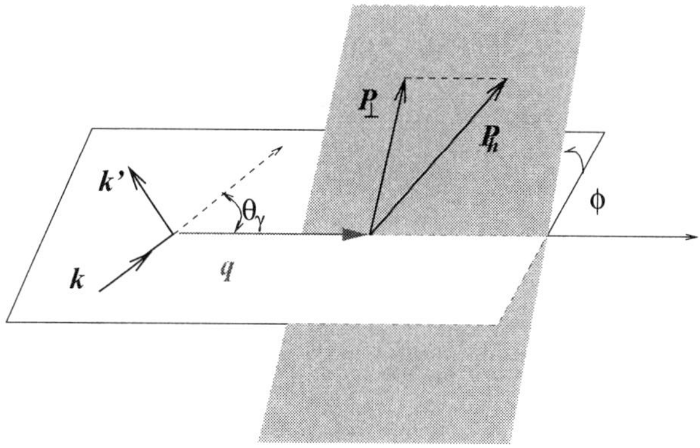

Figure 4. Kinematics for pion electroproduction. The 4-momenta of the incoming (k) and outgoing ($k\prime$) leptons and of the exchange photon (q) define the lepton scattering plane. The 4-momentum of the produced hadron define the azimuthal angle Φ with respect the virtual photon direction.

$$A_{UL}^{\sin\Phi} = \frac{\frac{L^+}{L_P^+}\int_0^{2\pi}\sin\Phi dN_+ - \frac{L^-}{L_P^-}\int_0^{2\pi}\sin\Phi dN_-}{1/2[\int_0^{2\pi}dN_+ - \int_0^{2\pi}dN_-]}$$

Here $\pm$ refer to the target spin state, $L^\pm$ are the dead-time corrected luminosities, and $L_P^\pm$ the luminosities weighted by the absolute value of the average target polarization. The $\sin\Phi$ moments of the single spin azimuthal asymmetries for π^+, π^- and π^0 were evaluated according to previous equation using the polarized target data. Results are presented in fig. 5 as a function of z in the kinematic range $0.2 < z < 1.0$. No corrections for smearing due to the detector resolution and for the spectrometer acceptance were applied. The first three data points correspond to the already published SIDIS azimuthal asymmetry[7,8]. In the exclusive region of high z and low missing mass (nucleon missing mass corresponds to $z \sim 0.98$), $A_{UL}^{\sin\Phi}$ shows a large variation for π^+ and π^0 while the value for π^- is consistent with zero within errors over the whole z-range. In case of π^+ the sign of $A_{UL}^{\sin\Phi}$ is opposite to that in the SIDIS region. The asymmetry of the exclusive pion is related to the interference between longitudinal and transverse photon amplitudes[1].

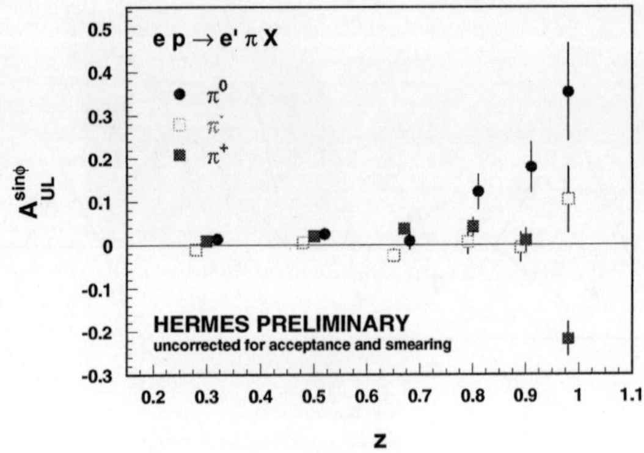

Figure 5. $A_{UL}^{\sin\Phi}$ for pion electroproduction as function of z. Only statistical errors are shown.

4 DVCS

Deeply Virtual Compton Scattering (DVCS) is the exclusive reaction with the simplest final state where a single photon is produced. As compared to exclusive meson production, DVCS has the advantage that the complication of a meson wave function convoluted with the final state of the target is avoided. However, DVCS has the identical final state as the Bethe-Heitler (BH) where the photon is radiated by the incoming or outgoing lepton. This make the DVCS difficult to isolate due to the large contribution of BH. Therefore at HERMES the Single Spin Azimuthal Asymmetry with respect to the helicity of the beam is used to isolate the interference term of the BH and DVCS processes, providing access to the DVCS contribution. Photons from DVCS and BH processes were identified by the detection of an energy deposition in the calorimeter and preshower detector without any associated track. The recoil proton was not detected and therefore the requirement for the exclusivity was ensured by a cut in the M_X missing mass distribution for the reaction $e^+p \to e^+\gamma X$. The M_X distribution of this reaction is shown in fig. 6 where a clear signal due to the DVCS and BH processes is seen at M_X around the nucleon mass.

Also shown in fig. 6 is the azimuthal dependence of the single spin (beam) asymmetry. The data in the exclusive region of missing mass from 0.4 to 1.4

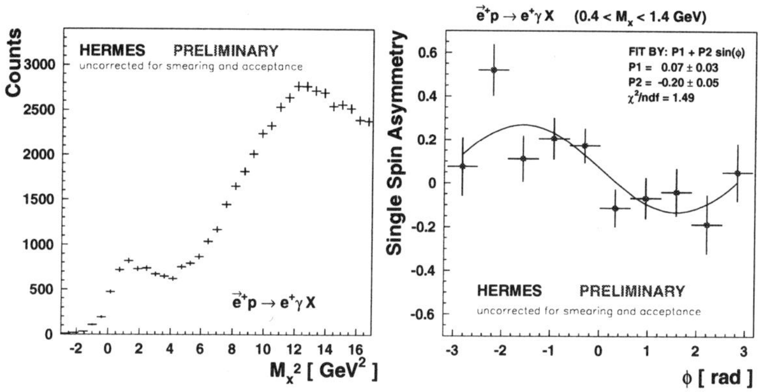

Figure 6. a) Missing mass distribution for the reaction $e^+p \to e^+\gamma X$. The enhancement at $M_X^2 \sim 1$ GeV^2 is the DVCS+BH contribution. b) Preliminary result for the single spin (beam) azimuthal asymmetry in the missing mass range $0.4 <$ GeV$M_X < 1.4$ GeV.

GeV, displayed in fig. 6, show a definite $\sin\phi$ behavior. Negligible single spin (beam) asymmetries were observed for higher values of M_X.

5 Conclusion and outlook

The first preliminary results on exclusive electroproduction at HERMES has been presented. The results are related to data collected in 96 and 97. Analysis of data collected in 98, 99 and 00 is under progress. In the year 02 and 03, HERMES will run with transversely polarized target. This will allow to precise measure the single spin (target) azimuthal asymmetry for exclusive π^+ which is expected to be large and to be related to the interference between the pion pole and non-pole terms. High luminosity runs with unpolarized target are also foreseen for cross section and single spin (beam) azimuthal asymmetry measurements. In addition a recoil detector project is under study which will greatly increase the capability on the measurement of DVCS and exclusive π^0 and η electroproduction.

References

1. For a recent review see K. Goeke *et al.*, hep-ph/0106012.
2. K. Ackerstaff *et al.*, *Phys. Lett.* B **464**, 123 (1999).
3. K. Ackerstaff *et al.*, *Nucl. Instrum. Methods* A **417**, 230 (1998).

4. A. Airapetian *et al.*, *Eur. Phys. Jour.* **C 17**, 389 (2000).
5. M. Vanderhaeghen *et al.*, *Phys. Rev.* D **60**, 094017 (1999).
6. A. Airapetian *et al.*, submitted to *Eur. Phys. Jour. C* and hep-ex/0104004.
7. A. Airapetian *et al.*, *Phys. Rev. Lett.* **84**, 4047 (2000).
8. A. Airapetian *et al.*, submitted to *Phys. Rev. D* and hep-ex/0104005.

SOFT PION PRODUCTION ASSOCIATED WITH DEEPLY VIRTUAL COMPTON SCATTERING

L. MOSSE

SPhN-DAPNIA, CEA/Saclay, F-91191 Gif-sur-Yvette, France
E-mail: lmosse@cea.fr

An evaluation of soft pion production associated with Deeply Virtual Compton Scattering is presented. A Low Energy Theorem is built to calculate the amplitudes, and its validity for soft pion production is discussed. Comparative results for typical experimental set-ups are shown.

Introduction

Exclusive Virtual Compton Scattering in the Deep inelastic regime (DVCS) is the simplest process which can be used to measure Generalized Parton Distributions (GPD)[1]. Because it is an exclusive reaction ($lp \to l'p'\gamma'$), a DVCS experiment is difficult to carry out, and special care must be taken of the competing reactions. Among the latter, the so-called Associated DVCS (AVCS), where the final state includes an additional low momentum pion, is very hard to disentangle from pure DVCS. It is then of major importance to compare AVCS and DVCS cross sections. In both DVCS and AVCS, the photon can be emitted either by the proton or by the leptonic line. The two contributions in the amplitude are referred to as pure VCS (Fig.1a, 1c) and Bethe-Heitler (Fig.1b:BH, 1d:ABH).

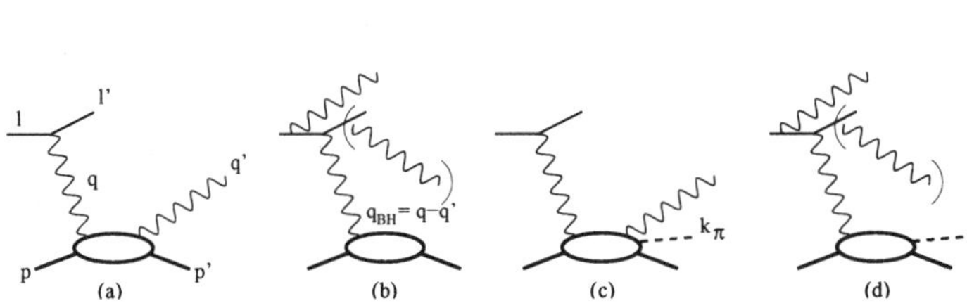

Figure 1. DVCS and AVCS contributions

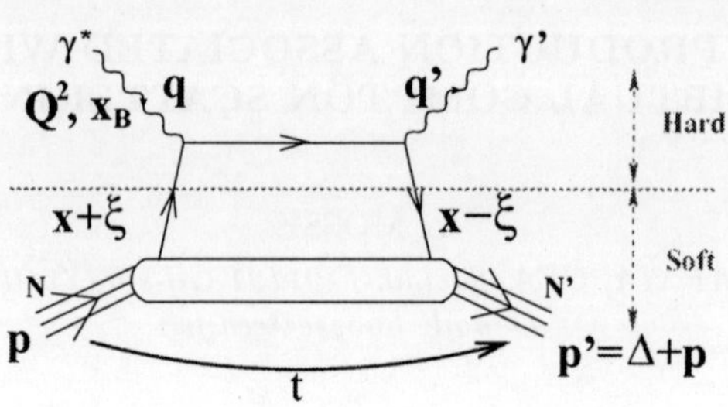

Figure 2. Handbag diagram for DVCS

1 Brief Review of DVCS

The BH amplitude is calculable in QED:

$$T_{BH} = -\frac{e^3}{t}\varepsilon_\nu'^\star L^{\nu\mu}\bar{u}(p')\left(F_1(t)\gamma_\mu + F_2(t)i\sigma_{\mu\nu}\frac{\Delta^\nu}{2M}\right)u(p) \tag{1}$$

where F_1 and F_2 are the Pauli and Dirac electromagnetic (em.) form factors, $\epsilon'(q')$ the photon polarization vector and $L^{\nu\mu}$ the leptonic tensor:

$$L_{\nu\mu} = \overline{u}(l')\left(\gamma_\nu\frac{\gamma\cdot(l'+q')+m_l}{(l'+q')^2-m_l^2}\gamma_\mu + \gamma_\mu\frac{\gamma\cdot(l-q')+m_l}{(l-q')^2-m_l^2}\gamma_\nu\right)u(l) \tag{2}$$

The DVCS amplitude, with lepton charge q_l, has a similar expression:

$$T_{DVCS} = q_l\frac{e^3}{Q^2}\varepsilon_\nu'^\star H^{\mu\nu}\bar{u}(l')\gamma_\mu u(l) \tag{3}$$

For a finite hadronic transfer $t = \Delta^2 = (p'-p)^2$, the DVCS hadronic tensor $H^{\mu\nu}$ factorizes [2,3] into a hard pQCD calculable part and a soft non perturbative part, leading to the so-called "handbag" diagram, see Fig.2. Four GPDs $H, \widetilde{H}, E, \widetilde{E}$, depending on the averaged momentum fraction x of both outgoing and incoming quarks, on the longitudinal fraction ξ of the momentum transfer and on the invariant transfer t, parametrise the soft part at leading twist. Introducing the operators $\Gamma^\mu_\pm$ depending on ξ and t:

$$\Gamma^\mu_\pm(\xi,t) = \int_{-1}^{1}dx\left(\frac{1}{x+\xi-i\epsilon}\pm\frac{1}{x-\xi+i\epsilon}\right)\left(H(x,\xi,t)\gamma^\mu + E(x,\xi,t)i\sigma^{\mu\nu}\frac{\Delta_\nu}{2M}\right)$$

$$\tilde{\Gamma}^\mu_\pm(\xi,t) = \int_{-1}^{1}dx\left(\frac{1}{x+\xi-i\epsilon}\pm\frac{1}{x-\xi+i\epsilon}\right)\left(\widetilde{H}(x,\xi,t)\gamma^5\gamma^\mu + \widetilde{E}(x,\xi,t)\gamma^5\frac{\Delta^\mu}{2M}\right)$$

and the Lorentz tensors:

$$S^{\mu\nu} = \frac{1}{2}\left(\widetilde{p}^{\mu} n^{\nu} + n^{\mu}\widetilde{p}^{\nu} - g^{\mu\nu}\right) \ , \ A^{\mu\nu} = \frac{1}{2} i\epsilon^{\mu\nu\rho\sigma}\widetilde{p}_{\rho} n_{\sigma} \tag{4}$$

where $\epsilon^{0123}=1$ and the Sudakov vectors n and $\widetilde{p}$ defined as in Ref.[4], the DVCS tensor follows as:

$$H^{\mu\nu} = i\int d^4 y e^{-iqy} \langle p' | T\{J^{\mu}_{em}(y) J^{\nu}_{em}(0)\} | p\rangle \tag{5}$$

$$= \bar{u}(p') \underbrace{\left(S^{\mu\nu} n\cdot\Gamma_{+}(\xi,t) + A^{\mu\nu} n\cdot\widetilde{\Gamma}_{-}(\xi,t)\right)}_{h^{\mu\nu}} u(p) \tag{6}$$

2 Associated DVCS

The ABH amplitude (Fig.1d) is:

$$T^{\alpha}_{ABH} = -\frac{e^3}{t_\gamma} \varepsilon^{,\nu\star} L_{\nu\mu} \langle p'\pi^{\alpha} | J^{\mu}_{em} | p\rangle \tag{7}$$

and the AVCS amplitude (Fig.1c) is

$$T^{\alpha}_{AVCS} = q_l \frac{e^3}{Q^2} \bar{u}(l')\gamma_\mu u(l)\, \varepsilon^{,\star}_{\nu}(q')\, i\int d^4 x\, e^{-iqx} \langle p'\,\pi^{\alpha} | T\{J^{\mu}_{em}(x) J^{\nu}_{em}(0)\} | p\rangle \tag{8}$$

The ABH hadronic matrix element is the same as the one for pion electroproduction. It can be calculated using a Low Energy Theorem (LET) for soft pion production[5]:

$$\begin{aligned}\langle p'\pi^{\alpha} | O | p\rangle &= \frac{i}{f_\pi} \langle p' | [Q^{\alpha}_5, O] | p\rangle \\ &+ \frac{ig_A}{2f_\pi}\left(\frac{1}{2p'\cdot k_\pi} \bar{u}(p')\gamma\cdot k_\pi \gamma_5 \tau^{\alpha} u(p') \langle p' | O | p\rangle \right. \\ &\left. - \langle p' | O | p\rangle \frac{1}{2p\cdot k_\pi} \bar{u}(p)\gamma\cdot k_\pi\gamma_5\tau^{\alpha} u(p)\right)\end{aligned} \tag{9}$$

where $f_\pi = 92.4 MeV$ is the pion decay constant, g_A the axial charge of the nucleon, Q^{α}_5 the axial charge operator and O a given operator. This expression is obtained in the *soft pion limit*: not only the pion is at rest, but its mass is also taken to be an expansion parameter, and only the first order terms are kept. m_π is must be smaller than any other scale in the process. Note that in Eq.(9) we have neglected a contribution where the pion is created from the vacuum by the operator O. When O is the em. current, this term is zero. In the VCS case it is suppressed if $t >> m^2_\pi$.

2.1 Soft pion electroproduction, ABH matrix element.

Although the LET has already been developed several times for soft pion electroproduction it is meaningful to do it once again here: hypotheses will be emphasized and the procedure will serve as a mock-up for the following AVCS development. The two terms on the right hand side of Eq.(9) are called the commutator term, $T_{com}^{\alpha,\mu}$, and the nucleon term, $T_{nucl}^{\alpha,\mu}$, respectively. The condition for writing a LET in the ABH case is

$$m_\pi \ll M,\ \sqrt{|t|},\ \sqrt{|q_{BH}^2|} \tag{10}$$

with $q_{BH} = q - q'$. Note that $q_{BH}^2 = t + O(m_\pi)$.

The commutator term: *Current Algebra* in isospin space allows one to write the commutator:

$$[Q_5^\alpha, J_{em}^\mu] = i\epsilon\epsilon^{\alpha 3\beta} A_\beta^\mu \tag{11}$$

where A_β^μ is the axial current. Using Eq.(11) in the commutator term of Eq.(9), assuming *pion pole dominance for the g_P form factor* and subtracting a term of order m_π/t in the latter, one gets:

$$T_{com}^{\alpha,\mu} = \bar{u}(p')\frac{[\tau^\alpha,\tau^3]}{2}\frac{ie}{2f_\pi}g_A(\Delta^2)\left(\gamma^\mu + \frac{(2k_\pi - q_{BH})^\mu}{\Delta^2 - m_\pi^2}\gamma\cdot\Delta\right)\gamma_5 u(p) \tag{12}$$

The nucleon term: Adding and subtracting terms of higher order in m_π in the propagators, the nucleon term can be written as

$$\begin{aligned} T_{nucl}^{\alpha,\mu} = \; & \frac{ieg_A}{2f_\pi}\bar{u}(p') \\ & \left[\gamma\cdot k_\pi\gamma_5\frac{\gamma\cdot(p'+k_\pi)+M}{(p'+k_\pi)^2 - M^2}\tau^\alpha\left(J_S^\mu + \tau^3 J_V^\mu\right)\right. \\ & \left. + \left(J_S^\mu + \tau^3 J_V^\mu\right)\tau^\alpha\frac{\gamma\cdot(p-k_\pi)+M}{(p-k_\pi)^2 - M^2}\gamma\cdot k_\pi\gamma_5\right] u(p) \end{aligned} \tag{13}$$

where $J_{S,V}^\mu = F_1^{S,V}(q_{BH}^2)\gamma^\mu + F_2^{S,V}(q_{BH}^2)i\sigma^{\mu\nu}q_{BH\nu}/2M$
and $F^{S,V} = (F^p \pm F^n)/2$.
The amplitude is not exactly gauge invariant because

$$q_{BH}\cdot T_{ABH}^\alpha = \frac{ie}{f_\pi}\bar{u}(p')\gamma\cdot k_\pi\gamma_5\left(g_A F_1^V(q_{BH}^2) - \frac{g_A(\Delta^2)}{2}\right)\frac{[\tau^\alpha,\tau^3]}{2}u(p)\ , \tag{14}$$

except if the same shape is assumed for F_1^V and g_A or if $q_{BH}^2 = 0$.
The amplitude (12) can be extended down to small t because the correction

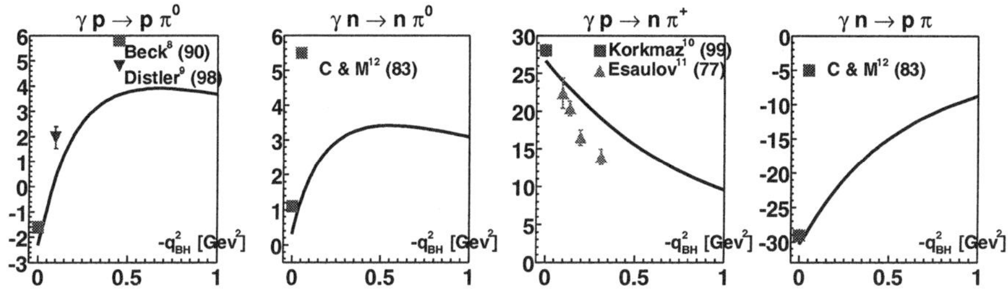

Figure 3. Q^2 dependency of threshold charged and neutral pion productions

terms contain no pion pole at the photon point. In a multipole development of the amplitude[6] only the first E_{0+} and L_{0+} multipoles survive at threshold. The comparison of LET predictions with data for E_{0+} allows one to check the validity of the method. Fig. 3 shows the plots of LET predictions and some experimental data. The charged pion photoproduction is dominated by commutator term and is well described by LET. On the other hand the neutral pion photoproduction predictions, relying on the Born term, display discrepancies with data: this is mainly due to pion loop contributions not included in the LET[7]. The neutral pion electroproduction point from Ref.[9] confirms the shape of the $-q_{BH}^2$ dependence for E_{0+}. Both charged and neutral pion production are contained in data points from Ref.[11], and their extraction is highly model dependent. However, the shape variation for charged pion electroproduction is governed by $g_A(-q_{BH}^2)$: the agreement at the photoproduction point also gives a measure of the agreement for electroproduction.

2.2 AVCS tensor

For the purpose of a first estimate, the LET technique will also be used for the calculation of the AVCS tensor. A first, but essential, remark must be made: although DVCS contains a hard scattering, the hard scale flows between the two photons, and $q - q' = q_{BH} \simeq \sqrt{(|t|)}$, finite, plays the "kinematical" role of the unique virtual photon of electroproduction. The direct application of LET with operator $O = i\int d^4x\, e^{-iqx} T\left\{J_{em}^{\mu}(x) J_{em}^{\nu}(0)\right\}$ leads to commutator and nucleon contributions.

The commutator term: In the soft limit, the axial charge is time independent up to terms of the order m_π^2. It is sensible to assume that terms of higher order in m_π remain so when integrated over. It is then possible to use the canonical commutation relation between the *constant axial charge* and quark em. current to obtain two T-products of axial and em. currents.

$$H_{com}^{\alpha,\mu\nu} = \frac{-ie}{f_\pi}\varepsilon^{\alpha 3\beta}\int d^4x e^{-iqx}\left\langle p'\left|T\left\{J_{em}^{\mu}(x)A^{\nu,\beta}(0)\right\} + T\left\{A^{\mu,\beta}(x)J_{em}^{\nu}(0)\right\}\right|p\right\rangle \tag{15}$$

A second hypothesis is the statement that the *handbag dominance* also stands for these T-products, see Fig. 2, with one of the currents being axial. Then a similar development as for DVCS[1,4] can be performed, with the following additional features: an extra γ^5 Dirac matrix in the Lorentz structure permutating the Lorentz structure association with Γ operators, and an extra quark isospin operator that can be translated into a nucleon isospin operator through the Wigner-Eckart theorem (with the assumption of isospin invariance, h and h' being the isospin of nucleons):

$$\langle p', h'\left|\bar{q}O^\alpha q\right|p, h\rangle = \langle p', \frac{1}{2}\left|\bar{q}O^3 q\right|p, \frac{1}{2}\rangle\langle p', h'\left|\tau_N^\alpha\right|p, h\rangle \tag{16}$$

Then, introducing the isovector operators $\Gamma^{(3)}$:

$$\Gamma^{(3)} = \Gamma^{u/p} - \Gamma^{d/p} \tag{17}$$

the commutator term takes the following form:

$$H_{com}^{\alpha,\mu\nu} = i\frac{e^2}{6f_\pi}\bar{u}(p')\left(S^{\mu\nu}n\cdot\widetilde{\Gamma}_{+}^{(3)}(\xi,t) + A^{\mu\nu}n\cdot\Gamma_{-}^{(3)}(\xi,t)\right)\frac{[\tau_N^\alpha,\tau_N^3]}{2}u(p) \tag{18}$$

The nucleon term is straightforward:

$$\begin{aligned} H_{\alpha,nucl}^{\mu\nu} = i&\frac{e^2 g_A}{2f_\pi}\bar{u}(p') \\ &\left[\gamma\cdot k_\pi\gamma_5\frac{\gamma\cdot(p'+k_\pi)+M}{(p'+k_\pi)^2-M^2}\tau^\alpha\left(h_S^{\mu\nu}+\tau^3 h_V^{\mu\nu}\right)\right. \\ &\left.+\left(h_S^{\mu\nu}+\tau^3 h_V^{\mu\nu}\right)\tau^\alpha\frac{\gamma\cdot(p-k_\pi)+M}{(p-k_\pi)^2-M^2}\gamma\cdot k_\pi\gamma_5\right]u(p) \end{aligned} \tag{19}$$

where $h_{S,V} = (h^p \pm h^n)/2$.

Because the perpendicular transfer $\Delta_\perp$ is not zero, gauge invariance for DVCS and equivalently for AVCS, is not insured at leading twist. It will be restored by hand, as in the early days for DVCS[4].

$$S^{\mu\nu} \to S^{\mu\nu} + \frac{\widetilde{p}^\nu}{\widetilde{p}\cdot q'}S^{\mu\rho}\Delta_{\perp\rho}\,,\; A^{\mu\nu} \to A^{\mu\nu} + \frac{\widetilde{p}^\nu}{\widetilde{p}\cdot q'}A^{\mu\rho}\Delta_{\perp\rho} \tag{20}$$

3 Results

In order to perform comparative studies with DVCS, the AVCS cross section has to be partially integrated over phase space. An additive pion in the final state increases the independent DVCS variable number from five to eight. The three new variables are chosen to be W, the final pion-nucleon center of mass energy and Ω^*_π, the pion solid angle in the pion-nucleon center of mass. The unpolarized differential AVCS cross section for a given lepton energy then reads:

$$\frac{d\sigma_{avcs}}{dQ^2 dx_{bj} dt_\gamma d\phi_\gamma dW d^2\Omega^*_\pi} = \frac{1}{32(2\pi)^4} \frac{x_{bj} y^2}{Q^4 \sqrt{1 + 4M^2 x_{bj}^2 / Q^2}} \times \frac{|\overrightarrow{p}*|}{2(2\pi)^3} |T_{ABH} + T_{AVCS}|^2 \tag{21}$$

with

$$y = \frac{q \cdot p}{l \cdot p}, \ |\overrightarrow{p}*| = \frac{\Lambda(W^2, M^2, m_\pi^2)}{2W} \tag{22}$$

$$\Lambda(x, y, z) = \sqrt{x^2 + y^2 + z^2 - 2xy - 2xz - 2yz} \tag{23}$$

The aim of the calculation is to evaluate the contribution of undetected soft pions. To achieve this goal, partial integration will be performed on the whole 4π pion solid angle and from threshold $W_{thr} = M + m_\pi$ up to a cut off value W_{max}. The choice of W_{max} is limited by the calculation hypothesis. As W goes close to M_Δ, the contribution of this resonance is no longer negligible[13]. As a result, W_{max} should not exceed $1.2\,GeV$. The cross section dependence on W_{max} is moderate. For four typical DVCS experimental set-ups, Fig. 4 compares the total DVCS and AVCS with neutral and charged pion for $W_{max} = 1.2\,GeV$.

Conclusion

This first evaluation of soft pion production associated with DVCS reveals a relative contribution of the order of one percent for the highest integration cut off. This has a major experimental importance when, for instance, the missing mass technique is applied for exclusivity check of DVCS. The next step to enlarge the scope of this calculation up to higher W_{max} is now to introduce the Δ resonance.

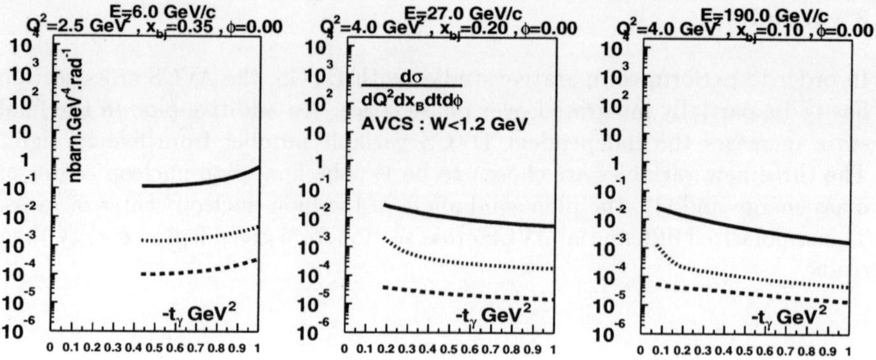

Figure 4. total cross sections for DVCS on proton (full line) and AVCS with neutral (dashed line) and charged (dotted line) pion for JLAB, HERMES and COMPASS typical kinematics

Acknowledgments

The author would like to acknowledge P.A.M. Guichon, who is at the origin of the calculation and M. Vanderhaeghen, for useful and extensive discussions. The author also thanks the CSSM in Adelaide for the kind invitation to present this work during the workshop.

References

1. X. Ji, *Phys. Rev. Lett.* **78**, 610 (1997); *Phys. Rev.* D **55**, 7114 (1997).
2. A.V. Radyushkin, *Phys. Lett.* B **385**, 333 (1996).
3. J.C. Collins, L. Frankfurt, and M. Strikman, *Phys. Rev.* D **56**, 2982 (1997).
4. M. Vanderhaeghen, P.A.M. Guichon and M. Guidal, *Phys. Rev.* D **60**, 094017 (1999).
5. B. de Alfaro and S. Fubini and G. Furlan and C. Rossetti, "Currents in hadron physics", North-Holland, Amsterdam, 1973
6. D. Drechsel and L. Tiator, *J. Phys.* G **18**, 449 (1992).
7. V. Bernard, N. Kaiser and U G Meissner, *Nucl. Phys.* A **607**, 379 (1996).
8. R. Beck *et al*, *Phys. Rev. Lett.* **65**, 1841 (1990).
9. M. O. Distler *et al*, *Phys. Rev. Lett.* **80**, 2294 (1998).
10. E. Korkmaz *et al*, *Phys. Rev. Lett.* **83**, 3609 (1999).
11. A. S. Esaulov *et al*, *Nucl. Phys.* B **136**, 511 (1978).
12. Crawford and Morton, *Nucl. Phys.* B **211**, 1 (1983).
13. K. Goeke, M.V. Polyakov, M. Vanderhaeghen hep-hp/0106012.

2. Spin Structure of Hadrons

AN EXPERIMENTAL REVIEW OF THE NUCLEON SPIN STRUCTURE FUNCTIONS

GERASSIMOS G. PETRATOS

Department of Physics, Kent State University, Kent, OH 44242, USA

E-mail: gpetrato@kent.edu

An experimental review of the nucleon spin structure functions g_1 and g_2, as measured in deep inelastic scattering of polarized leptons by polarized nucleons, is presented. Particular emphasis is given to the theoretical interpretation of the spin structure functions, to recent experiments at CERN, SLAC and DESY, and to the presentation and discussion of the available experimental data.

1 Theory Overview

This paper presents an experimental review of the spin structure functions g_1 and g_2 of the proton, neutron and deuteron. Recent experiments at CERN, SLAC and DESY with polarized lepton beams and polarized nucleon targets are described. A summary of all experimental data is provided together with their implications for the nucleon spin.

The spin structure functions G_1 and G_2 of the nucleons are determined by measuring the cross section asymmetries:[1]

$$A_{\|} = \frac{\sigma(\uparrow\Uparrow) - \sigma(\uparrow\Downarrow)}{\sigma(\uparrow\Uparrow) + \sigma(\uparrow\Downarrow)} = \frac{1-\epsilon}{(1+\epsilon R)W_1}\left[(E + E'\cos\theta)MG_1 - Q^2G_2\right]$$

$$A_{\perp} = \frac{\sigma(\uparrow\Rightarrow) - \sigma(\uparrow\Leftarrow)}{\sigma(\uparrow\Rightarrow) + \sigma(\uparrow\Leftarrow)} = \frac{(1-\epsilon)E'}{(1+\epsilon R)W_1}\left[(MG_1 + 2EG_2)\sin\theta\right]$$

in deep inelastic scattering (DIS) of polarized leptons from polarized nucleons. Here $\uparrow,\downarrow$ denotes the longitudinal spin of the incoming lepton (along or opposite the direction of its momentum) and $\Uparrow,\Downarrow$ or $\Leftarrow,\Rightarrow$ denotes the longitudinal or transverse spin of the target nucleon. The asymmetries are functions of the nucleon mass M, the incident lepton energy E, the scattered lepton energy E' and angle θ, the degree of the longitudinal polarization of the virtual photon exchanged in the scattering $\epsilon = [1+2(1+\nu^2/Q^2)\tan^2(\theta/2)]^{-1}$, and the spin-averaged structure functions W_1 and W_2 which are connected via $R = [(1+\nu^2/Q^2)W_2/W_1 - 1]$.

In the Bjorken scaling limit of large four-momentum $Q^2 = 4EE'\sin^2(\theta/2)$ and energy $\nu = E - E'$ transfers, the G_1 and G_2 spin structure functions are predicted to become functions only of the Bjorken scaling variable $x = Q^2/2M\nu$, which is the fraction of the nucleon momentum carried by the struck quark in the scattering: $M\nu^2 G_1(x, Q^2) \to g_1(x)$, $M\nu^2 G_2(x, Q^2) \to g_2(x)$.

In the quark-parton model, the spin structure function $g_1(x)$ of the nucleons is connected to the quark and anti-quark momentum distributions $q_i(x)$: $g_1(x) = \sum_i z_i^2 [q_i^{\uparrow}(x) - q_i^{\downarrow}(x)]$, where the arrow $\uparrow$ ($\downarrow$) represents quark spin parallel (anti-parallel) to the nucleon spin and the sum is over all quark flavors of charge z_i.

The spin structure functions g_1^p and g_1^n of the proton and neutron are related by the Bjorken sum rule:[2]

$$\int_0^1 [g_1^p(x) - g_1^n(x)]dx \simeq \frac{1}{6}\left|\frac{g_A}{g_V}\right|\left[1 - \frac{\alpha_s(Q^2)}{\pi}\right],$$

where g_A and g_V are the weak couplings of the nucleon beta decay and $\alpha_s(Q^2)$ is the coupling constant of quantum chromodynamics (QCD). This sum rule, first derived from current algebra and based on isospin symmetry, is a rigorous prediction of QCD.

The first moments of the spin structure functions have a simple interpretation in the quark-parton model:

$$\int_0^1 g_1^p(x)dx = \frac{1}{2}(\frac{4}{9}\Delta u + \frac{1}{9}\Delta d + \frac{1}{9}\Delta s), \quad \int_0^1 g_1^n(x)dx = \frac{1}{2}(\frac{1}{9}\Delta u + \frac{4}{9}\Delta d + \frac{1}{9}\Delta s),$$

where $\Delta q_i = \int_0^1 [q_i^{\uparrow}(x) - q_i^{\downarrow}(x)]dx$ represent the integrals over the momentum distributions of the up, down and strange quark of the nucleon. A measurement of the integrals of the spin structure functions in conjunction with the neutron beta decay relationship $\Delta u - \Delta d = F + D$ and the SU(3) relationship $\Delta d - \Delta s = F - D$ can be used to disentangle the $\Delta u, \Delta d$ and Δs quark spin distributions and determine their sum $\Delta\Sigma = \Delta u + \Delta d + \Delta s$. Here F and D are the SU(3) invariant matrix elements of the axial vector current.

The first experiment to extract the quark spin distributions was done by the EMC Collaboration,[3] which performed deep inelastic scattering of polarized muons by polarized protons at CERN. The EMC results confirmed the g_1 results of the pioneering SLAC experiments E80[4] and E130[5] that used polarized electrons and protons. The combined CERN/SLAC results showed that $\Delta\Sigma = 0.03 \pm 0.10$, indicating that the quarks carry only a small fraction of the proton spin ("proton spin crisis"). The results also indicated that a rather large and negative contribution to the proton spin should be attributed to the strange quark sea ($\Delta s = -0.22 \pm 0.06$).

The naive quark-parton model interpretation of the nucleon g_1 spin structure function neglects contributions to the nucleon spin from the gluon polarization and the orbital angular momentum L of the quarks and gluons. Inclusion of these two contributions results in the "spin sum rule": $\frac{1}{2} = \frac{1}{2}\Delta\Sigma + \Delta G + L$, where $\Delta G = \int_0^1 [G^{\uparrow}(x) - G^{\downarrow}(x)]dx$. Here the distributions $G^{\uparrow,\downarrow}$ have meaning similar to the $q^{\uparrow,\downarrow}$ ones. This gauge invariant and scheme independent sum rule[6] provides a possible resolution to the "spin crisis" but demands reliable measurements of ΔG and L. A detailed report on future measurements of ΔG is given in a recent review paper[7]. It has been suggested that the angular momentum contribution can be determined by means of deeply-virtual Compton scattering from the nucleon (see Ref. 6).

The g_2 spin structure function has no clear interpretation in the quark-parton model. It contains contributions from both longitudinal and transverse polarization distributions in the nucleon and it is sensitive to higher twist effects (effects suppressed by powers of $1/Q$) such as quark-gluon correlations and quark mass contributions. An expression for the twist-2 part of g_2 was derived by Wandzura and Wilczek (WW)[8] assuming that the twist-3 contributions are negligible: $g_2^{WW}(x, Q^2) = -g_1(x, Q^2) + \int_x^1 [g_1(x', Q^2)/x']dx'$. A comparison of measured values of g_2 with this prediction provides information about higher twist contributions to g_2. Burkhardt and Cottingham[9] have derived, using dispersion relations for virtual photon scattering, the sum rule: $\int_0^1 g_2(x)dx = 0$, which is valid to first order in perturbative QCD (pQCD). The validity of this sum rule, however, relies on g_2 obeying Regge theory at low-x, which may not be a good assumption.

2 Recent Experiments

The spin structure functions are determined experimentally by measuring the cross section asymmetries $A_{||}$, $A_{\perp}$ for lepton and nucleon/deuteron spins oriented parallel or transverse. Since the target and beam are not perfectly polarized, the asymmetries are diluted by the beam and target polarizations: $A_{meas} = p_b p_t f A_{true}$, where f, the dilution factor, represents the fraction of polarizable nucleons in the target. The factor f is further reduced when extraneous (unpolarized) target materials are in the path of the incident leptons. Lepton polarizations vary from 0.36 to 0.85, target polarizations from 0.33 to 0.85, and dilution factors from 0.11 to 1.

The Spin Muon Collaboration (SMC) experiment[10] used polarized muon beams with energy 100-190 GeV, polarized solid targets and a large acceptance forward spectrometer system. It is an evolution of the pioneering EMC experiment that first produced the surprising results on the g_1^p spin structure function. The strength (weakness) of the SMC experiment is its low-x

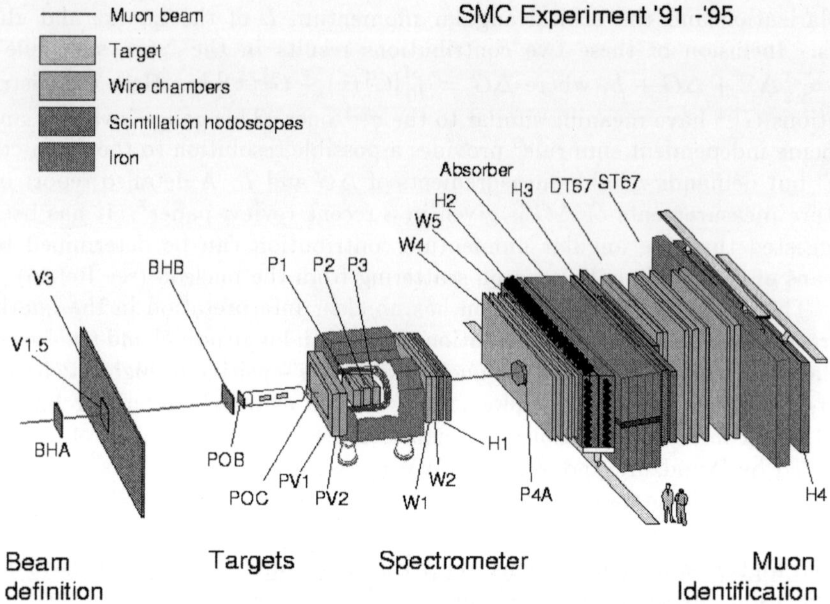

Figure 1. The CERN SMC experimental setup (see text).

coverage (statistical precision).

Polarized muons were produced by the decay in flight of pions and kaons. The muons are naturally polarized because of parity violation in the pion and kaon decays. Protons and deuterons were polarized by the Dynamic Nuclear Polarization (DNP) technique in butanol (C_4H_9OH), deuterated butanol (C_4D_9OD) and ammonia (NH_3) cryogenic targets. In the DNP technique, these solid targets are preirradiated to produce paramagnetic radical centers with quasifree electrons that are easily polarizable in a magnetic field of 2.5 T and at a temperature of 50 mK. The polarization of the electrons is transferred to the protons or deuterons by application of microwaves.

Scattered muons were detected in a spectrometer system based on a large aperture dipole magnet with $\int Bdl = 2.4\,\mathrm{Tm}$. Scattered particles were tracked with sets of proportional and drift chambers. A thick hadron absorber behind the chambers, followed by streamer tubes, provided muon identification. A schematic of the SMC experimental apparatus is shown in Figure 1.

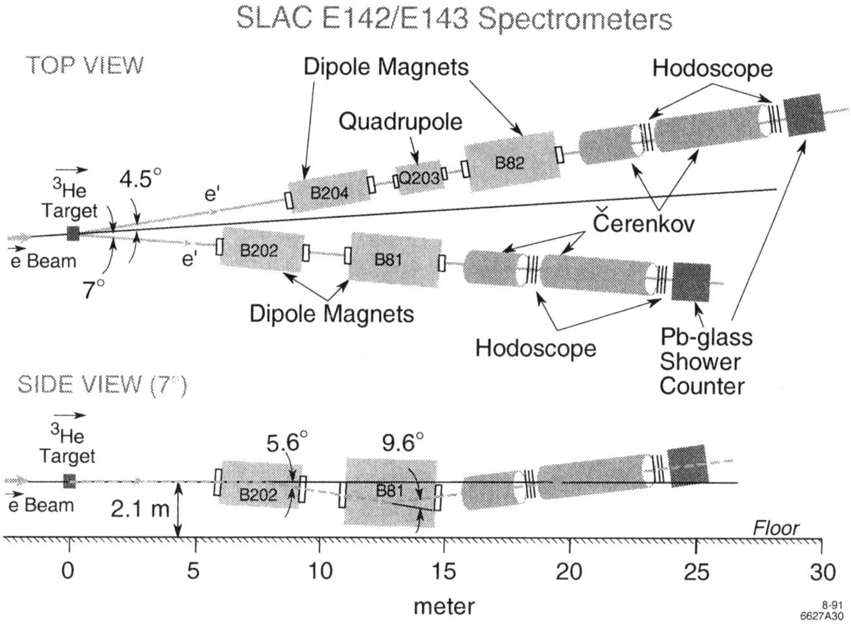

Figure 2. The SLAC E142 experimental setup (see text).

The SLAC E142[11], E143[12], E154[13] and E155[14] experiments used intense polarized electron beams with 19-48 GeV energy and fixed angle magnetic spectrometers in End Station A with large-x coverage. E142 and E154 used polarized gas ^{3}He targets; E143 and E155 used polarized solid targets. The strength (weakness) of the SLAC experiments is their statistical precision resulting from high luminosity (the limited low-x coverage).

Polarized electrons were produced by laser photoemission from an AlGaAs (strained AlGaAs) source for E142 (E143, E154 and E155). E142 and E154 used a polarized ^{3}He target as a source of polarized neutrons. The target consisted of a dual glass cell. Circularly polarized laser light polarized the valence electrons of Rb vapor in the upper cell. Spin exchange collisions transferred the polarization to ^{3}He nuclei. The lower target cell (free of Rb) was exposed to the beam. Both E143 and E155 used solid polarized targets based on DNP: NH_3 (E143 and E155), ND_3 (E143) and LiD (E155). The advantage of LiD over ND_3 is that Li can provide extra polarized deuterons as a cluster of an α-particle and a deuteron.

E142 and E143 used the same spectrometers. They were based on a DD

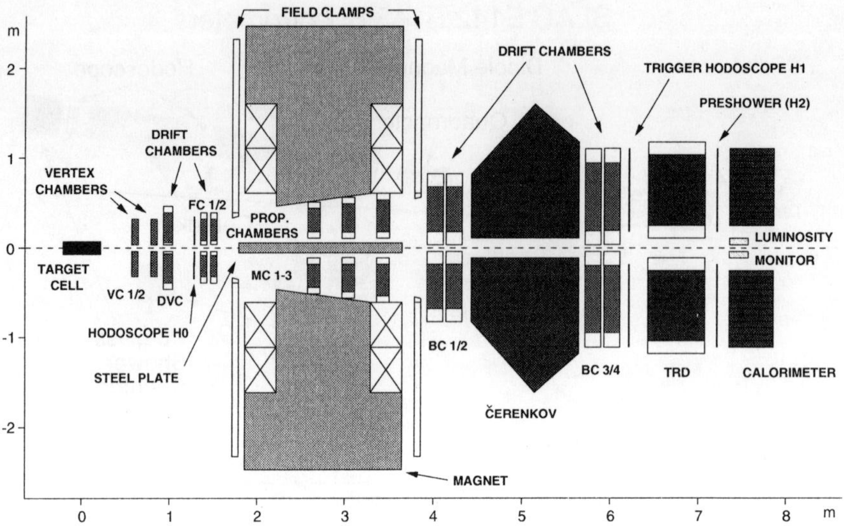

Figure 3. Side view of the DESY HERMES experimental setup (see text).

and DQD design that a) minimized a large photon background originating from the target and b) provided a very large momentum acceptance that minimized the experimental beam time. Scattered particles in each spectrometer were tracked with two highly segmented hodoscopes, and electron identification was provided by two gas threshold Čerenkov counters and a highly segmented Pb-glass calorimeter. A schematic of the E142/E143 setup is shown in Figure 2. For E154 and E155, the spectrometers were repositioned at smaller angles, the segmentation of the hodoscopes was increased, and the Čerenkov counters were elongated. In addition, E155 used another QDQ spectrometer for high-x coverage.

The HERMES experiment[15] uses 27.5 GeV polarized positron beams in the HERA ring and polarized internal targets. Scattered particles are detected in a forward magnetic spectrometer. The strength (weakness) of HERMES is the high statistical precision and the capability of semi-inclusive measurements (the limited x coverage).

The positron beam becomes transversely polarized in the emission of synchrotron radiation (Sokolov-Ternov effect). The polarization is rotated into the longitudinal direction by a series of dipole magnets. HERMES uses pure hydrogen (H), deuterium (D) and ^{3}He targets that flow into a windowless target cell in the beam, resulting in maximum dilution factor of 1 for H and

D and 1/3 for ^{3}He. Polarized protons and deuterons are produced by an atomic beam source based on the Stern-Gerlach effect to select specific spin states. Polarized ^{3}He nuclei are produced by spin exchange collisions with ^3S metastable ^{3}He atoms in a glass cell.

Scattered particles are detected in a forward spectrometer centered around a large aperture, 1.3 Tm, dipole magnet. The particle coordinates are determined by micro-strip gas chambers upstream of the target, drift chambers upstream/downstream of the dipole, and multi-wire proportional chambers inside the dipole. Particles are identified using a Pb-glass calorimeter, a transition radiation detector and a gas threshold Čerenkov counter (recently upgraded to a ring imaging counter). A schematic of the HERMES experimental setup is given in Figure 3.

3 Experimental Results

The results of the CERN, HERMES and SLAC experiments on the g_1 structure functions of the proton, neutron and deuteron are presented in Figure 4. There is a remarkable agreement between the many experiments. The precision of the existing data does not constrain the behavior of g_1 at low-x, necessitating the need for future measurements with higher projectile energies at a fixed target machine or a polarized electron-nucleus collider.

To calculate reliably the first moments of the g_1 nucleon spin structure functions, several groups[16] have performed a next-to-leading Order (NLO) perturbative QCD analysis of all available g_1 data. The NLO analyses use the factorization theorem of QCD[17] and are based on similar (although not identical) data sets. They differ mainly in the choice of parametrizations, definition of parameters, fitting procedure and handling of systematic errors.

For example, the SLAC E154 and E155 Collaborations expressed g_1 in terms of the polarized quark, antiquark, and gluon distributions $\Delta q(x)$, $\Delta\bar{q}(x)$, and ΔG: $g_1(x,Q^2) = \frac{1}{2}\sum_q^{N_f} {e_q}^2 \left[C_q \otimes (\Delta q + \Delta\bar{q}) + \frac{1}{N_f} C_G \otimes \Delta G\right]$, with the convolution $\otimes$ defined as: $(C \otimes q)(x,Q^2) = \int_x^1 C(x/z,\alpha_s) q(x,Q^2) dz/z$. The sum runs over all active quark flavors N_f and the coefficient functions $C_{q,G}(x,\alpha_s)$ correspond to the hard scattering photon-gluon cross sections (Wilson coefficients). They are calculated in perturbative QCD as an expression in powers of α_s.

A standard parametrization[18] was used to express the polarized parton distributions in terms of the unpolarized ones at the initial low scale $Q_\circ^2 = 0.34$ (GeV/c)^2: $\Delta f(x, {Q_\circ}^2) = A_f x^{\alpha_f}(1-x)^{\beta_f} f(x,{Q_\circ}^2)$. Here $\Delta f = \Delta u_v, \Delta d_v, \Delta\bar{Q}, \Delta G$ are the polarized valence (u and d), sea and gluon distri-

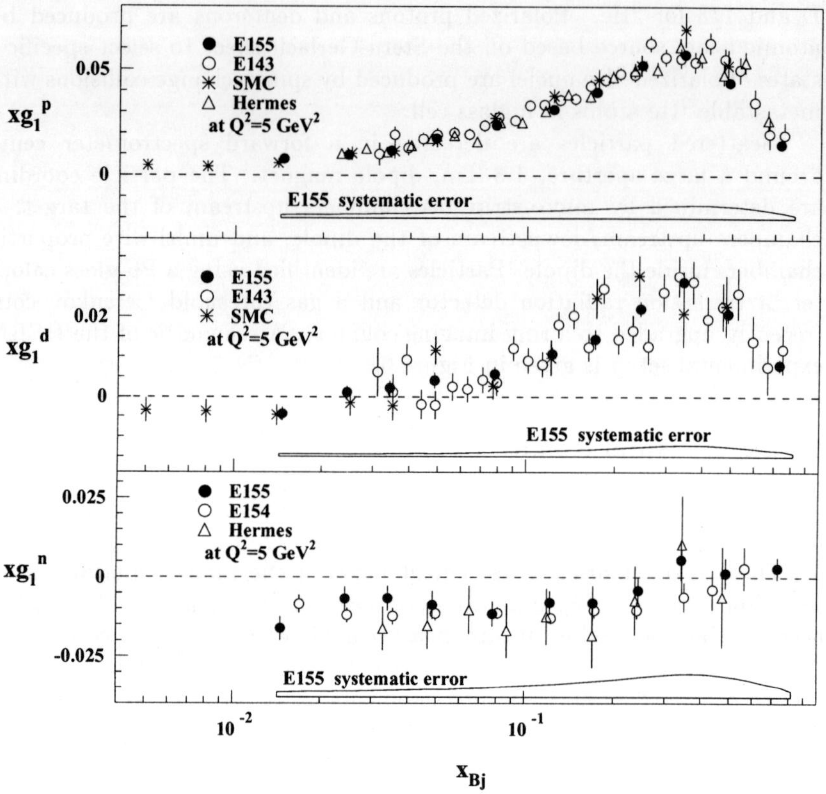

Figure 4. The g_1 structure function of the proton (top), deuteron (middle) and neutron (bottom) from SLAC, CERN and DESY Experiments versus the Bjorken scaling variable x (see text).

butions, and $f(x, Q_o^2)$ are the unpolarized parton distributions. The polarized parton distributions were evolved to the Q^2 values of the experimental measurements using the DGLAP[19] QCD evolution equations. A fit to all experimental data determined the free parameters of the parametrization.

All NLO analyses give similar results. For example, the E155 NLO analysis, which includes all available data to date finds the Bjorken sum rule to be 0.176 ± 0.008, at $Q^2 = 5$ (GeV/c)^2, where the error includes statistical and systematic but no theoretical uncertainties. This value is in agreement with the theoretical value 0.182 ± 0.005, which includes third order pQCD corrections, using $\alpha_s(M_Z^2) = 0.119 \pm 0.002$. The extraction of $\Delta\Sigma$ and ΔG, which is renormalization- and scheme-dependent, results, for the $\overline{\rm MS}$ scheme, to: $\Delta\Sigma = 0.23 \pm 0.04 \pm 0.06$ and $\Delta G = 1.6 \pm 0.8 \pm 1.1$, where the first errors are statistical, the second ones are systematic, and similarly sized theory errors are ignored.

The g_2 structure functions for the proton, neutron and deuteron have been measured[20] with less precision than their g_1 counterparts. All three data sets are consistent with g_2^{WW} indicating that twist-3 contributions to g_2 are not sizable. Within errors, each one of the g_2 data sets of the proton, neutron and deuteron satisfy the Burkhardt-Cottingham sum rule. To differentiate between $g_2 = g_2^{WW}$ and $g_2 = 0$, and to test theoretical model predictions, SLAC experiment E155 measured recently the proton and deuteron g_2 with a four-fold better statistical precision, but the results are not yet published.

4 Summary

Precision measurements at CERN, DESY and SLAC have provided a wealth of data on the spin structure functions g_1 and g_2 of the proton, neutron and deuteron. The g_1 results have confirmed the fundamental Bjorken sum rule of QCD. The quarks appear to account for $\sim 20\%$ of the nucleon spin with the remaining $\sim 80\%$ attributed to the polarization of the gluons and to the orbital angular momentum of the nucleon constituents. The g_2 results are consistent, within experimental errors, with the Wandzura-Wilczek calculation indicating that higher twist contributions are not sizable. Future low-x measurements of g_1, gluon ΔG measurements, and precision measurements of g_2 will be of great interest in improving our knowledge of the nucleon spin structure. Such measurements are within reach with existing facilities and future polarized electron-nucleus colliders or fixed target experiments.[7]

Acknowledgments

The author would like to acknowledge the support of the US National Science Foundation in this work and to thank the Workshop organizers for their kind invitation.

References

1. C. E. Carlson and W.-K. Tung, *Phys. Rev.* D **5**, 721 (1972); A.J.G. Hey and J.E. Mandula, *Phys. Rev.* D **5**, 2610 (1972).
2. J.D. Bjorken, *Phys. Rev.* **148**, 1467 (1966); *Phys. Rev.* D **1**, 1376 (1970).

3. EMC Collaboration, J. Ashman *et al.*, *Nucl. Phys.* B **328**, 1 (1989).
4. E80 Collaboration, M.J. Alguard *et al.*, *Phys. Rev. Lett.* **37**, 1261 (1976); *Phys. Rev. Lett.* **41**, 70 (1978).
5. E130 Collaboration, G. Baum *et al.*, *Phys. Rev. Lett.* **51**, 1135 (1983).
6. X. Ji, *Phys. Rev. Lett.* **78**, 610 (1997); and references therein.
7. B.W. Filippone and X. Ji, arXiv:hep-ph/0101224, Jan. 2001.
8. S. Wandzura and F. Wilczek, *Phys. Lett.* B **72**, 195 (1977).
9. H. Burkhardt and W.N. Cottingham, *Ann. Phys.* **56**, 453 (1970).
10. SMC Collaboration, B. Adeva *et al.*, *Phys. Rev.* D **58**, article 112001 (1998).
11. E142 Collaboration, P.L. Anthony *et al.*, *Phys. Rev.* D **54**, 6620 (1996).
12. E143 Collaboration, K. Abe *et al.*, *Phys. Rev.* D **58**, article 112003 (1998).
13. E154 Collaboration, K. Abe *et al.*, *Phys. Rev. Lett.* **79**, 26 (1997).
14. E155 Collaboration, P.L. Anthony *et al.*, *Phys. Lett.* B **463**, 339 (1999); P.L. Anthony *et al.*, *Phys. Lett.* B **493**, 19 (2000); G.S. Mitchell, *SLAC-R-540* (1999) (*Ph.D. Thesis*, U. Wisconsin, Madison).
15. HERMES Collaboration, K. Ackerstaff *et al.*, *Phys. Lett.* B **404**, 383 (1997); A. Airapetian *et al.*, *Phys. Lett.* B **442**, 484 (1998).
16. E154 Collaboration, K. Abe *et al.*, *Phys. Lett.* B **405**, 180 (1997); G. Altarelli *et al.*, *Nucl. Phys.* B **496**, 337 (1997); SMC Collaboration, B. Adeva *et al.*, *Phys. Rev.* D **58**, article 112002 (1998); E155 Collaboration, G.S. Mitchell *et al.*, *SLAC-PUB-8104* (1999).
17. H. Georgi and H.D. Politzer, *Phys. Rev.* D **9**, 416 (1974); D.J. Gross and F. Wilczek, *Phys. Rev.* D **9**, 980 (1974).
18. M. Gluck *et al.*, *Phys. Rev.* D **53**, 4775 (1996); T. Gehrmann and W.J. Stirling, *Phys. Rev.* D **53**, 6100 (1996).
19. V.N. Gribov and L.N. Lipatov, *Sov. J. Nucl. Phys.* **15**, 438 (1972); Y.L. Dokshitzer, *Sov. Phys. JETP* **46**, 461 (1977); G. Altarelli and G. Parisi, *Nucl. Phys.* B **126**, 298 (1977).
20. E154 Collaboration, K. Abe *et al.*, *Phys. Lett.* B **404**, 377 (1977); E155 Collaboration, P.L. Anthony *et al.*, *Phys. Lett.* B **458**, 529 (1999).

POLARIZED STRUCTURE FUNCTIONS IN QCD

J. KODAIRA

Department of Physics, Hiroshima University,
Higashi-Hiroshima 739-8526, JAPAN

We review the nucleon's polarized structure functions from the viewpoint of gauge invariant, nonlocal light-cone operators in QCD. We discuss a systematic treatment of the polarized structure functions and the corresponding parton distribution functions. We also address a question of what information on the structure of Nature will be obtained from the future polarized experiments. From this point of view, we will discuss the $W^{\pm}\gamma$ production at RHIC polarized experiment.

1 Introduction

In the last ten years, great progress has been made both theoretically and experimentally in hadron spin physics. Furthermore, in conjunction with new projects like the "RHIC spin project", "polarized HERA", etc., we are now in a position to obtain more information on the spin structure of nucleons. The spin dependent quantity is, in general, very sensitive to the structure of interactions among various particles. Therefore, we will be able to study the detailed structure of hadrons based on QCD. However the purpose of new experiments should not be limited to only the check of QCD. We also hope that we can find some clue to new physics beyond the standard model.

In this talk, we first review a systematic treatment of polarized structure functions and summarize the recent theoretical progress on the QCD evolutions [1]. Secondly, we address a question of what we will be able to learn from the future polarized experiments by considering the phenomena called Radiation Zeros (RAZ) [2] as an example. We reanalyze this phenomena at the realistic RHIC polarized collider. We will point out that the polarization of colliding protons will emphasize the RAZ phenomena in the cross section and a "moderate energy" machine is better than the extremely high energy machines to find this phenomena.

2 Classification of structure functions

To describe a variety of high-energy processes to which the factorization theorem can be applied, it is desirable to have a formulation based on a universal language in QCD. The traditional approach relies on the operator product expansion (OPE), but it can be applied only to a limited class of processes. This calls for an approach based on the factorization as a generalization of the OPE. As a result of factorization, the structure functions (cross section) are given as convolution of the short- and long- distance parts. The former contains all the dependence on the hard scale and the latter is controlled by the nonperturbative dynamics of QCD.

For the long-distance part, the parton distribution functions (PDFs) are introduced and it is now standard to define them as the nucleon's matrix elements of nonlocal light-cone operators in QCD[3]. The momentum of nucleon (mass M) P will be written in terms of two auxiliary light-like vectors p^μ and w^μ ($p^2 = w^2 = 0$, $p \cdot w = 1$), $P^\mu = p^\mu + \frac{M^2}{2} w^\mu$. The (quark) PDF is defined by the Fourier transform of the nucleon's matrix element of the nonlocal light-cone quark operator as,

$$\int_{-\infty}^{+\infty} \frac{d\lambda}{2\pi} e^{i\lambda x} \langle PS|\bar{\psi}(0)[0, \lambda w]\Gamma\psi(\lambda w)|PS\rangle \, . \tag{1}$$

Here $|PS\rangle$ is the nucleon state with spin S which is decomposed into the longitudinal and transverse parts with respect to $p^\mu(w^\mu)$ direction, $S^\mu = S^\mu_{||} + S^\mu_{\perp}$. The variable x can be interpreted as the momentum fraction of parton (Bjorken x) since $\lambda x = (xP) \cdot (\lambda w)$. Γ is a generic Dirac matrix and the link operator $[0, \lambda w]$ which makes the operators gauge invariant is given by,

$$[y, z] = \mathrm{Pexp}\left(ig \int_0^1 dt (y - z)_\mu A^\mu (ty + (1-t)z) \right) \, .$$

Possible choices for Γ and spin states which depend on the processes considered in Eq.(1) lead to various PDFs.

2.1 Quark distributions

Let us first consider the quark distributions. An important observation made in Ref.[4] is that one can generate all quark distribution functions up to twist-4 by substituting all the possible Γ. By decomposing Eq.(1) into independent tensor structures, one finds nine independent quark distribution functions associating with each tensor structure[1]. Their spin, twist and chiral classifications are listed in Table 1. The distributions in the first row are spin-independent, while those in the second and third rows correspond to the longitudinally ($S_{||}$) and transversely ($S_\perp$) polarized nucleons. Each column refers to the twist. The distributions marked with "$\star$" are referred to as chiral-odd, because they correspond to chirality-violating Dirac matrix structures $\Gamma = \{\sigma_{\mu\nu} i\gamma_5, 1\}$. The other distributions are chiral-even, because of the chirality-conserving structures $\Gamma = \{\gamma_\mu, \gamma_\mu\gamma_5\}$. In the massless quark limit, chirality is conserved through the propagation of a quark. This means that in the DIS one can measure only the chiral-even distributions up to tiny quark mass corrections because the perturbative quark-gluon and quark-photon couplings conserve the chiralities. On the other hand, in the Drell-Yan and certain other processes, both chiral-odd and chiral-even distribution functions can be measured [5] because the chiralities of the quark lines originating in a single nucleon are uncorrelated.

Table 1: Spin,twist and chiral classification of quark distributions.

Twist	2	3	4
S ave.	$q(x)$	$e(x)^\star$	$f_4(x)$
$S_\parallel$	$\Delta q(x)$	$h_L(x)^\star$	$g_3(x)$
$S_\perp$	$\delta q(x)^\star$	$g_T(x)$	$h_3(x)^\star$

2.2 *Gluon distributions*

The gluon distribution functions can be defined in the similar way as for the quark distributions. The gauge-invariant definition of the gluon distribution functions is provided by [3,6],

$$\frac{2}{x}\int \frac{d\lambda}{2\pi} e^{i\lambda x}\langle PS|w_\alpha G^{\alpha\mu}(0)[0,\lambda w]w_\beta G^{\beta\nu}(\lambda w)|PS\rangle \,.$$

Corresponding to independent tensor structures, we have four gluon distribution functions summarized in Table 2. The gluon distributions mix through renormalization with the flavor singlet chiral-even quark distributions. On the other hand, there exists no gluon distributions that mix with the chiral-odd quark distributions.

Table 2: Spin and twist classification of gluon distribution functions.

Twist	2	3	4
S ave.	$\mathcal{G}(x)$		$\mathcal{G}_4(x)$
$S_\parallel$	$\Delta\mathcal{G}(x)$		
$S_\perp$		$\mathcal{G}_{3T}(x)$	

2.3 *Twist-3 three particle distributions*

Coherent many-particle contents of the nucleon are described by multi PDFs. In this talk, we mention only the twist-3 quark-gluon correlation functions,

$$\int \frac{d\lambda}{2\pi}\frac{d\zeta}{2\pi} e^{i\lambda x+i\zeta(x'-x)}\langle PS|\bar\psi(0)\Gamma[0,\zeta w]g\, G^{\mu\nu}(\zeta w)[\zeta w,\lambda w]\psi(\lambda w)|PS\rangle.$$

Similarly to the quark distributions, one can define the multiparton distributions by considering possible Dirac matrices for Γ which are listed in Table 3.

Table 3: Spin and chiral classification of quark-gluon correlations at twist-3.

S ave.	$\Phi(x,x')^*$
$S_\parallel$	$\widetilde{\Phi}(x,x')^*$
$S_\perp$	$\Psi(x,x')$, $\widetilde{\Psi}(x,x')$

The treatment here can be extended to the case of three-gluon correlation functions, which are relevant to the (singlet) quark distribution $g_T(x)$ and the gluon distribution $\mathcal{G}_{3T}(x)$. The details have been discussed in Refs.[6,7].

3 QCD evolution

In the case of the lowest twist, their QCD evolution can be easily estimated using the standard techniques. The calculation for the higher twist terms is, however, generally very complicated due to the presence of multiparton distributions. Although it is possible to generalize the DGLAP approach to the three-body case, one is forced to use some particular techniques, e.g. the use of the light-like axial gauge etc. [7,9] On the other hand, it has been known that one can construct a set of "local composite operators" from the nonlocal one by taking the moment of PDFs. Therefore we can take an approach based on these local composite operators [10]. The advantage of this approach is that all the relevant steps can be worked out based on the standard and familiar field theory techniques in any gauge.

The common feature for the higher twist operators is that the number of a set of local operators with the same quantum numbers increases with spin (moment) and these operators are related through the QCD equation of motion [11,12]. However, once a complete set of operators is identified, the anomalous dimensions which derive QCD evolution is easily calculated following the theorem [13] explained now. This theorem tells us that in non-abelian gauge theory, three kinds of operators will mix under renormalization of gauge invariant operators. (I)the gauge invariant operator itself. (II)the BRST invariant operators. (III)the operators which are proportional to the equation of motion (EOM). The EOM operators involve both BRST "invariant" and "variant" ones [10,14]. Although the physical matrix elements of the BRST invariant and EOM operators vanish [1], it is necessary to consider these operators to complete the renormalization. At the lowest twist level, these complexities do not come into play because there exists neither EOM nor BRST invariant operator of twist-2. The EOM as well as the BRST invariant operators always have smaller spin by at least one unit than the possible highest spin operators of the same dimension.

The explicit calculations at the one-loop level appear in Refs.[10,8,15] for g_T and in Ref.[16] for h_L. Much more recent progress on the twist-3 parton densities can be found in Ref.[17].

4 $W^{\pm}\gamma$ production at RHIC

The polarized PDFs will be measured precisely by the future polarized experiments and we will be able to check if the QCD predictions are consistent or not. However, we believe that the polarized experiments will provide us with a great store of knowledge on the structure of all interactions. From this point of view, the final part of this talk is devoted to a discussion on the radiative weak boson production $pp \to W^{\pm}\gamma$ at RHIC polarized experiment. We point out two reasons why we reanalyze this process. (1) Due to the $V - A$ structure of the W boson interaction, only the initial quarks which have definite helicities can participate in the process. Therefore, an experiment with the polarized beams will be more efficient to study this process [18] and an information on $\Delta q(x)$ is relevant. (2) Many works so far assumed rather high energy collisions [19]. However, a realistic experiment with the polarized beams becomes available at RHIC whose center of mass energy is around $\sqrt{s} \sim 500$ GeV.

4.1 RAZ at the partonic level

Radiative weak boson ($W\gamma$) production in hadronic collisions has been the subject of much theoretical interests since this process contains the gauge boson trilinear coupling and develops the so-called radiation zero (RAZ) [2]. The RAZ is a typical example which is sensitive to the structure of electroweak interaction.

The RAZ is a phenomena that the cross section (amplitude) for some process develops zero in the some point of the phase space. To understand this phenomena in the simplest way, although not rigorous, let us use the soft photon approximation. Since the soft photon factorizes as the eikonal factor, the amplitude which contains one photon can be written as, $M_\gamma \simeq e\, J \cdot \epsilon(k) M$ where $J^\mu = \sum_i Q_i \eta_i \frac{p_i^\mu}{p_i \cdot k}$ and $\epsilon^\mu(k)$ is the polarization vector of photon. Q_i is the charge of i-the particle and $\eta_i = +(-)$ for the incoming (outgoing) particle. The sum is taken over all external particles. If $\frac{Q_i}{p_i \cdot k} =$ const. for all i, the energy-momentum conservation $\sum_i \eta_i p_i^\mu = 0$ implies $J^\mu = M_\gamma = 0$. The well-known example is the process, $u(p_1) + \bar{d}(p_2) \to W^+(p_3) + \gamma(k)$. In this case, the identity $\frac{2}{3}\frac{1}{p_1 \cdot k} = \frac{1}{3}\frac{1}{p_2 \cdot k} = \frac{1}{p_3 \cdot k}$ is satisfied at $\cos\theta_\gamma = -\frac{1}{3}$.

4.2 Hadronic cross section

To obtain a realistic (hadronic) cross section, we must convolute the partonic cross section with the PDFs [20]. Fig.1 shows the cross section in Pb for the various proton's spin configurations at $\sqrt{s} = 500$GeV. Denoting the helicity of initial protons by $p(\pm)$, Figs.(1.a1),(1.a2),(1.b1) and (1.b2) correspond to

$p(-)p(+)$, $p(+)p(-)$, $p(+)p(+)$ and $p(-)p(-)$ respectively. We have chosen the minimum cut-off energy for the photon to be 5 GeV.

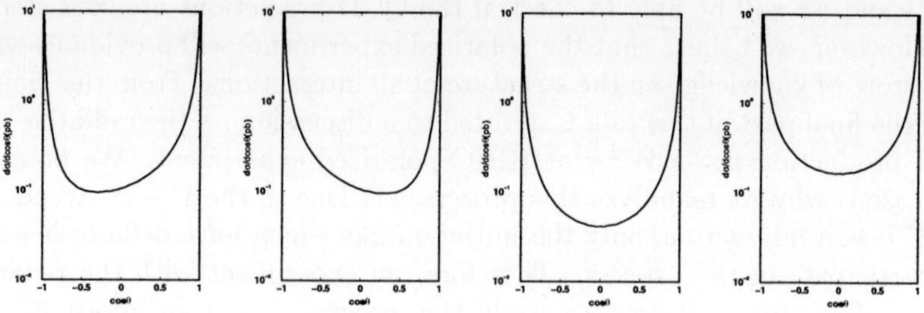

(a1) (a2) (b1) (b2)

Figure 1: Distribution of $\cos\theta_\gamma$ in the laboratory frame for the reaction $pp \to W^+\gamma + X$.

One can see a rather clear dip in the cross section when the initial proton's helicities are parallel each other. This is because the helicity distributions of quarks depend on the spin of the parent protons. Note that if the proton contains the equal parton densities of both helicity states as for the unpolarized case, the convolution completely smears out the RAZ. What is the energy dependence of this smearing effect coming from the convolution? It is easily supposed that the dip will be more smeared as the energy becomes higher. It is because the small x partons start to participate in the process at higher energy and those sea partons carry less information of the parent proton's spin. Namely the contribution from the small x partons is almost the same for the polarized and unpolarized protons.

We also define an asymmetry by,

$$A = \frac{d\sigma\,(p(-)p(+) \to W^+\gamma) - d\sigma\,(p(+)p(-) \to W^+\gamma)}{d\sigma\,(p(-)p(+) \to W^+\gamma) + d\sigma\,(p(+)p(-) \to W^+\gamma)}\,,$$

and plot it in Fig.2. This asymmetry amounts to 40 %.

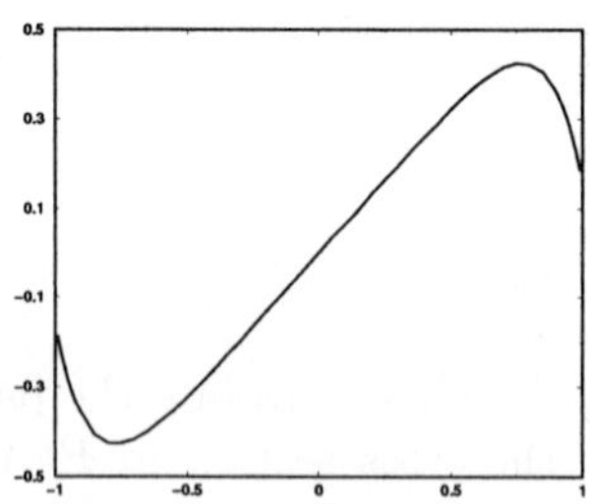

Figure 2; Asymmetry w.r.t. $\cos\theta_\gamma$.

A comment is in order concerning the higher order corrections. Since the RAZ occurs only at the partonic tree level, the higher order (QCD) corrections might give important effects and it is possible that RAZ is completely smeared out in the physical hadronic cross sections. There has been much effort to estimate higher order corrections in the standard model and phenomenological analyses [21]. The QCD radiative corrections can be classified into three effects: (1) virtual corrections and soft gluon emissions (2) hard gluon emissions (3) gluon initiated processes $q_1 g \to V\gamma q_2$ and $g\bar{q}_2 \to V\gamma\bar{q}_1$. Among these effects, the first one leaves the RAZ intact. Its main effect is an almost constant K factor. The second and third effects, on the other hand, will completely wash out the RAZ phenomena. These effects, however, have been known to be important at high energies. It is, therefore, expected that in the RHIC energy region, it is sufficient to include only the first effect which will not change the shape of the cross section from the tree level one except for some multiplicative enhancement. In particular, the asymmetry, Fig.2, remains the same.

5 Conclusions

We have surveyed the polarized structure functions from a viewpoint of gauge invariant nonlocal operators. We explained an approach to obtain the QCD evolution which preserves maximal (BRST and Lorentz) symmetries of the theory at every step of investigation. We also presented the radiative weak boson production at RHIC which strongly depends on the helicity distribution of quarks inside the proton. We have pointed out that the experiments in the RHIC energy region will be very efficient to study this process.

We hope that various kinds of new experiments and theoretical investigations will be able to clarify not only perturbative and nonperturbative aspects of QCD but also the full structure of all interactions in Nature.

Acknowledgments

The author would like to thank the organizers for their hospitality and inviting him to this stimulating workshop. He also thanks H. Kawamura, K. Morii and K. Tanaka for collaborations. This work is supported in part by the Monbu-kagaku-sho Grant-in-Aid for Scientific Research No. C-13640289.

References

1. J. Kodaira and K. Tanaka, *Prog. Theor. Phys.* 101, 191 (1999).
2. R.W. Brown and K.O. Mikaelian, *Phys. Rev.* D **19**, 922 (1979); R.W. Brown, K.O. Mikaelian and D. Shadev, *Phys. Rev.* D **20**, 1164 (1979); K.O. Mikaelian, M. A. Samuel and D. Shadev, *Phys. Rev. Lett.* **43**, 746 (1979).
3. J. C. Collins and D. E. Soper, *Nucl. Phys.* B **194**, 445 (1982).
4. R. L. Jaffe and X. Ji, *Nucl. Phys.* B **375**, 527 (1992).

5. J. Ralston and D. E. Soper, *Nucl. Phys.* B **152**, 109 (1979); J. L. Cortes, B. Pire and J. P. Ralston, *Z. Phys.* C **55**, 409 (1992).
6. A. V. Manohar, *Phys. Rev. Lett.* **65**, 2511 (1990); **66**, 289 (1991).
7. A. P. Bukhvostov, E. A. Kuraev and L. N. Lipatov, *Sov. J. Nucl. Phys.* **38**, 263 (1983); **39**, 121 (1984); *JETP Letters* **37**, 483 (1984); *Sov. Phys. JETP* **60**, 22 (1984).
8. J. Kodaira, T. Nasuno, H. Tochimura, K. Tanaka and Y. Yasui, *Prog. Theor. Phys.* **99**, 315 (1998); *in* Proc. of 2nd Topical Workshop on Deep Inelastic Scattering off Polarized Targets: Theory Meets Experiment (DESY-Zeuthen, 1997), DESY 97-200, p. 210.
9. B. Geyer, D. Müller and D. Robaschik, *Nucl. Phys. Proc. Suppl.* **51C**, 106 (1996); D. Müller, *Phys. Lett.* B **407**, 314 (1997).
10. J. Kodaira, Y. Yasui and T. Uematsu, *Phys. Lett.* B **344**, 348 (1995); J. Kodaira, Y. Yasui, K. Tanaka and T. Uematsu, *Phys. Lett.* B **387**, 855 (1996).
11. H. D. Politzer, *Nucl. Phys.* B **172**, 349 (1980).
12. E. V. Shuryak and A. I. Vainshtein, *Nucl. Phys.* B **199**, 951 (1982); *Nucl. Phys.* B **201**, 141 (1982).
13. S.D. Joglekar and B.W. Lee, *Ann. of Phys.* **97**, 160 (1976).
14. H. Kawamura, T. Uematsu, J. Kodaira and Y. Yasui, *Mod. Phys. Lett.* A **12**, 135 (1997).
15. Y. Koike, N. Nishiyama and K. Tanaka, *Phys. Lett.* B **437**, 153 (1998).
16. Y. Koike and K. Tanaka, *Phys. Rev.* D **51**, 6125 (1995).
17. V. M. Braun, G. P. Korchemsky and A. N. Manshov, hep-ph/0010128; hep-ph/0102313; *Phys. Lett.* B **476**, 455 (2000); X. Ji, W. Lu, J. Osborne and X. Song, *Phys. Rev.* D **62**, 094016 (2000); A. Belitsky, X. Ji, W. Lu, and J. Osborne, hep-ph/0007305.
18. M. Wiest, D. R. Stump, D. O. Carlson and C. -P. Yuan, *Phys. Rev.* D **52**, 2724 (1995).
19. for a recent article, see D. de Florian and A. Signer, hep-ph/0002138.
20. M. Hellmund and G. Ranft, *Z. Phys.* C **12**, 333 (1982).
21. V. Berger, T. Han, D. Zeppenfeld and J. Ohnemus, *Phys. Rev.* D **41**, 2782 (1990); J. Smith, W. L. van Neerven and J. A. M. Vermaseren, *Z. Phys.* C **30**, 621 (1986); J. Smith, D. Thomas and W. L. van Neerven, *Z. Phys.* C **44**, 267 (1989); S. Mendoza, J. Smith and W. L. van Neerven, *Phys. Rev.* D **47**, 3913 (1993); J. Ohnemus, *Phys. Rev.* D **47**, 940 (1993); U. Baur, T. Han, N. Kauer, R. Sobey and D. Zeppenfeld, *Phys. Rev.* D **56**, 140 (1997); U. Baur, T. Han and J. Ohnemus, *Phys. Rev.* D **48**, 5140 (1993).

SINGLE SPIN ASYMMETRIES AND QUARK FRAGMENTATION

M. ANSELMINO

Dipartimento di Fisica Teorica, Università di Torino and INFN, Sezione di Torino, Via P. Giuria 1, I-10125 Torino, Italy

D. BOER

RIKEN-BNL Research Center Brookhaven National Laboratory, Upton, NY 11973, USA

U. D'ALESIO and F. MURGIA

Istituto Nazionale di Fisica Nucleare, Sezione di Cagliari and Dipartimento di Fisica, Università di Cagliari C.P. 170, I-09042 Monserrato (CA), Italy

We study the fragmentation of a transversely polarized quark into a non collinear ($\boldsymbol{k}_\perp \neq 0$) spinless hadron and the fragmentation of an unpolarized quark into a non collinear transversely polarized spin 1/2 baryon. These nonperturbative properties are described by new spin and $\boldsymbol{k}_\perp$ dependent fragmentation functions and are revealed in the observation of single spin asymmetries. Recent data on the production of pions in polarized semi-inclusive DIS and older data on Λ polarization in unpolarized p–N processes are considered: these new fragmentation functions can describe the experimental results and the single spin effects in the quark fragmentation turn out to be surprisingly large.

1 Introduction

Several large and puzzling single spin asymmetries in high energy inclusive processes are experimentally well known since a long time and new ones have just been or are being measured. These effects are absent in massless perturbative QCD dynamics and they depend on new and interesting aspects of nonperturbative QCD; as such, they deserve a careful study, both theoretically and experimentally.

We consider here spin properties of quark fragmentation processes, which have been suggested in the literature [1,2,3], and address the question of whether or not they might explain some single spin asymmetries observed in inclusive processes: in particular we look at the recently observed [4,5] azimuthal dependence of the number of pions produced in polarized semi-inclusive DIS, $\ell p^\uparrow \to \ell \pi X$, and at the longstanding problem of the polarization of Λ's produced in unpolarized p–p and p–n interactions, $pN \to \Lambda^\uparrow X$ [6]. Both these unexpected spin dependences should originate in the fragmentation process of a quark, polarized in the first case ($q^\uparrow \to \pi X$) and unpolarized in the second one ($q \to \Lambda^\uparrow X$).

2 $q^\uparrow \to h\,X$ - Collins effect

The inclusive production of hadrons in DIS with transversely polarized nucleons, $\ell N^\uparrow \to \ell h X$, is the ideal process to study the so-called Collins effect, *i.e.* the spin and $\boldsymbol{k}_\perp$ dependence of the fragmentation process of a transversely polarized quark, $q^\uparrow \to hX$. In such a case, in fact, possible effects in quark distribution functions[7], which require initial state interactions[8,9], are expected to be negligible.

If one looks at the $\gamma^* N^\uparrow \to hX$ process in the γ^*–N c.m. frame, the elementary interaction is simply a γ^* hitting head on a transversely polarized quark, which bounces back and fragments into a jet containing the detected hadron. The hadron $\boldsymbol{p}_T$ in this case coincides with its $\boldsymbol{k}_\perp$ inside the jet; the fragmenting quark polarization can be computed from the initial quark one.

The spin and $\boldsymbol{k}_\perp$ dependent fragmentation function for a quark with momentum $\boldsymbol{p}_q$ and a *transverse* polarization vector $\boldsymbol{P}_q$ ($\boldsymbol{p}_q \cdot \boldsymbol{P}_q = 0$) which fragments into a hadron with momentum $\boldsymbol{p}_h = z\boldsymbol{p}_q + \boldsymbol{p}_T$ ($\boldsymbol{p}_q \cdot \boldsymbol{p}_T = 0$) can be written as:

$$D_{h/q}(\boldsymbol{p}_q, \boldsymbol{P}_q; z, \boldsymbol{p}_T) = \hat{D}_{h/q}(z, p_T) + \frac{1}{2}\,\Delta^N D_{h/q^\uparrow}(z, p_T)\,\frac{\boldsymbol{P}_q \cdot (\boldsymbol{p}_q \times \boldsymbol{p}_T)}{|\boldsymbol{p}_q \times \boldsymbol{p}_T|} \qquad (1)$$

where $\hat{D}_{h/q}(z, p_T)$ is the unpolarized fragmentation function. Notice that – as required by parity invariance – the only component of the polarization vector which contributes to the spin dependent part of D is that perpendicular to the $q - h$ plane; in general one has:

$$\boldsymbol{P}_q \cdot \frac{\boldsymbol{p}_q \times \boldsymbol{p}_T}{|\boldsymbol{p}_q \times \boldsymbol{p}_T|} = P_q \sin\Phi_C\,, \qquad (2)$$

where $P_q = |\boldsymbol{P}_q|$ and we have defined the *Collins angle* Φ_C.

When studying single spin asymmetries one considers differences of cross-sections with opposite transverse spins; by reversing the nucleon spin all polarization vectors, including those of quarks, change sign and the quantity which eventually contributes to single spin asymmetries is:

$$D_{h/q}(\boldsymbol{p}_q, \boldsymbol{P}_q; z, \boldsymbol{p}_T) - D_{h/q}(\boldsymbol{p}_q, -\boldsymbol{P}_q; z, \boldsymbol{p}_T) = \Delta^N D_{h/q^\uparrow}(z, p_T)\,\frac{\boldsymbol{P}_q \cdot (\boldsymbol{p}_q \times \boldsymbol{p}_T)}{|\boldsymbol{p}_q \times \boldsymbol{p}_T|} \qquad (3)$$

which implies the existence of a *quark analysing power* for the fragmentation process $q \to h\,X$:

$$A_q^h(\boldsymbol{p}_q, \boldsymbol{P}_q; z, \boldsymbol{p}_T) = \frac{D_{h/q}(\boldsymbol{p}_q, \boldsymbol{P}_q; z, \boldsymbol{p}_T) - D_{h/q}(\boldsymbol{p}_q, -\boldsymbol{P}_q; z, \boldsymbol{p}_T)}{D_{h/q}(\boldsymbol{p}_q, \boldsymbol{P}_q; z, \boldsymbol{p}_T) + D_{h/q}(\boldsymbol{p}_q, -\boldsymbol{P}_q; z, \boldsymbol{p}_T)} \tag{4}$$

$$= \frac{\Delta^N D_{h/q^\uparrow}(z, p_T)}{2\,\hat{D}_{h/q}(z, p_T)}\, \frac{\boldsymbol{P}_q \cdot (\boldsymbol{p}_q \times \boldsymbol{p}_T)}{|\boldsymbol{p}_q \times \boldsymbol{p}_T|} \equiv A_q^h(z, p_T)\, \frac{\boldsymbol{P}_q \cdot (\boldsymbol{p}_q \times \boldsymbol{p}_T)}{|\boldsymbol{p}_q \times \boldsymbol{p}_T|}\,.$$

This results in a single spin asymmetry [10]:

$$A_N^h(x, y, z, \Phi_C, p_T) = \frac{d\sigma^{\ell+p,\boldsymbol{P}\to\ell+h+X} - d\sigma^{\ell+p,-\boldsymbol{P}\to\ell+h+X}}{d\sigma^{\ell+p,\boldsymbol{P}\to\ell+h+X} + d\sigma^{\ell+p,-\boldsymbol{P}\to\ell+h+X}}$$

$$= \frac{\sum_q e_q^2\, h_{1q}(x)\, \Delta^N D_{h/q^\uparrow}(z, p_T)}{2\sum_q e_q^2\, f_{q/p}(x)\, \hat{D}_{h/q}(z, p_T)}\, \frac{2(1-y)}{1+(1-y)^2}\, P\, \sin\Phi_C\,, \tag{5}$$

where P is the transverse (with respect to the γ^* direction) proton polarization.

We wonder how large the quark analysing power $A_q^h(z, p_T)$ can be. Such a question has been addressed in Ref. [10], where recent data on A_N^π [4,5] were considered. We refer to that paper for all the details and only outline the main procedure here. Under some realistic assumptions and using isospin and charge conjugation invariance, Eq. (5) gives ($i = +, -, 0$):

$$A_N^{\pi^i}(x, y, z, \Phi_C, p_T) = \frac{h_i(x)}{f_i(x)}\, A_q^\pi(z, p_T)\, \frac{2(1-y)}{1+(1-y)^2}\, P\, \sin\Phi_C \tag{6}$$

where:

$$i = + : \quad h_+ = 4h_{1u} \qquad f_+ = 4f_{u/p} + f_{\bar{d}/p} \tag{7}$$

$$i = - : \quad h_- = h_{1d} \qquad f_- = f_{d/p} + 4f_{\bar{u}/p} \tag{8}$$

$$i = 0 : \quad h_0 = 4h_{1u} + h_{1d} \qquad f_0 = 4f_{u/p} + f_{d/p} + 4f_{\bar{u}/p} + f_{\bar{d}/p}\,. \tag{9}$$

The f's are the unpolarized distribution functions and the h_1's are the transversity distributions. Notice that the above equations imply – at large x values – $A_N^{\pi^+} \simeq A_N^{\pi^0}$ as observed by HERMES [11].

We bound the unknown transversity distributions saturating Soffer's inequality [12]

$$|h_{1q}| \le \frac{1}{2}\,(f_{q/p} + \Delta q)\,, \tag{10}$$

and, by comparing with SMC data [5]

$$A_N^{\pi^+} \simeq -(0.10 \pm 0.06)\, \sin\Phi_C\,, \tag{11}$$

we obtain the significant lower bound for pion valence quarks:

$$|A_q^\pi(\langle z\rangle, \langle p_T\rangle)| \gtrsim (0.24 \pm 0.15) \qquad \langle z\rangle \simeq 0.45\,, \quad \langle p_T\rangle \simeq 0.65\,\mathrm{GeV}/c\,. \quad (12)$$

A similar result is obtained by using HERMES data [4], although their proton transverse polarization is much smaller. These lower bounds of the quark analysing power are remarkably large.

The Collins mechanism might be (at least partly) responsible for other observed spin asymmetries: in particular the E704 data [13] on single transverse spin asymmetries in the inclusive process $p^\uparrow p \to \pi X$ can be related to the Collins function [8,14]:

$$\frac{E_\pi d^3\sigma^{p^\uparrow p\to\pi X}}{d^3\boldsymbol{p}_\pi} - \frac{E_\pi d^3\sigma^{p^\downarrow p\to\pi X}}{d^3\boldsymbol{p}_\pi} = \sum_{a,b,c,d} \int \frac{dx_a\, dx_b}{\pi z} \int d^2\boldsymbol{k}_\perp$$

$$\times\, h_{1a}(x_a)\, f_{b/p}(x_b)\, \Delta_{NN}\hat{\sigma}^{ab\to cd}(x_a, x_b, \boldsymbol{k}_\perp)\, \Delta^N D_{\pi/c^\uparrow}(z, \boldsymbol{k}_\perp) \qquad (13)$$

where $\Delta_{NN}\hat{\sigma}$ is a double spin elementary asymmetry, given by pQCD dynamics:

$$\Delta_{NN}\hat{\sigma}^{ab\to cd} \equiv \frac{d\hat{\sigma}^{a^\uparrow b\to c^\uparrow d} - d\hat{\sigma}^{a^\uparrow b\to c^\downarrow d}}{d\hat{\sigma}^{a^\uparrow b\to c^\uparrow d} + d\hat{\sigma}^{a^\uparrow b\to c^\downarrow d}}\,. \qquad (14)$$

Eq. 13 allows to obtain, with a simple parameterization of the Collins function, good fits of the E704 results [8,14].

3 $q \to h^\uparrow X$ - Λ polarization

We consider now the possibility that an unpolarized quark fragments into a transversely polarized hadron [2,3]; in analogy to Eq. (1) we write

$$\hat{D}_{h^\uparrow/q}(z, \boldsymbol{k}_\perp) = \frac{1}{2}\,\hat{D}_{h/q}(z, k_\perp) + \frac{1}{2}\,\Delta^N D_{h^\uparrow/q}(z, k_\perp)\, \frac{\hat{\boldsymbol{P}}_h \cdot (\boldsymbol{p}_q \times \boldsymbol{k}_\perp)}{|\boldsymbol{p}_q \times \boldsymbol{k}_\perp|} \qquad (15)$$

for an unpolarized quark with momentum $\boldsymbol{p}_q$ which fragments into a spin 1/2 hadron h with momentum $\boldsymbol{p}_h = z\boldsymbol{p}_q + \boldsymbol{k}_\perp$ and polarization vector along the $\uparrow = \hat{\boldsymbol{P}}_h$ direction. $\Delta^N D_{h^\uparrow/q}(z, k_\perp)$ (denoted by $D_{1T}^\perp$ in Ref. [2]) is a new *polarizing fragmentation function.*

This reflects into a possible polarization of hadrons inclusively produced in the high energy interaction of unpolarized nucleons. Indeed, it is well known since a long time that Λ hyperons produced with $x_F \gtrsim 0.2$ and $p_T \gtrsim 1$ GeV/c in the collision of two unpolarized nucleons are polarized perpendicularly to

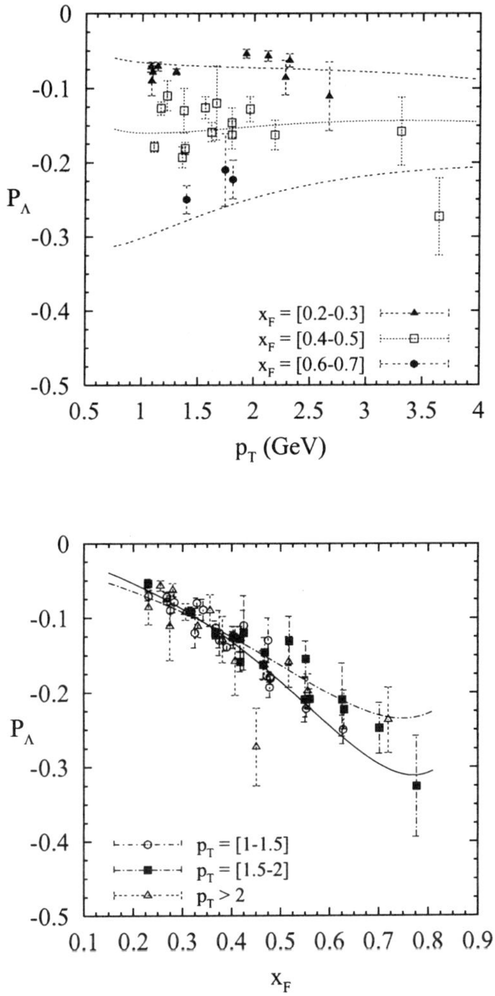

Figure 1: Our best fit to P_Λ data from p–Be reactions (a partial collection from Ref. [6]) as a function of p_T (above) and of x_F (below). For each x_F-bin, the corresponding theoretical curve is evaluated at the mean x_F value in the bin. The two theoretical curves, in the lower plot, correspond to $p_T = 1.5$ GeV/c (solid) and $p_T = 3$ GeV/c (dot-dashed).

the production plane, as allowed by parity invariance; a huge amount of experimental information, for a wide energy range of the unpolarized beams, is available on such single spin asymmetries[6]:

$$P_\Lambda = \frac{d\sigma^{pN\to\Lambda^\uparrow X} - d\sigma^{pN\to\Lambda^\downarrow X}}{d\sigma^{pN\to\Lambda^\uparrow X} + d\sigma^{pN\to\Lambda^\downarrow X}} \,. \tag{16}$$

By taking into account intrinsic $\boldsymbol{k}_\perp$ in the hadronization process, and assuming that the factorization theorem holds also when $\boldsymbol{k}_\perp$'s are included[1], one obtains

$$\frac{E_\Lambda \, d^3\sigma^{pN\to\Lambda X}}{d^3\boldsymbol{p}_\Lambda} \, P_\Lambda = \sum_{a,b,c,d} \int \frac{dx_a \, dx_b \, dz}{\pi z^2} \, d^2\boldsymbol{k}_\perp \, f_{a/p}(x_a) \, f_{b/N}(x_b)$$

$$\times \, \hat{s} \, \delta(\hat{s}+\hat{t}+\hat{u}) \, \frac{d\hat{\sigma}^{ab\to cd}}{d\hat{t}}(x_a, x_b, \boldsymbol{k}_\perp) \, \Delta^N D_{\Lambda^\uparrow/c}(z, \boldsymbol{k}_\perp) \,. \tag{17}$$

We use the above equation, together with a simple parameterization of the polarizing fragmentation functions, to see whether or not one can fit the experimental data on Λ and $\bar{\Lambda}$ polarization. All details can be found in Ref. [3] and some results are shown here in Fig. 1.

The data can be described with remarkable accuracy in all their features: the large negative values of the Λ polarization, the increase of its magnitude with x_F, the puzzling flat $p_T \gtrsim 1$ GeV/c dependence and the $\sqrt{s}$ apparent independence; data from p–p processes are in agreement with data from p–Be interactions and also the tiny or zero values of $\bar{\Lambda}$ polarization are well reproduced. The resulting functions $\Delta^N D_{\Lambda^\uparrow/q}$ are very reasonable and realistic.

We conclude by stressing that a systematic phenomenological approach towards the description and prediction of single transverse spin asymmetries, based on perturbative QCD dynamics and nonperturbative quark properties, is now possible and worth being developed.

Acknowledgments

One of us (M.A.) would like to thank the special research Centre for the Subatomic Structure of Matter (CSSM) and its Director, Prof. A. Thomas, for support and for providing working facilities and a stimulating atmosphere during the whole Workshop.

References

1. J.C. Collins, *Nucl. Phys.* B **396**, 161 (1993).

2. P.J. Mulders and R.D. Tangerman, *Nucl. Phys.* B **461**, 197 (1996); *Nucl. Phys.* B **484**, 538 (1997) (E).
3. M. Anselmino, D. Boer, U. D'Alesio and F. Murgia, *Phys. Rev.* D **63**, 054029 (2001).
4. H. Avakian (on behalf of the HERMES collaboration), *Nucl. Phys.* B **79**, 523 (1999)(Proc. Suppl.); HERMES Collaboration, A. Airapetian *et al.*, *Phys. Rev. Lett.* **84**, 4047 (2000).
5. A. Bravar (on behalf of the SMC collaboration), *Nucl. Phys.* B **79**, 520 (1999) (Proc. Suppl.).
6. For a review of data see, *e.g.*, K. Heller, in Proceedings of Spin 96, C.W. de Jager, T.J. Ketel and P. Mulders, Eds., World Scientific (1997); or A.D. Panagiotou, *Int. J. Mod. Phys.* A **5**, 1197 (1990).
7. D. Sivers, *Phys. Rev.* D **41**, 83 (1990); *Phys. Rev.* D **43**, 261 (1991).
8. M. Anselmino, M. Boglione and F. Murgia, *Phys. Rev.* D **60**, 054027 (1999).
9. M. Anselmino, M. Boglione, J. Hansson and F. Murgia, *Eur. Phys. J.* C **13**, 519 (2000).
10. M. Anselmino and F. Murgia, *Phys. Lett.* B **483**, 74 (2000).
11. HERMES Collaboration, A. Airapetian *et al.*, e-Print Archive: hep-ex/0104005.
12. J. Soffer, *Phys. Rev. Lett.* **74**, 1292 (1995).
13. D.L. Adams *et al.*, *Phys. Lett.* B **264**, 462 (1991); A. Bravar *et al.*, *Phys. Rev. Lett.* **77**, 2626 (1996).
14. M. Boglione and E. Leader, *Phys. Rev.* D **61**, 114001 (2000).

DETERMINING ΔG WITH POLARIZED PHOTO- AND HADROPRODUCTION OF HEAVY QUARKS

I. BOJAK
CSSM, Adelaide University, SA 5005, Australia
E-mail: ibojak@physics.adelaide.edu.au

We present the first calculation of the complete NLO QCD corrections to the production of heavy flavors with longitudinally polarized hadrons. Predictions for the extraction of the gluon helicity density at RHIC using this reaction are made and their theoretical uncertainties are briefly discussed. We also mention the potential of polarized photoproduction of heavy flavors at the EIC project.

1 Introduction

The gluon helicity density $\Delta g(x, \mu_f^2) \equiv g_+(x, \mu_f^2) - g_-(x, \mu_f^2)$ continues to be only weakly determined by experiments [1,2]. Hence the gluonic contribution to the nucleon spin, given by the first Mellin moment $\Delta g_{n=1}(\mu_f^2)$, also remains uncertain. All data are compatible with a vanishing $\Delta g \equiv 0$ at a low input scale $\mu_{f0}^2 = 0.4$ GeV^2, but we cannot yet completely exclude full saturation $g_\pm \equiv 0$ either [1]. With $\Delta g_{n=1}(5 \text{ GeV}^2)$ somewhere between -0.8 and 1.7, improvements are urgently needed. Of course gluons can be probed only indirectly, but there are three additional reasons why Δg is much less constrained than its unpolarized counterpart $g \equiv g_+ + g_-$. First, polarized DIS data come only from fixed target experiments. These data have a severely restricted kinematical range in which approximate "scaling" prevents pinning down the gluon. Second, the momentum sum rule $g_{n=2} + \Sigma_{n=2} = 1$ constrains the unpolarized gluon distribution. But the helicity sum rule $\Delta g_{n=1} + \Delta\Sigma_{n=1} + L_z = \frac{1}{2}$ is currently useless, since the angular momentum of the partons L_z remains unknown. Third, data from exclusive processes is still missing in the polarized parton fits[a], but play a major rôle in the unpolarized fits.

Two polarized experiments will collect such data for open heavy quarks soon: COMPASS (photoproduction) and RHIC (hadroproduction) [3]. In both cases, we have provided the first complete calculation in NLO QCD. The older photoproduction analysis included predictions for COMPASS [4,5], and has later been confirmed by a second calculation [6]. As an update, we estimate here what the EIC project, which would add a polarized electron beam to RHIC, could achieve with the photoproduction of charm quarks in about one month of

[a]Semi-inclusive DIS data has been used, but does not constrain the gluon helicity density.

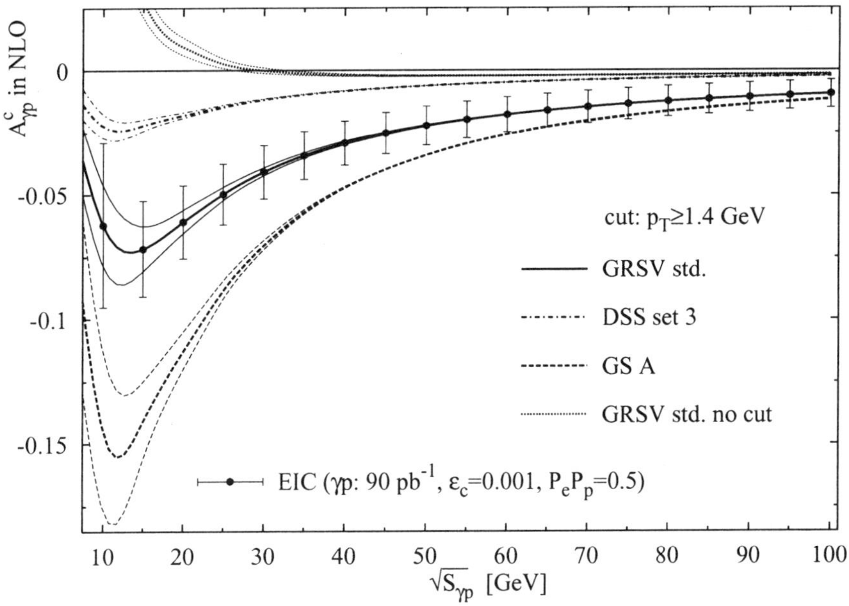

Figure 1. NLO $\bar{c}$ asymmetry with a cut $p_T \geq 1.4$ GeV using different helicity densities [2]. The bands correspond to $m_c = 1.4 - 1.6$ GeV and the EIC error is for one month running.

running, see Fig. 1. Different helicity density sets [2] could be distinguished very easily. In the following we will discuss the recently completed hadroproduction calculation and make first predictions for RHIC in NLO QCD. Note that we could also calculate *resolved* photon contributions in NLO QCD now.

2 Partonic Results for Hadroproduction

The calculation methods we have used are explained in great detail in a previous publication [5]. Here we remind the reader of just a few important details. The partonic reactions that have to be considered in NLO are

Born & virtual loops	real contributions	
$\vec{g} + \vec{g} \to Q + \overline{Q}$,	$\vec{g} + \vec{g} \to Q + \overline{Q} + g$,	(1)
$\vec{q} + \vec{\overline{q}} \to Q + \overline{Q}$,	$\vec{q} + \vec{\overline{q}} \to Q + \overline{Q} + g$,	(2)
	$\vec{g} + \vec{q}/\vec{\overline{q}} \to Q + \overline{Q} + q/\overline{q}$.	(3)

Spurious singularities are regularized by calculating in $n = 4 + \varepsilon$ dimensions. Since the helicity projections involve the purely four-dimensional γ_5 and $\epsilon_{\mu\nu\rho\sigma}$, we employ the HVBM [7] prescription to extend these objects to arbitrary dimensions. This adds two complications as compared to the unpolarized case: First, we have to integrate over non-vanishing scalar products in the ε-dimensional subspace. Second, we have to cure the unphysical violation of helicity conservation at the qqg-vertex by subtracting a finite piece in mass factorization.

We keep the full helicity dependence of the squared matrix element $|M|^2(\lambda_1, \lambda_2)$. Thanks to parity conservation, then

$$|M|^2(\lambda_1,\lambda_2) = \overline{|M|^2} + \lambda_1\lambda_2\,\Delta|M|^2 \,. \tag{4}$$

Thus we *simultaneously* obtain the unpolarized $\overline{|M|^2}$ and polarized $\Delta|M|^2$ squared matrix elements. Hence we can indirectly check our polarized results against previously published unpolarized calculations of the same reactions. For renormalization we use a modified $\overline{\text{MS}}$ scheme: light flavors are treated by the standard $\overline{\text{MS}}$ prescription, but the produced heavy flavor is renormalized on-shell and its loop contribution to the gluon self-energy is subtracted. This explicitly decouples the heavy flavor at low energies, and of the $n_f = n_{lf} + 1$ flavors, only the light flavors n_{lf} contribute in the running of the coupling. For mass factorization we have also used the $\overline{\text{MS}}$ scheme.

Since we are observing a massive particle in the final state, we can obtain a finite *total* partonic cross section[b][c]

$$\tilde{\hat{\sigma}} = \frac{\alpha_s^2(\mu_r^2)}{m^2}\left\{\tilde{f}^{(0)}(\eta) + 4\pi\alpha_s(\mu_r^2)\left[\tilde{f}^{(1)}(\eta) + \tilde{\bar{f}}^{(1)}(\eta)\ln\frac{\mu_f^2}{m^2}\right]\right\}\,, \tag{5}$$

where $\eta \equiv \frac{s-4m^2}{4m^2}$. The scaling functions $\tilde{f}(\eta)$ are convenient for displaying the partonic results. Most of them have been shown before [5], so Fig. 2 contains the recently completed gluon-gluon ones. In the polarized case we observe, as has been known for the unpolarized case, that the NLO corrections are sizable everywhere. However, this does not indicate a breakdown in the perturbation series, since the large deviations stem from Feynman graph topologies occurring for the first time in NLO. For example, a Coulomb singularity leads to a constant NLO cross section at threshold, whereas LO vanishes simply because the responsible gluon interaction is of higher order. In NNLO we do not expect similar qualitative changes, hence the conclusion is that NLO rather than LO is the first trustworthy order for this process.

[b] $\tilde{\hat{\sigma}}$ can be either the unpolarized $\hat{\sigma}$ or the polarized $\Delta\hat{\sigma}$, and the $\tilde{f}$ match accordingly.
[c] Set $\alpha_s(\mu_r^2) \to \alpha_s(\mu_f^2)[1 + \alpha_s(\mu_f^2)(\beta_0/4\pi)\ln(\mu_r^2/\mu_f^2)]$ in Eq. (5), if μ_r does not equal μ_f.

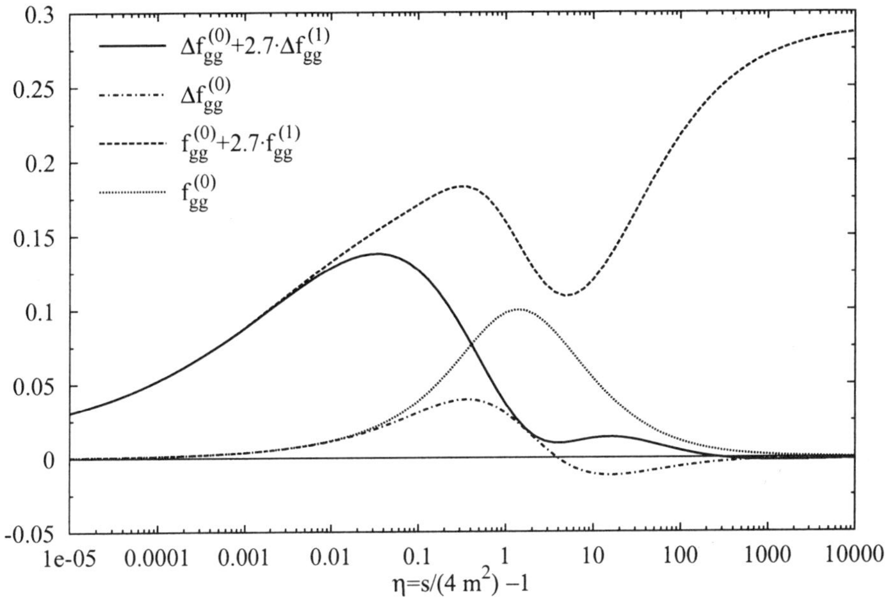

Figure 2. The NLO and LO, polarized and unpolarized, gluon-gluon scaling functions in the (modified) $\overline{\text{MS}}$ scheme. We have set $4\pi\alpha_s \simeq 2.7$, as appropriate for charm.

3 Hadron Level Results

By convoluting our partonic results with parton densities, we can obtain the cross sections, and from them the double spin asymmetry $A = (d)\Delta\sigma/(d)\sigma$. In this ratio systematic errors will mostly cancel and the absolute normalization is not needed. However, we stress that in light of the lack of precision for charm and mismatch for bottom of *unpolarized* theory vs. experiment, one *must* measure $\Delta\sigma$ and σ separately here! With that caveat in mind, let us turn to the stability of the asymmetry prediction displayed in Fig. 3. We see that there has been a drastic improvement in going from LO to NLO with respect to the variation of $\mu_f(=\mu_r)$ and m_c: the prediction varies by -90% to 130% in LO, but only by -20% to 20% in NLO. Note furthermore that the uncertainty in the NLO case stems almost completely from the m_c variation.

Encouraged by the strongly improved stability in NLO, we now present predictions for the charm asymmetry at RHIC in Fig. 4. We show the asymmetry as a function of $x_T^{\min}$, which means p_T-cuts: $p_T \geq p_T^{\min} = x_T^{\min} \cdot p_T^{\max}$.

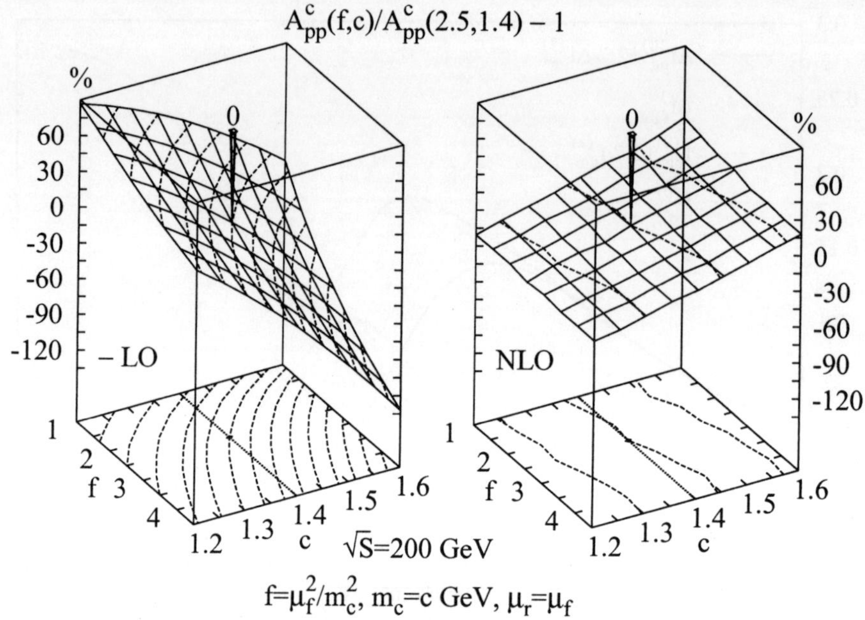

Figure 3. Deviation in % of the asymmetry from its value at $\mu_f^2 = 2.5\, m_c^2$ and $m_c = 1.4$ GeV as the scales $\mu_f^2 (= \mu_r^2)$ and m_c are varied. The GRSV'00 std. helicity densities [1] are used.

We see that a wide range of current and older helicity densities [1,2] can be clearly distinguished. However, there is a marked difference between the LO and NLO predictions, hence a LO calculation is not a good approximation even from a practical point of view. We conclude that at RHIC one should be able to determine Δg well with this process. However, polarized and unpolarized, charm and bottom cross sections should all be measured separately, with other Δg measurements at RHIC providing essential consistency checks. Then spin-dependent data may even help solving the enigma why theory describes the production of charm better than the production of bottom.

Acknowledgements

I thank M. Stratmann for the fruitful collaboration on this project. This work has been supported in part by the 'Bundesministerium für Bildung, Wissenschaft, Forschung und Technologie' and the Australian Research Council.

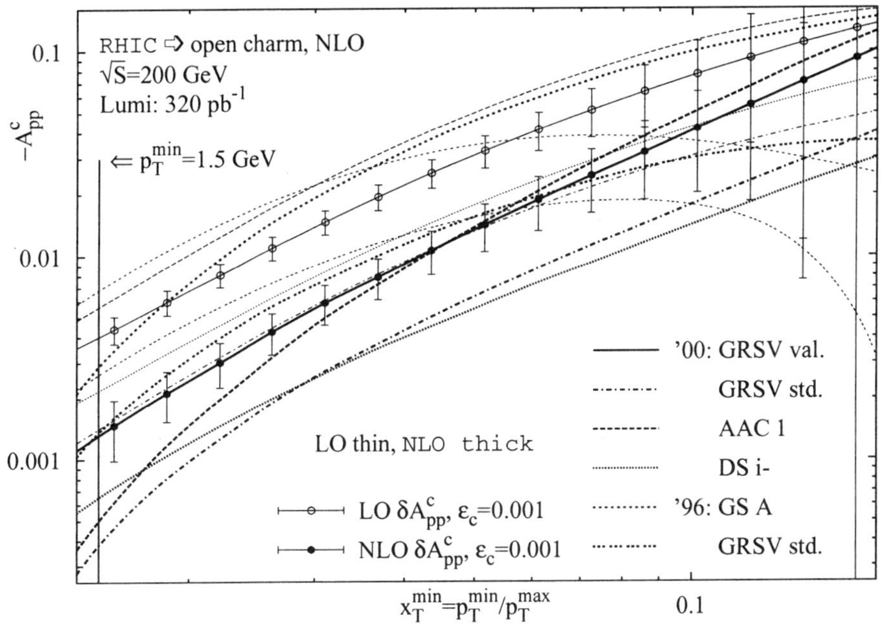

Figure 4. NLO and LO predictions for the anticharm asymmetry at RHIC using different sets of helicity densities [1,2]. $x_T^{\min}$ represents a p_T-cut: $p_T \geq p_T^{\min} = x_T^{\min} \cdot p_T^{\max}$.

References

1. M. Glück *et al.*, *Phys. Rev.* D **63**, 094005 (2001).
2. M. Glück *et al.*, *Phys. Rev.* D **53**, 4775 (1996); T. Gehrmann and W.J. Stirling, *Phys. Rev.* D **53**, 6100 (1996); D. de Florian *et al.*, *Phys. Rev.* D **57**, 5803 (1998); Y. Goto *et al.*, AA Collab., *Phys. Rev.* D **62**, 034017 (2000); D. de Florian and R. Sassot, *Phys. Rev.* D **62**, 094025 (2000).
3. F. Bradamante, *Prog. Part. Nucl. Phys.* **44**, 339 (2000); G. Bunce *et al.*, *Ann. Rev. Nucl. Part. Sci.* **50**, 525 (2000).
4. I. Bojak and M. Stratmann, *Phys. Lett.* B **433**, 411 (1998), *Nucl. Phys.* B **540**, 345 (1999) and Erratum *Nucl. Phys.* B **569**, 694 (2000).
5. I. Bojak, Ph.D. Thesis, Universität Dortmund, 2000, hep-ph/0005120.
6. A.P. Contogouris *et al.*, *Phys. Lett.* B **482**, 93 (2000) and *Phys. Rev.* D **62**, 114509 (2000).
7. G. 't Hooft and M. Veltman, *Nucl. Phys.* B **44**, 189 (1972); P. Breitenlohner and D. Maison, *Comm. Math. Phys.* **52**, 11 (1977).

FLAVOUR SYMMETRY BREAKING IN THE POLARIZED NUCLEON SEA

FU-GUANG CAO and A. I. SIGNAL
Institute of Fundamental Sciences, Massey University
Palmerston North, New Zealand

After a brief review on flavour symmetry breaking (FSB) in the unpolarized nucleon sea, we discuss theoretical predications for FSB in the polarized nucleon sea from meson cloud and 'Pauli blocking'.

1 Flavour symmetry breaking in the unpolarized nucleon sea

The possible breaking of parton model symmetries by the nucleon's quark distribution functions has been a topic of great interest since the experimental discoveries that the Ellis-Jaffe and Gottfried sum rules are violated. In particular, the flavour asymmetry in the nucleon sea ($\bar{d} > \bar{u}$) has been confirmed by several experiments [1]. This asymmetry can be naturally explained in the meson cloud model [2], in which the physical nucleon can be viewed as a bare nucleon plus some meson-baryon Fock states which result from the fluctuation $N \to MB$,

$$|p\rangle_{\rm phys.} = |p\rangle_{\rm bare} + |\pi N\rangle + |\pi\Delta\rangle + \cdots. \tag{1}$$

The valence anti-quark in the meson contributes (via a convolution) to the anti-quark distributions in the proton sea. Since the probability of the Fock state $|n\pi^+\rangle$ is larger than that of the $|\Delta^{++}\pi^-\rangle$ state in the proton wave function, the asymmetry $\bar{d} > \bar{u}$ emerges naturally in the proton sea. Another possible source for this asymmetry is that the bare nucleon, $|p\rangle_{\rm bare}$ may have an intrinsic asymmetry associated with it. According to the Pauli exclusion principle, the $d\bar{d}$ is more likely to be created than the $u\bar{u}$ pair since there are two valence u quarks and only one valence d quark in the proton. So there is a small excess of $d\bar{d}$ pairs over $u\bar{u}$ pairs. This asymmetry has also been studied in the chiral quark model [3], the chiral quark-soliton model [4], and the instanton model [5]. It was shown by Melnitchouk, Speth and Thomas [6] that by using the meson cloud model together with the Pauli blocking, the data for both the ratio $\bar{d}(x)/\bar{u}(x)$ and difference $\bar{d}(x) - \bar{u}(x)$ can be described reasonably well, while using one of these effects will not (see Fig. 1). About half of the asymmetry can be attributed to the meson cloud and the other half to the Pauli blocking. We would like to point out that the data that were compared with is the E866 data in 1998. Recently E866 collaboration reported its improved measurements [7]

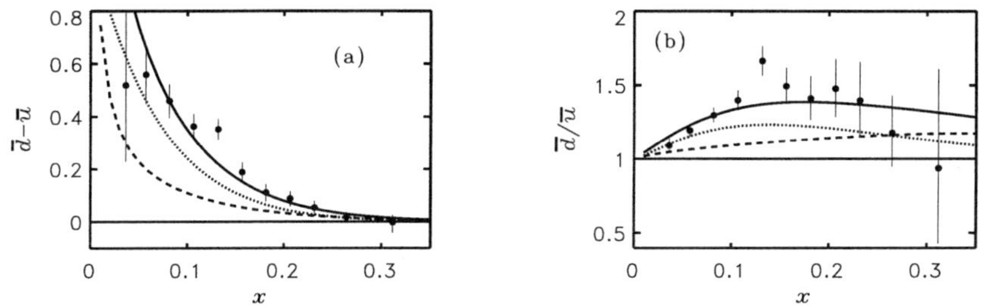

Figure 1: Contributions from pions with $\Lambda_{\pi N} = 1$ GeV and $\Lambda_{\pi\Delta} = 1.3$ GeV (dashed) and from antisymmetrization (dotted) to the (a) $\bar{d} - \bar{u}$ difference and (b) $\bar{d}/\bar{u}$ ratio, and the combined effect (solid). Taken from Phys. Rev. D **59**, 014003 (1998).

in which the statistics are improved and the measured x-range is extended to lower x (from 0.036 to 0.026). An interesting point is that comparing with the previous results, the ratio for x being 0.315 is pushed down from about 0.9 to about 0.4 and the difference $\bar{d} - \bar{u}$ becomes negative, while the other data remain nearly unchanged. So for the last data point in large x in Fig. 1 the theoretical predications will be well outside the error bars. More studies are needed in theoretical calculations and experimental measurements.

2 Flavour symmetry breaking in the polarized nucleon sea

Recently there has been increasing interest in the question of whether this asymmetry extends also to the polarized sea distributions *i.e.* $\Delta\bar{d}(x) \neq \Delta\bar{u}(x)$? Such a polarized sea asymmetry would make a direct contribution to the Bjorken sum rule. Although well established experimental evidence for a polarized sea asymmetry is still lacking, some experimental studies have been done [8]. Moreover several parameterizations [9] for the polarized parton distributions arising from fits of the world data from polarized experiments leave open the possibility of this asymmetry. There have also been some theoretical studies on this asymmetry. A much larger asymmetry in the polarized sea distributions than that in the unpolarized sea distributions is predicted in the chiral quark-soliton model (using the large-N_C limit) [10]. Such sizeable asymmetries would make an important contribution (around 20%) to the Bjorken sum rule. This asymmetry has also been studied by considering the ρ meson cloud in the meson cloud model [11]. The prediction for $\Delta\bar{d}(x) - \Delta\bar{u}(x)$ is more than one order of magnitude smaller than the result from the chiral quark-soliton model.

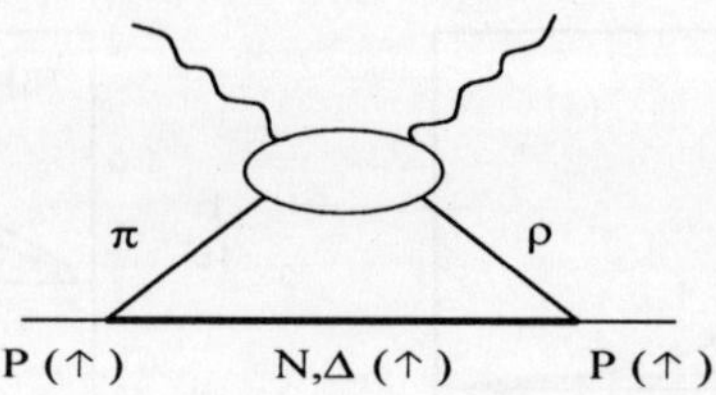

Figure 2: Schematic illustration of interference contributions to the polarized anti-quark distributions.

More theoretical calculations can be found in reference [12]. Here we report a study [13] on the flavour asymmetry of the non-strange polarized anti-quarks using the meson cloud model and 'Pauli blocking'.

2.1 FSB in the meson cloud model

It was assumed in the meson cloud model (MCM) that the lifetime of a virtual baryon-meson Fock state is much larger than the interaction time in the deep inelastic or Drell-Yan process, thus the quark and anti-quark in the virtual meson-baryon Fock states can contribute to the parton distributions of the nucleon. For polarised parton distributions in the model it is necessary to include all the terms which can lead to the same final state [14]. This allows the possibility of interference terms between different terms in the nucleon wavefunction Eq. (1). For polarised anti-quark distributions the interference will be between terms with different mesons and the same baryon e.g. $N\pi$ and $N\rho$ (see Fig. 2). We will consider the fluctuations $p \to N\pi, N\rho, N\omega$ and $p \to \Delta\pi, \Delta\rho$. The fluctuation $p \to \Delta\omega$ is neglected because of isospin conservation. The flavour asymmetry is studied by calculating the difference $x(\Delta\bar{d} - \Delta\bar{u})$ which turns out to be

$$\begin{aligned} x(\Delta\bar{d} - \Delta\bar{u}) = & \ [\frac{2}{3}\Delta f_{\rho N/N} - \frac{1}{3}\Delta f_{\rho\Delta/N}] \otimes \Delta v_\rho \\ & +[-\Delta f_{(\rho^0\omega)p/p} + \frac{2}{3} f^0_{(\pi\rho)N/N} - \frac{1}{3} f^0_{(\pi\rho)\Delta/N} - f^0_{(\pi^0\omega)p/p}] \otimes \Delta v_\rho \\ = & \ \Delta f_\rho \otimes \Delta v_\rho + \Delta f_{int} \otimes \Delta v_\rho . \end{aligned} \quad (2)$$

where Δv_ρ is the polarized parton distribution of the ρ meson, $\Delta f_{(V_1V_2)B/N} = f^1_{(V_1V_2)B/N} - f^{-1}_{(V_1V_2)B/N}$ is the polarized fluctuation function and $f^\lambda_{(M_1M_2)B/N} = \sum_{\lambda'} \int_0^\infty dk_\perp^2 \phi^{\lambda\lambda'}_{M_1B}(y, k_\perp^2) \phi^{*\,\lambda\lambda'}_{M_2B}(y, k_\perp^2)$ is the helicity dependent fluctuation function. $\phi^{\lambda\lambda'}_{MB}(y, k_\perp^2)$ is the wave function of the Fock state containing a meson

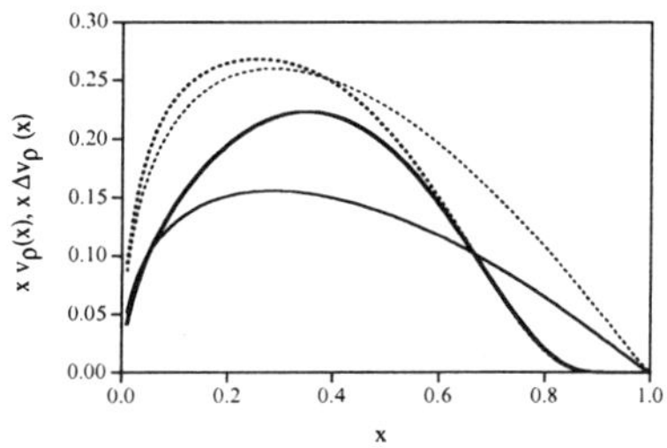

Figure 3: The polarized and unpolarized valence quark distributions of the ρ meson at $Q^2 = 4$ GeV^2. See the text for more explanations.

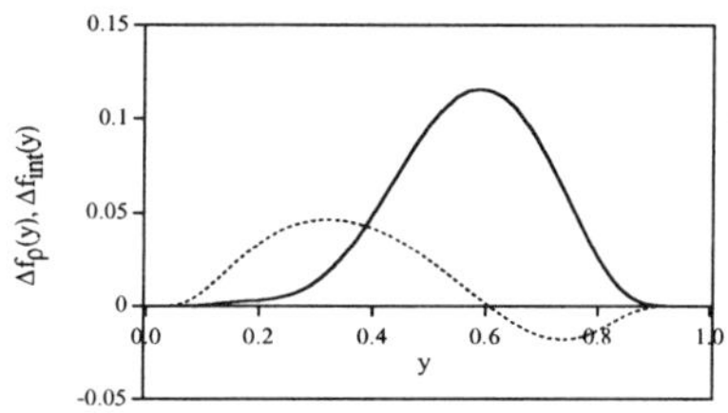

Figure 4: The polarized fluctuation functions Δf_ρ (the solid curve) and the interference term Δf_{int} (the dashed curve). $\Lambda_{oct} =$ 1.08 GeV, $\Lambda_{dec} = 0.98$ GeV.

(M) with longitudinal momentum fraction y, transverse momentum $\mathbf{k}_\perp$, and helicity λ, and a baryon (B) with momentum fraction $1-y$, transverse momentum $-\mathbf{k}_\perp$, and helicity λ'. The first term in Eq. (2) is the same as the result given in [11]. The second term in Eq. (2) is the interference contribution. We note that there are no contributions directly from the ω meson due to its charge structure.

We adopt two prescriptions for Δv_ρ (see Fig. 3): (i) employing the MIT bag model calculation, $\Delta v_\rho^{MIT}(x)$ (the thick solid curve) and (ii) adopting the ansatz $\Delta v_\rho(x) = 0.6 v_\pi(x)$ (the thin solid curve) as in [11]. The parameters of the MIT bag model calculation are fixed by fitting the calculated unpolarized parton distribution of the ρ meson (the thick dashed curve) to the Glück-Reya-Schienbein parameterization [15] for the valence parton distribution of the pion (the thin dashed curve). The first moment of $\Delta v_\rho^{MIT}(x)$ is found to be about 0.60 at $Q^2 = 4$ GeV^2, which is in agreement with the lattice value [16] of 0.60. It can be seen that the distribution $0.6\, x v_\pi(x)$ is smaller than $x \Delta v_\rho^{MIT}(x)$ in the intermediate x region, although both distributions satisfy the same normalization condition. Also the bag model calculated parton distribution has a different x-dependence from the unpolarized distribution.

The fluctuation functions Δf_ρ and Δf_{int} in Eq.(2) are calculated by using time-ordered perturbation theory in the infinite momentum frame [13,17]. (see Fig. 4. Λ is a cut-off parameter in the phenomenological form factor introduced to describe the unknown dynamics in the fluctuation $N \to MB$.) It can be seen that the maximum of Δf_{int} is about 40% that of Δf_ρ. So the interference contribution to $x(\Delta\bar{d} - \Delta\bar{u})$ will not be negligible, although Δf_{int} changes sign from positive to negative at about $y = 0.6$.

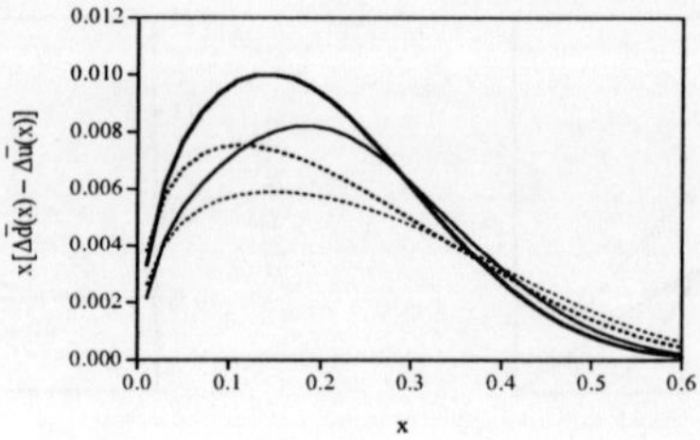

Figure 5: The flavour asymmetry of the anti-quark in the proton. The solid curves are the predictions using $x\Delta v_\rho^{MIT}$, while the dashed curves are obtained by using $0.6\,xv_\pi(x)$. The thin curves are the results without interference contribution while the thick curves are the results with interference contribution.

The results for $x(\Delta\bar{d}-\Delta\bar{u})$ are shown in Fig. 5. The interference effect increases sizably the predictions for the flavour asymmetry, and pushes the curves towards the small x region due to Δf_{int} being peaked at smaller y ($y_{max} \simeq 0.3$) than the Δf_ρ ($y_{max} \simeq 0.60$). Also the calculations with $x\Delta v_\rho^{MIT}(x)$ are larger than that with $0.6\,xv_\pi(x)$ in the intermediate x region, and have their maxima at larger x.

The integral

$$I_\Delta = \int_0^1 dx[\Delta\bar{d}(x) - \Delta\bar{u}(x)]$$
$$= \int_0^1 dx \Delta v_\rho(x) \int_0^1 dy[\Delta f_\rho(y) + \Delta f_{int}(y)] \quad (3)$$

will be the same for both models for the polarized parton distribution of the ρ as they have the same first moment for the polarized distribution. We find the integral to be 0.023 (0.031) without (with) the interference terms for $\Lambda_{oct} = 1.08$ GeV and $\Lambda_{dec} = 0.98$ GeV. The interference effect increases the integral by about 30%. The prediction for the integral I_Δ has a strong dependence on the cut-off parameters Λ_{oct} and Λ_{dec}. For example, the results with (without) interference contribution vary from 0.0043 (0.0027) to 0.033 (0.020) for the cut-off parameters changing from $\Lambda_{oct} = \Lambda_{dec} = 0.8$ GeV to 1.10 GeV. Clearly these values obtained using the meson cloud model are very different from those obtained using the chiral quark-soliton model[10] which have a magnitude of around 0.3. It is interesting that both models agree well with the experimental data for the unpolarized asymmetry, yet predict very different results for the polarized asymmetry. As the magnitude of the predicted polarized asymmetry appears to be fairly natural in each of these models, experimental data will provide a valuable test of these models, and give insight into the relation between helicity dependent and helicity independent observables in quark models.

2.2 FSB from 'Pauli blocking'

Now we consider the contribution to the asymmetry arising from 'Pauli blocking' effects[6,18,19]. In a model such as the bag model, where the valence quarks are confined by a scalar field, the vacuum inside a hadron is different from the vacuum outside. This manifests itself as a distortion in the negative energy Dirac sea, which is full for the outside (or free) vacuum, whereas there will be empty states in the Dirac sea of the bag. To an external probe this change in vacuum structure appears as an intrinsic, non-perturbative sea of $q\bar{q}$ pairs. This change in the vacuum is similar to the change in the Fermi-Dirac distribution when the temperature is raised above absolute zero. Now when a quark is put into the ground state of the bag it will have the effect of filling some of the empty negative energy states in the sea of the bag vacuum. The reason for this is that the ground state wavefunction can be written as a wavepacket in terms of plane wave states of positive and negative energy, with the energy distribution of the wavepacket centred at the ground state energy eigenvalue, but with non-zero contributions from negative energy plane waves. Hence the presence of a quark in the bag ground state lowers the probability of a negative energy state being empty, which is the same as lowering the probability of finding a positive energy antiquark. As the proton consists of two up quarks and one down quark, the probability of finding a $\bar{u}$ antiquark is reduced more than the probability of finding a $\bar{d}$ antiquark *i.e.* $\bar{d} > \bar{u}$.

When we include spin in the analysis of Pauli blocking, we find that putting a spin up quark into the bag ground state has the effect of filling some of the negative energy spin up quark states in the bag vacuum, which is equivalent to lowering the probability of finding a positive energy spin down antiquark. As the $SU(6)$ wavefunction of the spin up proton is dominated by terms with the two up quarks having spin parallel to the proton spin and the down quark having spin anti-parallel, Pauli blocking predicts that the probabilities of finding spin down $\bar{u}$ antiquarks and spin up $\bar{d}$ antiquarks are reduced *i.e.* $\bar{u}^{\uparrow} > \bar{u}^{\downarrow}$, $\bar{d}^{\downarrow} > \bar{d}^{\uparrow}$ or $\Delta\bar{u}(x) \geq 0$, $\Delta\bar{d}(x) \leq 0$.

We estimate the contribution of the Pauli blocking effect to the polarized asymmetry using the Adelaide group's argument for calculating parton distributions in the bag model. It was found

$$\bar{d}(x) - \bar{u}(x) = F_{(4)}(x), \quad \Delta\bar{d}(x) - \Delta\bar{u}(x) = -\frac{5}{3}G_{(4)}(x) \tag{4}$$

where $F_{(4)}(x)$ and $G_{(4)}(x)$ are the spin independent and spin dependent kinematic integrals over the momentum of the intermediate four quark state. As $G_{(4)}(x)$ is positive at all x, Pauli blocking gives a negative contribution to the spin dependent flavour asymmetry in the sea, whereas the meson cloud contribution tended to be positive. Also noting that as $F_{(4)}(x) \geq G_{(4)}(x)$, we can integrate over all x and then obtain an upper limit for the size of the Pauli blocking contribution to the spin dependent asymmetry in terms of the contribution to the spin independent asymmetry:

$$-\int_0^1 dx[\Delta\bar{d}(x) - \Delta\bar{u}(x)] \leq \frac{5}{3}\int_0^1 dx[\bar{d}(x) - \bar{u}(x)]. \tag{5}$$

As an estimate for the integral on the rhs of Eq. (5) we may use the value of 0.07 given by the analysis of reference [6]. This then gives an upper limit of about 0.12 for the magnitude of the integral over the polarized asymmetry. In the bag model, the ratio $G_{(4)}(x)/F_{(4)}(x)$ varies from about 0.8 at low x to unity at large x, which gives us a value of about -0.09 for the integrated polarized asymmetry. While these values are calculated at some scale appropriate to the bag model, the values of the integrals are not much affected by evolution up to experimental scales, so we expect the relation between polarized and unpolarized sea asymmetries to be approximately correct at all scales. The value of the Pauli blocking contribution to the integrated polarized asymmetry is much larger than that we have calculated in the meson cloud model, in contrast to approximate equality in the unpolarized case. Thus the experimental observation of any asymmetry in the polarized sea distributions is much more a test of the Pauli blocking hypothesis than of the meson cloud model. We estimate that the contribution to the Bjorken sum rule from Pauli blocking plus meson cloud effects is about 5-10% of the value of the sum rule.

3 Summary

We report a study on the flavour asymmetry of the non-strange polarized anti-quarks using the meson cloud model and 'Pauli blocking'. In the meson cloud model, we have included the contributions from both the vector meson cloud and the interference terms between pseudoscalar and vector mesons. It was found that the interference terms can provide sizable contribute to the asymmetry in the intermediate x region. We have also discussed the effect of 'Pauli blocking' on the asymmetry, and have seen that this effect gives a larger contribution to the asymmetry than meson cloud effects, in contrast to the unpolarized case.

Acknowledgments

We would like to thank Thomas, Melnitchouk, Schreiber and Tsushima for helpful discussions. This work was partially supported by the Science and Technology Postdoctoral Fellowship of the Foundation for Research Science and Technology, and the Marsden Fund of the Royal Society of New Zealand.

1. A. Baldit et al., NA51 Collaboration, *Phys. Lett.* **B332**, 244 (1994); E. A. Hawker et al., E866/NuSea Collaboration, *Phys. Rev. Lett.* **80**, 3715 (1998).
2. A. W. Thomas, *Phys. Lett.* B **126**, 97 (1983).
3. A. Szczurek, A. J. Buchmann and A. Faessler, *J. Phys. G* **22**, 1741 (1996).
4. P. V. Pobylitsa, M. V. Polyakov, K. Goeke, T. Watabe, and C. Weiss; *Phys. Rev.* D **59**, 03402 (1999).
5. A. E. Dorokhov and N. I. Kochelev, *Phys. Lett.* B **259**, 335 (1991).
6. W. Melnitchouk, J. Speth, and A. W. Thomas, *Phys. Rev.* D **59**, 014033 (1998).
7. E. A. Hawker et al., E866/NuSea Collaboration, **hep-ex/0103030**.
8. B. Adeva, SMC Collaboration, *Phys. Lett.* B **420**, 180 (1998); K. Ackerstaff, HERMES Collaboration, *Phys. Lett.* B **464**, 123 (1999).
9. see *e.g.* M. Glück, E. Reya, M. Stratmann, and W. Vogelsang, *Phys. Rev.* D **53**, 4775 (1996).
10. D. I. Diakonov et. al., *Phys. Rev.* D **56**, 4069 (1997).
11. R. J. Fries and A. Schäfer, *Phys. Lett.* B **443**, 40 (1998).
12. A. E. Dorokhov and N. I. Kochelev, *Phys. Lett.* B **304**, 167 (1993); K. G. Boreskov and A. B. Kaidalov, *Eur. Phys. J.* C **10**, 143 (1999); R. S. Bhalerao *Phys. Rev.* C **63**, 025208 (2001).
13. F.-G. Cao and A. I. Signal, **hep-ph/0103113**.
14. A. W. Schreiber, P. Mulders, A. I. Signal and A. W. Thomas, *Phys. Rev.* D **45**, 3069 (1992).
15. M. Glück, E. Reya, and I. Schienbein, *Eur. Phys. J.* C **10**, 313 (1999).
16. C. Best et. al., *Phys. Rev.* D **56**, 2743 (1997).
17. H. Holtmann, A. Szczurek, and J. Speth, *Nucl. Phys. A* **569**, 631 (1996).
18. A. I. Signal and A. W. Thomas, *Phys. Rev.* D **40**, 2832 (1989).
19. R. D. Field and R. P. Feynman, *Phys. Rev.* D **15**, 2590 (1977).

3. Perturbative – Nonperturbative QCD Transition

LEPTON SCATTERING AND QUARK-HADRON DUALITY STUDIES AT JLAB

ROLF ENT
Jefferson Lab, 12000 Jefferson Avenue, Newport News VA 23602, USA
E-mail: ent@jlab.org

At high enough energies asymptotic freedom guarantuees the deep inelastic scattering cross sections to be calculated as nearly free electron-quark scattering. However, confinement guarantees that the experimentally observed final states particles are hadrons. Low-energy quark-hadron duality suggests that hadronic cross sections, when averaged over an appropriate energy range, nevertheless coincide with the naive leading-twist quark-gluon calculations. Deep inelastic inclusive scattering shows that scaling at modest Q^2 and ν already arises from very few resonance channels. This is reflected by the striking agreement ($< 10\%$) between data in the nucleon resonance region and the deep inelastic ($W^2 > 4$ GeV^2) region for $Q^2 > 0.5$ (GeV/c)^2, known as Bloom-Gilman duality. Electron-hadron scattering allows for further investigation of quark-hadron duality by virtue of its ability to select resonances, by tagging with either spin or flavor.

1 Introduction

In QCD, the quark and gluon degrees of freedom are confined, and therefore the transition from quark-gluon to hadron degrees of freedom should in principle just be a matter of convenience. This is referred to as quark-hadron duality. At high energies, where the interactions between quarks and gluons become weak and quarks can be considered asymptotically free, an efficient description of nuclear and particle physics is possible in terms of quarks. At low energies, where the effects of confinement make QCD highly nonperturbative, it is more convenient to work in terms of the physical mesons and baryons. Quark-hadron duality is therefore an expression of the relationship between confinement and asymptotic freedom, and is intimately related to the nature of the transition from non-perturbative to perturbative QCD.

A global kind of quark-hadron duality is well established: low-energy resonance production can be shown to be related to the high-energy behavior of hadron-hadron scattering [1]; the familiar ratio of $e^+e^- \rightarrow$ hadrons over $e^+e^- \rightarrow$ muons uses duality to relate the hadron production to the sum of the squared charges of the quarks [2]; rigorous calculations exist in the infinite number of colors limit in leptonic heavy-quark decays [3].

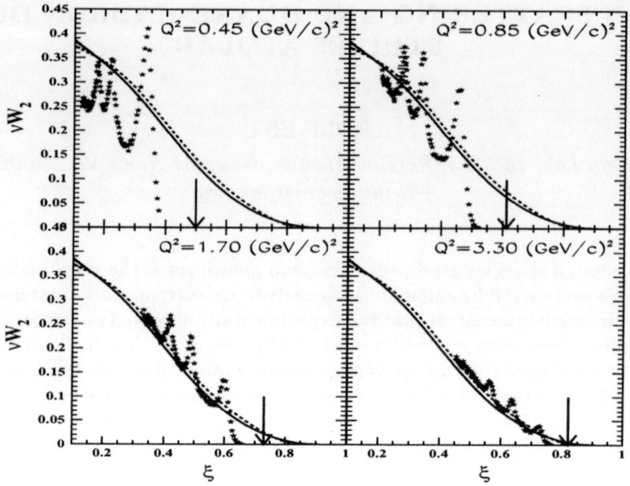

Figure 1. Sample hydrogen νW_2 structure function spectra obtained at Q^2 = 0.45, 0.85, 1.70, and 3.30 (GeV/c)^2 and plotted as a function of the Nachtmann scaling variable ξ $(= 2x/(1+\sqrt{1+4M^2x^2/Q^2}))$. Arrows indicate elastic kinematics. The solid (dashed) line represents the NMC fit of deep inelastic structure function data at $Q^2 = 10$ (GeV/c)^2 ($Q^2 = 5$ (GeV/c)^2).

2 Bloom-Gilman Duality

One of the more intriguing examples, initially observed three decades ago, is in inclusive electron-nucleon scattering. In studying inelastic electron scattering in the resonance region and the onset of scaling behavior, Bloom and Gilman [4] found that the inclusive F_2 structure function at low W generally follows a global scaling curve which describes high W data, to which the resonance structure function averages. Recently, high precision data on the F_2 structure function from Hall C at Jefferson Lab [5] have quantified these earlier observations, and demonstrated that duality works to better than 10% for both the total nucleon resonance region and each of the separate low-lying nucleon resonance regions, for $Q^2 \geq 0.5$ (GeV/c)^2. This is illustrated in Fig. 1, where the nucleon resonance data for various Q^2 are compared to a parameterizations of deep inelastic scattering data at constant $Q^2 = 5$ and 10 (GeV/c)^2 [6]. Such behavior shows that the distinction between the nucleon resonance region and the deep inelastic region is spurious; if properly averaged, the nucleon resonance regions closely mimic the deep inelastic region.

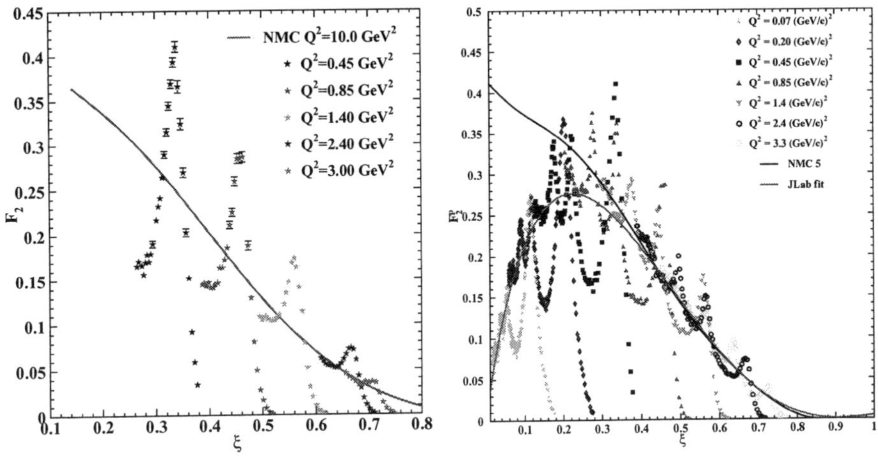

Figure 2. (left) Sample hydrogen νW_2 structure function spectra obtained at various Q^2, for the $N-\Delta$ transition region, defined as $1.2 < W^2 < 1.9$ GeV^2, only, and plotted as a function of the Nachtmann scaling variable ξ. The solid line represents the NMC fit of deep inelastic structure function data at $Q^2 = 10$ (GeV/c)^2. (right) Sample hydrogen νW_2 structure function spectra obtained at various Q^2 and plotted as a function of the Nachtmann scaling variable ξ. The data at $Q^2 = 0.07$ and 0.20 (GeV/c)^2 are from older SLAC experiments. The solid line represents the NMC fit of deep inelastic structure function data at $Q^2 = 5$ (GeV/c)^2. The light solid line represents a fit of the various nucleon resonance spectra.

To emphasize that this, at least for the F_2 structure function, also works in a localized region, we show in Fig. 2 (left) the region of $1.2 < W^2 < 1.9$ (GeV/c)^2, the "N-Δ" region, for various values of Q^2, in comparison to the NMC parameterization at $Q^2 = 10$ (GeV/c)^2. Recently, Close and Isgur [7] argued that in the quark model one expects duality in the F_2 structure function to work at low Q^2 (≈ 0.5 (GeV/c)^2), and to start at $Q^2 \approx 0$ for magnetic contributions only. It is also argued that for neutron targets, in contrast, duality only works if one integrates over the full nucleon resonance region, up to $W^2 \approx 3.3$ (GeV/c)^2.

Experiments investigating such behavior have started. In Hall A a $^3\vec{\mathrm{He}}$(e,e′) experiment is being analyzed [8], whereas in Hall B the results of a $^1\vec{\mathrm{H}}$(e,e′) experiment are under investigation [9]. Results are anticipated for Q^2 up to 1-2 (GeV/c)^2. Initial results of the Hall A experiment [10] indicate impressive agreement between g_1 data in the nucleon resonance region and deep inelastic region at $Q^2 \approx 1$ (GeV/c)^2 (for the nucleon resonance region), with the exception of the region of the $N - \Delta$ transition that, at this Q^2, still renders negative values as opposed to the positive values of the deep inelastic region. A dedicated experiment to investigate this issue up to $Q^2 = 5$ $(\mathrm{GeV/c})^2$ has recently been approved [11]. In addition, duality results on the longitudinal structure function F_L are forthcoming from a Hall C experiment [12]. There are predictions [13] and some scant experimental evidence [14] that F_L exhibits duality.

The Hall C experiments additionally showed that the nucleon resonance region data *at all* Q^2 oscillate around a single smooth curve, as shown in Fig. 2 (right). This curve coincides with the deep inelastic scaling curve at $Q^2 > 0.5$ $(\mathrm{GeV/c})^2$, consistent with Bloom-Gilman duality, and resembles neutrino/anti-neutrino xF_3 data or a valence-like sensitivity only [15] below $Q^2 \approx 0.5$ $(\mathrm{GeV/c})^2$. This is perhaps not too surprising in the quark model where the nucleon resonances act as valence quark transitions, while at low Q^2 not many sea quarks can be "seen" yet. However, it is surprising that all the strongly-interacting nucleon resonances shuffle their strength around to closely follow a single scaling curve. Do we see duality down to the lowest values of Q^2?

A simple quantum-mechanical model assuming two confined, relativistic, valence scalar quarks, with one of the quarks infinitely heavy, is able to reproduce all qualitative features of Bloom-Gilman duality [16] described above. Such modeling is instrumental to understand the physical mechanism that causes the large cancellations, on average, of the higher-twist processes responsible for the nucleon resonance transitions, in inclusive electron-proton scattering. Previous work on understanding Bloom-Gilman duality was done in terms of the operator-product expansion [17]. This is the subject of the next Section.

3 Moments of the F_2 Structure Function at Low Q^2

An analysis of the resonance region at smaller W^2 and Q^2 in terms of QCD was first presented by DeRujula, Georgi, and Politzer [17]. The integrals of the structure function, performed by Bloom and Gilman [4] over the energy transfer ν, were translated into integrals over the variable x (or Nachtmann $\xi = 2x/(1+\sqrt{1+4M^2x^2/Q^2})$, in order to account for finite target mass effects). Bloom-Gilman duality was translated into a correspondence between the $n=2$ moment of the F_2 structure function in the low Q^2 region, characterized by resonances, and in the high Q^2 scaling region, respectively. The fall of the resonances along a smooth scaling curve with increasing Q^2 was to be attributed [17] to the fact that there exist only small changes in the low n moments of the F_2 structure function due to power corrections in addition to the predicted perturbative ones. The appearence of power corrections is interpreted as a signal of deviations of the inclusive cross section from perturbative predictions, which one can envisage as due to the increasing importance of interactions between the quark struck in the electron-nucleon hard scattering process and the other quarks in the nucleon. Such effects are inversely proportional to Q^2, and can therefore be large at small Q^2. If they are not, averages of the F_2 structure function over a sufficient range in x at moderate Q^2 are approximately the same as at high Q^2. Recently, similar arguments were presented for the case of polarized lepton-nucleon scattering [18].

We constructed the experimental moments of the structure function, F_2, for the Q^2 range up to 5 $(\mathrm{GeV/c})^2$ [19]. The Cornwall-Norton moments are defined as

$$M_n(Q^2) = \int_0^{x_{thr}} dx x^{n-2} F_2(x, Q^2), \tag{1}$$

and the Nachtmann moments as

$$M_n(Q^2) = \int_0^{x_{thr}} dx \frac{\xi^{n+1}}{x^3} \left[\frac{3 + 3(n+1)r + n(n+2)r^2}{(n+2)(n+3)}\right] \nu W_2(x, Q^2). \tag{2}$$

Here, $r = (1 + 4M^2\, x^2/Q^2)^{1/2}$, and x_{thr} is Bjorken x for pion threshold. We add to these integrals the elastic contribution, at $x = 1$.

We show the values for the second, fourth, sixth, and eigth Cornwall-Norton (top) and Nachtmann (bottom) moments of the proton, extracted from the world's data, including deep inelastic, nucleon resonance, and elastic data, as described above, in Fig. 3. The elastic contribution dominates the moments at the lowest Q^2. Note that the Cornwall-Norton moments will become unity, i.e. the proton charge squared, at $Q^2 = 0$, whereas the Nachtmann moments will vanish at $Q^2 = 0$.

We find that the moments of F_2 show a smooth transition from the deep inelastic limit down to $Q^2 \approx 0$ $(\mathrm{GeV/c})^2$, and that the nucleon resonances tend to oscillate around one smooth curve, supporting the findings [20,21] that higher-twist effects are small if one looks at the low-Q^2 behavior of F_2 for $Q^2 \simeq 1$ $(\mathrm{GeV/c})^2$. The dynamical process of local duality dictates minimal Q^2 dependence of F_2 at small Q^2; in terms of pQCD this can be explained if the higher-twist effects are reduced *on average* in the nucleon resonance region. Nonetheless, higher-twist effects must be responsible for the nucleon resonances themselves.

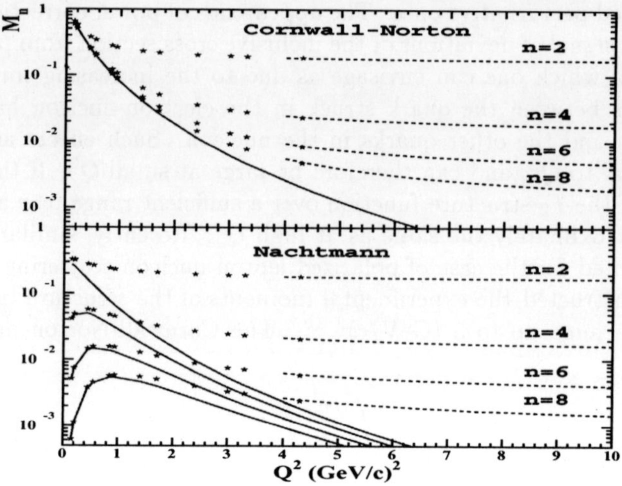

Figure 3. Cornwall-Norton moments (top) and Nachtmann moments (bottom) extracted from the world's electron-proton scattering data, for n = 2, 4, 6, and 8. The solid lines indicate the elastic contribution. At low Q^2 ($\leq$ 4.3 (GeV/c)^2) the moments (stars) are directly constructed from the world's electron-proton F_2 database. At larger Q^2, the moments have been extracted from appropriate fits to the world's data on inclusive scattering to both the nucleon resonance and deep inelastic regions (dashed lines).

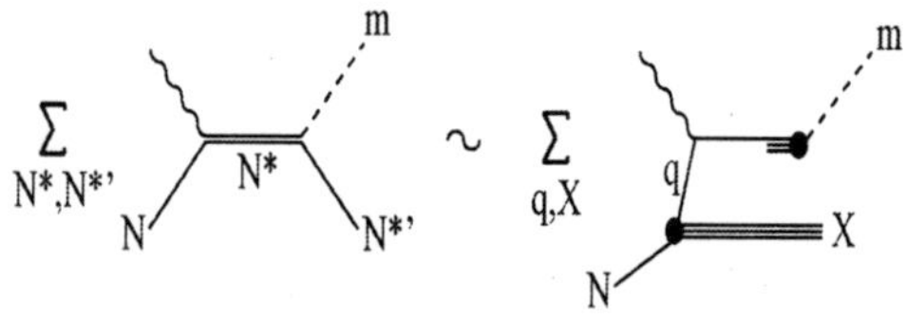

Figure 4. Duality between descriptions of semi-inclusive meson production in terms of quark (right) and nucleon resonance (left) degrees of freedom.

4 Fragmentation Duality

A largely unexplored domain with potentially broad applications is the production of mesons (M) in semi-inclusive electron scattering, $eN \to e'MX$ [22]. At high energy the scattering and production mechanisms factorize, with the cross section at leading order in QCD given by a simple product of the structure function, $\sim \sum_q e_q^2\, q(x)$, and a quark $\to$ meson fragmentation function, $D_{q\to M}$, for a given elasticity $z = E_M/\nu$, as in Fig. 4.

In terms of hadronic variables, the same process can be described through the excitation of nucleon resonances, N^*, where the $\gamma^* N \to N^*$ transition form factor, $F_{\gamma^* N \to N^*}$, depends on the mass of the virtual photon and the excited nucleon, $W = M_{N^*}$, and a function $\mathcal{D}_{N^* \to N^{*'} M}$ describing their subsequent decays into mesons and lower lying resonances, $N^{*'}$, with W' the invariant mass of the final state $N^{*'}$:

$$\sum_{N^*, N^{*'}} F_{\gamma^* N \to N^*}(Q^2, W^2)\, \mathcal{D}_{N^* \to N^{*'} M}(W^2, W'^2) \sim \sum_q e_q^2\, q(x)\, D_{q \to M}(z)\,.$$

The hadronic description is rather elaborate, as the production of a fast outgoing meson in the current fragmentation region at high energy requires non-trivial cancellations of the angular distributions from various decay channels [16]. Such a cancellation may not be unlikely if one realizes that duality has also been observed in hadronic τ decays [23].

For a kinematics region that mimics single-quark scattering, in analogy with the inclusive scattering case ($W^2 > 4$ GeV^2 and $Q^2 > 1$ (GeV/c)^2), the question here is whether the remaining part of the process can be described in terms of a struck quark hadronizing into the detected meson. Similar as in the inclusive case where the nucleon resonances average at low energies to a scaling curve, here the nucleon resonances left in the final state after removing a fast meson may average to a fragmentation function. Such a signature of "truncated" or "tagged" duality has never been investigated yet, and could be related to the open question to what extent factorization applies at lower than asymptotic energy. Recently, it has been suggested that one may expect factorization and approximate duality to work at $Q^2 < 3$ (GeV/c)^2 and relatively low W^2 [7]. In addition, hints of a scaling behavior in older Cornell data can be observed [24]. Lastly, we show in Fig. 5 a comparison between $\pi^\pm$ electroproduction data at low energy (beam energy of 5.5 GeV, $W^2 = 5.3$ GeV^2, $W'^2 < 4$ GeV^2) and a next-to-leading order fragmentation function fit to high energy data. Experiment E00-108 has recently been approved at JLab to investigate fragmentation duality and the onset of factorization in more detail [25].

5 Conclusion

If one understands duality it may be used as a tool. It can give guidance to cuts typically used to select "hard scattering" regions. E.g., it shows that the $W^2 > 4$ cut to select deep inelastic events is spurious. One can access the very large x region [26], where, without escape, one encounters the nucleon resonance region. This could provide us with data for parton distribution functions in

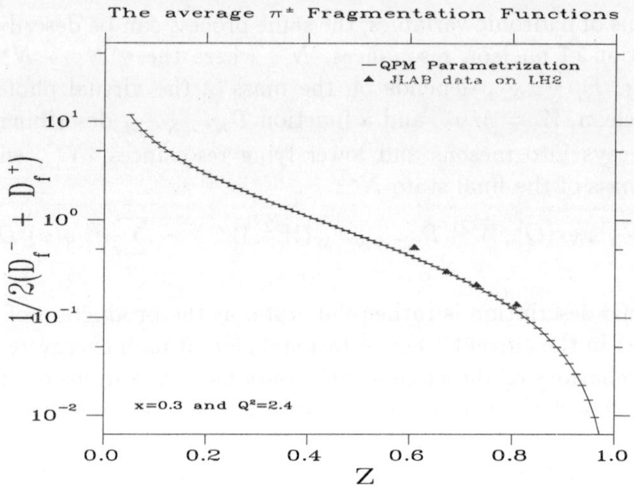

Figure 5. A comparison between $\pi^{\pm}$ electroproduction data at low energy and a next-to-leading order (NLO) fragmentation function fit to high energy data (solid line). The curve corresponds to the average of positive and negative fragmentation functions, which is compared to the average of π^{+} and π^{-} data points. The data shown here were limited to pion momenta above 2.4 GeV/c in order to minimize $\pi^{\pm}N$ interactions in the final state.

the strict valence region, and allow investigations of the Q^2 evolution of large-x parton distribution functions. Initial studies utilizing the fermi broadening in nuclear targets as "averaging tool" are underway [27,28].

Additionally, it is worth stressing that confirmation of factorization and truncated duality would open the way to an enormously rich semi-inclusive program, allowing unprecedented spin and flavor decompositions of quark distributions.

Acknowledgments

The author thanks the JLab Theory Group for stimulating discussions. The Southeastern Universities Research Association (SURA) operates the Thomas Jefferson National Accelerator Facility for the United States Department of Energy under contract DE-AC05-84ER40150.

References

1. P.D.B. Collins, *An Introduction to Regge Theory and High Energy Physics*, Cambridge University Press, Cambridge, 1977.
2. E.C. Poggio, H.R. Quin and S. Weinberg, *Phys. Rev.* D **13**, 1958 (1976).
3. M.B. Voloshin and M. Shifman, *Sov. J. Nucl. Phys.* **47**, 511 (1988); N. Isgur, *Phys. Rev.* D **40**, 101 (1989); *Phys. Lett.* B **448**, 111 (1999).
4. E.D. Bloom and F.J. Gilman, *Phys. Rev.* D **4**, 2901 (1971); *Phys. Rev. Lett.* **25**, 1140 (1970).
5. I. Niculescu *et al.*, *Phys. Rev. Lett.* **85**, 1186 (2000).
6. M. Arneodo *et al.*, *Phys. Lett.* B **364**, 107 (1995).
7. F.E. Close and N. Isgur, *Phys. Lett.* B **509**, 81 (2001).
8. Z.E. Meziani, G. Cates and J.P. Chen, JLab experiment E94-010.
9. V. Burkert, D. Crabb and R. Minehart, JLab experiment E91-023.
10. N. Liyanage, private communications (2001).
11. N. Liyanage, J. P. Chen and S. Choi, JLab experiment E01-012.
12. C.E. Keppel, JLab experiment E94-110.
13. C.E. Carlson and N. Mukhopadhyay, *Phys. Rev. Lett.* **74**, 1288 (1995).
14. R. Ent, C.E. Keppel and I. Niculescu, *Phys. Rev.* D **62**, 73008 (2000).
15. I Niculescu *et al.*, *Phys. Rev. Lett.* **85**, 1182 (2000).
16. N. Isgur, S. Jeschonnek, W. Melnitchouk and J. Van Orden, *hep-ph/0104022* (2001).
17. A. DeRujula, H. Georgi and H.D. Politzer, *Annals of Phys.* **103**, 315 (1977).
18. J. Edelmann, G. Piller, N. Kaiser and W. Weise, *Nucl. Phys.* A **665**, 125 (2000).
19. C.S. Armstrong *et al.*, *Phys. Rev.* D **63**, 094008 (2001).
20. U.K. Yang and A. Bodek, *Phys. Rev. Lett.* **82**, 2467 (1999).
21. I. Alekhin and A.L. Kataev, *Nucl. Phys.* A **666-667**, 179 (2000).
22. A. Afanasev, C.E. Carlson and C. Wahlquist, *Phys. Rev.* D **62**, 074011 (2000).
23. M. Shifman, *hep-ph/0009131* (2000).
24. A. Calogeracos, N. Dombey and G.B. West, *Phys. Rev.* D **51**, 6075 (1995).
25. R. Ent, H. Mkrtchyan and G. Niculescu, JLab experiment E00-108.
26. W. Melnitchouk, *Phys. Rev. Lett.* **86**, 35 (2001).
27. J. Arrington *et al.*, *Phys. Rev.* C **64**, 014602 (2001); J. Arrington, *et al.*, in preparation.
28. J. Arrington, JLab experiment E00-101.

QUARK-HADRON DUALITY IN INCLUSIVE ELECTRON-HADRON SCATTERING

W. MELNITCHOUK

Special Research Centre for the Subatomic Structure of Matter, Adelaide University, 5005, Australia, and
Jefferson Lab, 12000 Jefferson Avenue, Newport News, VA 23606, USA

We discuss the origin of Bloom-Gilman duality and the relationship between resonances and scaling in deep-inelastic scattering. A simple quantum mechanical model is used to illustrate the essential features of Bloom-Gilman duality at low Q^2. As an application of local duality, we present model-independent relations between structure functions at $x \sim 1$ and elastic electromagnetic form factors.

1 Introduction

Quark-hadron duality addresses some of the most fundamental issues in strong interaction physics, in particular the nature of the transition from the perturbative to the nonperturbative region of QCD. In its broadest form, it postulates that calculations of physical quantities performed in either a quark-gluon or hadronic basis should yield identical results. Although duality is in practice almost never realized exactly, there are rare cases where the average of hadronic observables can be described to good accuracy within perturbative QCD.

The correspondence between hadronic and partonic descriptions has in fact been observed in a number of processes, such as e^+e^- annihilation into hadrons, heavy quark decays, and inclusive electron-hadron scattering. It also forms the basis for theoretical approaches such as QCD sum rules. However, despite many years of studying duality, a deep understanding of its origins remains elusive. In this review, I shall discuss some recent progress made in unraveling the origin of duality in electron scattering, first observed some 30 years ago by Bloom and Gilman.

2 Bloom-Gilman Duality

In studying inelastic electron scattering in the resonance region and the onset of scaling, Bloom and Gilman found[1] that the inclusive F_2 structure function at low W generally follows a global scaling curve which describes high W data, to which the resonance structure function averages. More recently, this behavior was confirmed in high precision measurements of the F_2 structure function in the resonance region at Jefferson Lab[2]. The similarity between the averaged resonance and scaling structure functions implies that the lowest moment of F_2 is approximately independent of Q^2. Furthermore, the data clearly demonstrate that duality works remarkably well for each of the low-lying resonances, including the elastic, to rather low values of Q^2 (~ 0.5 GeV^2), as Fig. 1 illustrates.

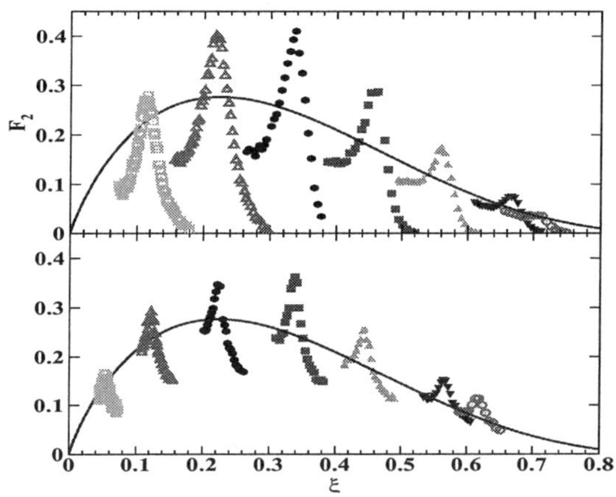

Figure 1. Proton F_2 structure function versus the Nachtmann scaling variable, ξ, in the first (upper panel) and second (lower panel) resonance regions, for various Q^2 between 0.07 GeV^2 (smallest ξ) and 3 GeV^2 (largest ξ). The solid line is the global scaling curve determined from all nucleon resonance data[3].

Before the advent of QCD, Bloom-Gilman duality was interpreted in the context of finite-energy sum rules[4], in analogy with the s and t channel duality observed in hadron-hadron scattering. In QCD, Bloom-Gilman duality can be reformulated[5] in the language of the operator product expansion, in which moments of structure functions are organized in powers of $1/Q^2$. The leading terms are associated with free quark scattering, and are responsible for the scaling, while the $1/Q^2$ terms involve interactions between quarks and gluons. The weak Q^2 dependence of the low moments of the structure function is then interpreted as indicating that the non-leading, $1/Q^2$-suppressed, interaction terms do not play a major role even at low Q^2.

An important consequence of duality is that the strict distinction between the resonance and deep-inelastic regions becomes entirely artificial. To illustrate this, consider that at $Q^2 = 1$ GeV^2 around 2/3 of the total cross section comes from the resonance region, $W < 2$ GeV. However, the resonances and the deep-inelastic continuum conspire to produce only about a 10% correction to the lowest moment of the scaling F_2 structure function at the same Q^2. Even though each resonance is built up from a multitude of twists, when combined the resonances interfere in such a way that they resemble the leading twist component.

This can be even more dramatically illustrated by considering QCD in the large N_c limit, where the hadron (or more specifically, meson) spectrum consists of infinitely narrow, noninteracting resonances. Since the quark level calculation still yields a smooth scaling curve, one sees that an averaging over hadrons must be invoked even in the scaling limit[6]. In the next section we demonstrate this duality explicitly in a simple model of large N_c QCD.

3 Understanding the Origin of Duality

The essential features of the dynamics behind Bloom-Gilman duality can be exposed with the help of a simple model in which the hadronic spectrum is dominated by infinitely narrow resonances made of valence quarks. To strip away irrelevant details which may complicate the illustration, we consider scattering from a relativistic scalar quark confined to an infinitely massive core by an oscillator-like potential. Although a model with these features will not give a realistic description of data, it will allow us to study the critical questions of when and why duality holds. Furthermore, it affords exact solutions for the complete spectrum of excited states.

If all the excited states are infinitely narrow resonances, the structure function is given entirely by a sum of squares of transition form factors[6],

$$\mathcal{S}(\nu,\vec{q}^{\,2}) = \sum_{N=0}^{N_{\max}} \frac{|\vec{q}|}{4E_0E_N} \; |F_{0N}(\vec{q})|^2 \; \delta(E_N - E_0 - \nu)\,, \tag{1}$$

where E_0 and E_N are the energies of the ground state and N-th excited state, respectively, and F_{0N} is the transition form factor. The sum over N is restricted to a maximum value $N_{\max}$ allowed by the available energy transfer, ν. Note that for a scalar probe, the structure function has dimensions of $(\text{mass})^{-2}$. The corresponding scaling variable[6],

$$u = \frac{1}{2m}\left(\sqrt{\nu^2+Q^2}-\nu\right)\left(1+\sqrt{1+\frac{4m^2}{Q^2}}\right)\,, \tag{2}$$

is defined with respect to the quark mass, m (rather than the target mass, M, which is infinite), and takes into account nonzero mass effects at finite Q^2. In the limit $Q^2 \to \infty$, the variable $u \to (m/M)x$.

In Fig. 2 we show the onset of scaling for the structure function $\mathcal{S}$ as a function of u (the energy-dependent δ-function has been smoothed out by a Breit-Wigner shape with some width in order to display the u dependence). With increasing Q^2 the resonances are seen to move out towards higher u, as observed in the physical spectrum in Fig. 1. Furthermore, the area under the curves remains approximately constant, indicating that global duality is reproduced by the model. Remarkably, the resonance "spikes" at lower Q^2 also tend to oscillate (at least qualitatively) around the scaling curve, reminiscent of the local duality observed in the proton F_2 data. The curve at $Q^2 = 5$ GeV^2 is already close to the asymptotic scaling function which is obtained by taking the continuum limit for the energy, and taking the sum over all excited states[6].

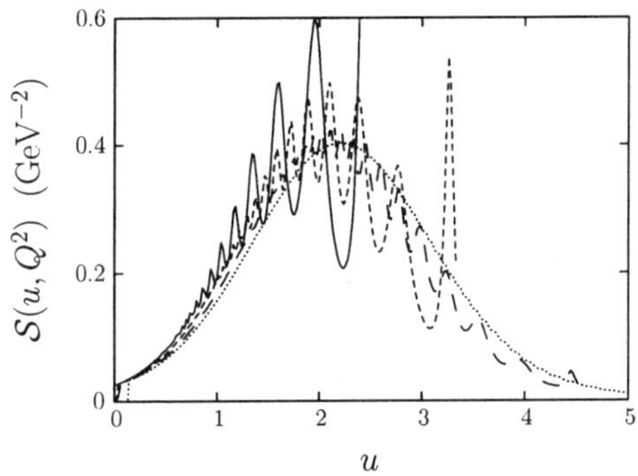

Figure 2. Scalar structure function versus the scaling variable u, for $Q^2 = 0.5$ (solid), 1 (short-dashed), 2 (long-dashed) and 5 GeV^2 (dotted).

Note that the curves in Fig. 2 are at fixed Q^2, but sweep over all ν. Since $\vec{q}^{\,2} = Q^2 + \nu^2$, as ν is increased, more and more thresholds for creating excited states are crossed, so that the density of states per $\vec{q}^{\,2}$ interval increases correspondingly. In fact, the correct density of states is crucial to compensate for the falling off with $\vec{q}^{\,2}$ of each individual form factor, $F_{0N} \sim \vec{q}^{\,N} \exp(-\vec{q}^{\,2}/4\,\beta^2)$. Had the spacing between the excited states not been Regge-like[7], $E_N \sim \sqrt{N}$, but linear as for the nonrelativistic case, the density of states would not have been sufficient to produce duality. While it remains an interesting exercise to see whether potentials other than a harmonic oscillator give rise to similar behavior, this simple example illustrates that Bloom-Gilman duality appears a natural consequence of even the simplest quantum mechanical systems, with the only requirements being confinement and a correct treatment of kinematics.

4 Applications of Local Duality

If the inclusive–exclusive connection via local duality is taken seriously, one can relate structure functions measured in the resonance region to electromagnetic transition form factors[1,5]. Isolating an individual resonance contribution to the inclusive structure function is problematic, however, since the separation of the resonance from the nonresonant background is model-dependent. For the extreme case of elastic scattering, on the other hand, there is no background below the pion production threshold, so the extraction of the elastic form factors from the inclusive structure function data (and vice versa) is free of this ambiguity.

The elastic magnetic form factor of the proton has in fact been extracted[2] from the recent Jefferson Lab data, and found to agree to within 30% with the experimental form factor for $Q^2 < 5$ GeV^2. Conversely, empirical electromagnetic form factors at large Q^2 can be used to predict the $x \to 1$ behavior of deep-inelastic structure functions[1]. Knowledge of structure functions at large x is vital for several reasons — the $x \to 1$ behavior, for instance, is very sensitive to mechanisms for spin-flavor SU(6) symmetry breaking, for which there are nonperturbative and perturbative QCD predictions[8,9].

Of particular interest is the polarization asymmetry, A_1, which at large Q^2 is given by the ratio of spin-dependent to spin-averaged structure functions, $A_1 \approx g_1/F_1$. Assuming that the area under the elastic peak is the same as the area under the scaling function (at much larger Q^2) when integrated from the pion threshold to the elastic point[1], the polarization asymmetry at threshold can be written as[10]:

$$A_1(x_{\rm th}) = \left\{ \frac{G_M\,(G_M - G_E)}{4M^2(1+\tau)^2} + \frac{1}{1+\tau}\left(\frac{d(G_E G_M)}{dQ^2} + \tau \frac{dG_M^2}{dQ^2} \right) \right\} \Big/ \frac{dG_M^2}{dQ^2} \quad (3)$$

where $x_{\rm th} = Q^2/(2m_\pi M + m_\pi^2 + Q^2)$, and $\tau = Q^2/4M^2$. In the limit $x_{\rm th} \to 1$, corresponding to $Q^2 \to \infty$, both F_1 and $g_1 \sim dG_M^2/dQ^2$, so that $A_1^{p,n} \to 1$. Note that SU(6) symmetry predicts that for valence quarks $A_1 = 5/9$ for the proton and 0 for the neutron, while the perturbative QCD expectation based on helicity conservation is $A_1^{p,n} \to 1$ as $x \to 1$.

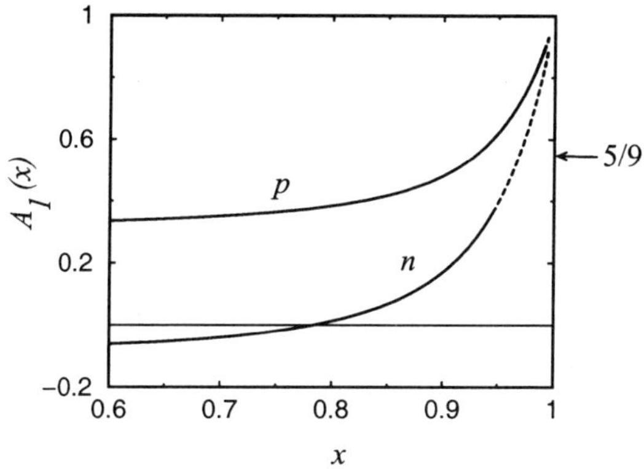

Figure 3. Polarization asymmetries A_1 for the proton and neutron at large x. The SU(6) predictions are 5/9 for p and 0 for n. The dashed extensions represent asymmetries calculated from extrapolations of form factors beyond the currently measured regions of Q^2.

Using parameterizations of global form factor data, we show in Fig. 3 the proton and neutron asymmetries A_1 as a function of $x = x_{\rm th}$. The solid curves represent the asymmetry calculated from actual form factor data, while the dashed extensions at larger x illustrate the extrapolation of the form factors beyond the currently measured regions of Q^2. Unfortunately the current data on A_1 extend only out to an average $\langle x \rangle \sim 0.5$, and are inconclusive about the $x \to 1$ behavior. While the proton A_1 data do indicate a rise at $x \sim 0.5 - 0.6$, the neutron asymmetry is, within errors, consistent with zero over the measured range. It will be of great interest in future to observe whether, and at which x and Q^2, the A_1 asymmetries start to approach unity.

Expressions similar to (3) can be derived also for other structure functions[10]. The ratios of the neutron to proton F_1, F_2 and g_1 structure functions are shown in Fig. 4 as a function of x, with x again evaluated at $x_{\rm th}$. Asymptotically, each of the structure functions approaches dG_M^2/dQ^2, so that in the dipole approximation the n/p ratios $\sim \mu_n^2/\mu_p^2$. Also indicated in Fig. 4 are some leading twist model predictions for the F_2^n/F_2^p, namely 2/3 from SU(6) symmetry, 3/7 from broken SU(6) with helicity conservation, and 1/4 from broken SU(6) with scalar diquark dominance[9]. Note, however, that the structure functions predicted from the duality relations contain both leading twist and higher twist contributions (for a discussion of the conditions under which the form factors can yield leading twist structure functions see Close & Isgur[11]).

The reliability of the duality predictions is of course only as good as

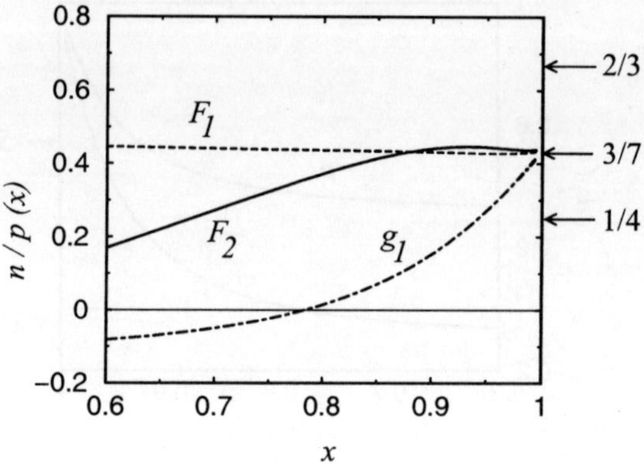

Figure 4. Neutron to proton ratio for the F_1 (dashed), F_2 (solid) and g_1 (dot-dashed) structure functions at large x. Several leading twist model predictions for F_2 in the $x \to 1$ limit are indicated by the arrows: 2/3 from SU(6), 3/7 from SU(6) breaking via helicity retention, and 1/4 from SU(6) breaking through d quark suppression.

the quality of the empirical data on the electromagnetic form factors and resonance structure functions. While the duality relations are expected to be progressively more accurate with increasing Q^2, the difficulty in measuring form factors at large Q^2 also increases. More data on form factors at larger Q^2 would allow more accurate predictions for the $x \to 1$ structure functions, and new experiments at Jefferson Lab, Mainz and elsewhere will provide valuable constraints.

5 Conclusion

Quark-hadron duality offers the prospect of addressing the physics of the transition from the strong to weak coupling limits of QCD, where neither perturbative QCD nor effective descriptions such as chiral perturbation theory are applicable. While considerable insight into quark-hadron duality has been gained from recent theoretical studies, it will be important in future to understand more quantitatively the features of the electron scattering data in the resonance region and the phenomenological N^* spectrum in terms of realistic models of QCD.

On the experimental side, the spin and flavor dependence of duality can be most readily accessed through semi-inclusive scattering, which requires both high luminosity and a high duty factor. An energy upgraded Jefferson Lab would be an ideal facility to study meson production in the current fragmentation region at moderate Q^2, allowing the onset of scaling to be tracked in the pre-asymptotic regime. This would shed considerable light on the relationship between incoherent (single quark) and coherent (multi-quark) processes, and on the nature of the quark $\rightarrow$ hadron transition in QCD.

Acknowledgments

This work was supported by the Australian Research Council, the U.S. Department of Energy contract DE-AC05-84ER40150, under which the Southeastern Universities Research Association (SURA) operates the Thomas Jefferson National Accelerator Facility (Jefferson Lab), and cooperative research agreement DE-FC02-94ER40818.

References

1. E.D. Bloom and F.J. Gilman, *Phys. Rev. Lett.* **16**, 1140 (1970).
2. I. Niculescu et al., *Phys. Rev. Lett.* **85**, 1182 (2000).
3. C.S. Armstrong et al., *Phys. Rev.* D **63**, 094008 (2001).
4. R. Dolen, D. Horn and C. Schmid, *Phys. Rev. Lett.* **19**, 402 (1967).
5. A. de Rújula, H. Georgi and H.D. Politzer, *Ann. Phys.* **103**, 315 (1975).
6. N. Isgur, S. Jeschonnek, W. Melnitchouk and J.W. Van Orden, *Phys. Rev.* D, in print, hep-ph/0104022.
7. G. Domokos, S. Koveni-Domokos and E. Schonberg, *Phys. Rev.* D **3**, 1184 (1971).
8. W. Melnitchouk and A.W. Thomas, *Phys. Lett.* B **377**, 11 (1996).
9. N. Isgur, *Phys. Rev.* D **59**, 034013 (1999).
10. W. Melnitchouk, *Phys. Rev. Lett.* **86**, 35 (2001).
11. F.E. Close and N. Isgur, hep-ph/0102067.

SIGNALS OF LOCAL DUALITY FROM A PERTURBATIVE QCD ANALYSIS OF INCLUSIVE ep SCATTERING.

S. LIUTI

Physics Department, University of Virginia, Charlottesville Virginia, USA
and
I.N.F.N. Sezione Roma III, Università di Roma III, Italy,
E-mail: sl4y@virginia.edu

We study the Q^2 dependence of large x data on F_2 from Jefferson Lab, SLAC, and BCDMS, with the aim of providing a perturbative-QCD (pQCD) based, quantitative verification of Bloom and Gilman's local duality relation. We find that most of the data in the resonance region (with invariant mass $W^2 = Q^2(1/x-1)+M^2 \leq 4$ GeV^2) can be described within a pQCD based picture, similarly to the large W^2 data, including large x resummation contributions, target mass corrections, and relatively small size *residual* power corrections. We uncover however a possible breakdown of the twist expansion in the $N-\Delta$ transition region which, by affecting the initial evolution of parton distributions, bears consequences on pQCD analyses at large x and very large Q^2. A dynamical model for local duality is proposed which accounts for the Q^2 dependence unravelled in the first part of our work.

1 Introduction

An overwhelming number of observations shows that QCD has manifestly a *dual* nature. One of the key problems of the theory is in fact in formulating a connection between the description of the hard or short-distance scattering processes, which can be calculated in terms of quarks and gluons degrees of freedom using perturbative methods, and the physical asymptotic states, the hadrons, whose properties are in principle only remotely related to parton dynamics and are therefore not calculable within perturbative-QCD (pQCD) [1].

Experimental information on the transition between the perturbative and non-perturbative regimes in QCD is often elusive and not easily extracted from available data. The general approach is based on QCD factorization theorems according to which the cross section for any hard process is factorized into an elementary hard subprocess at a scale given by the four-momentum transfer, Q, and a *hadronization* process, by which partons become the constituents of hadrons. Partons in the initial hadron also emit other partons with maximum virtuality given by Q^2, producing the perturbative evolution of the initial hadron's structure function. Similar parton showers are produced in the final state. Perturbative QCD does not provide information on a confinement mechanism that converts the partons into final hadrons, or on how constituent partons form the initial hadrons. However, an idea seeded in pQCD is that a preconfinement phase might precede the formation of hadrons, in which at a scale Q_o^2 where α_S is still small, as a consequence of forced gluon splitting ending the parton shower, color-neutral clusters are formed from pairs of adjacent color connected quarks and anti-quarks, with a mass distribution which is damped at high mass and is independent of Q^2 [2]. This property has been confirmed in jet fragmentation studies where the shape of hadron spectra, or fragmentation functions, has been found to be proportional to the parton spectra (calculable within pQCD), with Q^2 independent proportionality constants. The sharp transition between partonic and hadronic degrees of freedom, defined as the Local Parton-Hadron Duality hypothesis [3], is interpreted as an indication that hadrons are formed from partons locally in phase space. Further investigations of hadron jets at e^+e^- [4], ep [5], and $p\bar{p}$ [6] colliders show that several infrared-sensitive features of the inclusive hadron distributions can be reproduced by a pQCD description of the parton shower down to $Q_o \approx \Lambda_{QCD}$, Q_o being an effective cut-off defining the onset of the non-perturbative regime [a]. A similar signature has been also found in Deep Inelastic Scattering (DIS) experiments at very low Bjorken x that support the idea that pQCD can describe the structure function F_2 down to $Q^2 \approx 1\,\text{GeV}^2$, where a smooth and fast transition to the non-perturbative regime occurs, leading eventually to the expected behavior in the real photon scattering limit ($F_2(x, Q^2) \rightarrow 0$ for $Q^2 \rightarrow 0$).

Motivated by these studies we have addressed yet another set of experiments where a transition between the perturbative and non-perturbative regimes of QCD can be studied in detail, namely inclusive ep and eD scattering in the region of low final state invariant mass, or the resonance region. The resonance region is of particular interest because of the observation of Bloom and Gilman (BG) duality [8], according to which the resonances contribution is equivalent in average to the DIS one, obtained at a much larger Q^2. BG duality turns out to be a consequence of pQCD evolution in the range $1\ \text{GeV}^2 \leq Q^2 \leq 30\ \text{GeV}^2$, where the moments of the structure functions in the resonance and in the deep inelastic regions are, in fact, approximately equivalent to one another, modulo perturbative corrections and additional power corrections [9]. In this language, it is possible to redefine local BG duality as the "averaging of the resonances (in given intervals of W^2) to a smooth curve which, once evolved according to pQCD, will coincide with the fits to the data at much larger Q^2". Here, the standard expectation is that a series of inverse Q^2 power terms (formally a

[a] Notice, however, that detailed angular correlation measurements have been found to be in disagreement with available pQCD estimates [7].

twist expansion) due to multi parton "final state interaction" processes, which occur at large distances, is eventually responsible for the production of low invariant mass states or resonances. These corrections have been observed to approximately cancel one another in the resonance region (see *e.g.* Figure 2 in [10]), therefore generating the duality phenomenon. In this contribution we suggest a possible mechanism for such cancellation. We present a two step program: In order to confirm whether the idea of duality embodies the right physics and/or whether a breakdown of the twist expansion is most likely to happen, we first determine a breakpoint where pQCD no longer applies and a transition occurs, similar to what is observed *e.g.* in the low x regime. Our pQCD analysis is presented in Section 2. In Section 3, we propose a dynamical model to interpret this situation.

2 Perturbative QCD analysis of the resonance region.

We study the question of the limit of applicability of pQCD in the resonance region. We perform our study directly in x space at variance with previous analyses performed by using the moments of the structure function [9,11,12],

$$M_n(Q^2) = \int_0^1 dx x^{n-2} F_2(x, Q^2). \tag{1}$$

Although moments, by definition, circumvent the problem of dealing with the complicated structure of resonances by providing a natural averaging procedure [9], at the Q^2 values of the available data [10,14,15] they are dominated by elastic scattering, as shown in detail in [13]. We notice however that the high accuracy of the recent data allows us to perform a fit of F_2 in the resonance region, to a smooth curve [16]. The fitting procedure, can be applied separately to bins with fixed invariant mass, W^2, with a $\chi^2/d.f.$ varying from 0.8 to 1.1. The curves and uncertainties in each W^2 bin are represented by the hatched area in Figure 1. Notice that the spectra at fixed W^2 have Q^2 varying for each spectrum as an increasing function of x, $Q^2 \equiv Q^2(x) = (W^2 - M^2)x/(1-x)$. The choice of variables, (x, W^2), is a natural one in the resonance region where peaks appear as a function of the invariant mass, and it is useful practically because it allows for a more accurate determination of the scale dependence of the data which are binned directly in W^2. We assume that only valence quarks contribute to F_2 at low $W^2 \leq 10\,\text{GeV}^2$ and $x \geq 0.3$.

PQCD evolution to $Q^2 \equiv Q^2(x)$ or at fixed W^2, for the Non-Singlet (NS)

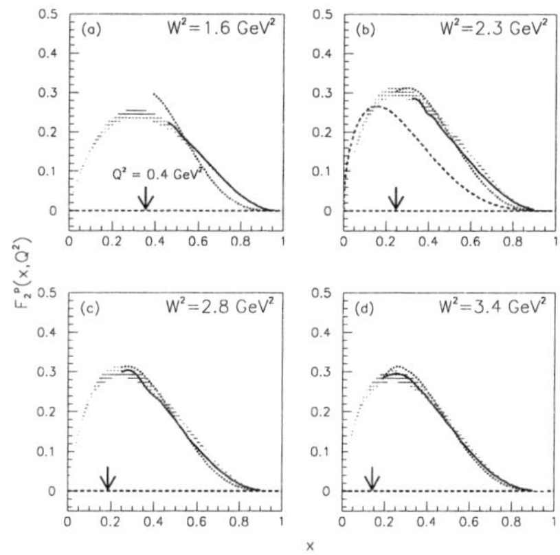

Figure 1: Comparison of NLO pQCD calculations (dotted lines), and NLO pQCD+TMC (full lines) with the data on F_2^p (hatched areas). The dashed curve, shown for comparison in (b), is the pQCD evaluation at $Q^2 = 200$ GeV^2. The data are averaged with the procedure described in the text. The arrows correspond to: $Q_o^2 = 0.4$ GeV^2. Below these values the observation of duality is no longer tracked by pQCD.

or valence distributions at next-to-leading-order (NLO) is given by:

$$q_i^{(-)}(x, Q^2) = \int_{Q_o^2}^{Q^2(x)} \frac{dQ'^2}{Q'^2} \frac{\alpha_S(Q'^2)}{2\pi} \int_x^1 \frac{dy}{y} P_{qq}\left(\frac{x}{y}, \alpha_S(Q'^2)\right) q_i^{(-)}(y, Q'^2), \tag{2}$$

where $q_i^{(-)} = q_i - \bar{q}_i \equiv q_i^v$, $i = u, d$ and the expressions for the splitting function $P_{qq}\left(z, \alpha_S(Q^2)\right)$, and for the corresponding endpoints at NLO in $\overline{MS}$ scheme can be found in [17]. The structure function, F_2, is then obtained by convoluting with the quark coefficient function, B_2^{NS} [17]. At large x this convolution is dominated by $\ln(1-z)$, $z = x/y$. Large logarithms enhance the higher order terms, they are expected to hamper the perturbative prediction, and it is necessary to re-sum them to all orders. We performed the resummation in x space by replacing the Q^2 scale with a z-dependent one, $\widetilde{W} = Q^2(1-z)/z$. It is well known that such a procedure introduces an ambiguity in the evaluation of the running coupling at low values of the scale $\widetilde{W}$, *i.e.* as $z \to 1-\Lambda^2/Q^2$ ([18] and references therein). This ambiguity is however under control in the kinematical region of present low W^2 data (with the exclusion data at $W^2 \leq 1.3$ GeV^2), so long as we consider large x values which, at fixed W^2, are attained at large enough Q^2. Moreover, our results for NS evolution are in good agreement with

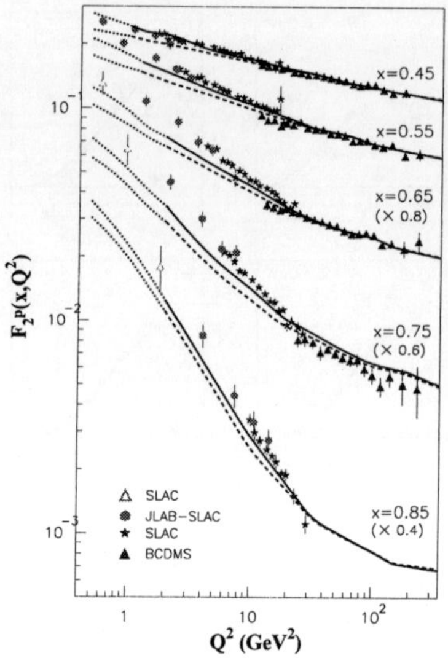

Figure 2: Comparison of pQCD+TMC calculations at NLO (dashed lines) and with resummation (full lines), with current large x data. The full dots are in the resonance region, $1.3 \leq W^2 \leq 3.4$ GeV^2; the open triangles correspond to $W^2 \leq 1.3$ GeV^2. The dotted lines represent the regions where TMC contributions are uncertain.

more recent resummation calculations performed in n space and anti-Mellin transformed as in [19].

Finally, we take into account Target Mass Corrections (TMC). We use the expression in [9], obtained by a formal inversion of the Nachtmann moments. Ambiguities in this procedure are expected to arise because of the neglect of higher order and higher twist corrections. To safely disregard the effect of such corrections on TMC, we use a "minimal criterion" that only the kinematical points (x, Q^2), yielding low values of the parameter $\propto x^2M^2/Q^2$, in the TMC expansion [9] are kept. Comparisons of pQCD+TMC predictions with the data are shown in Figure 1 and in Figure 2.

Although resummation does improve the agreement with the data, substantial discrepancies remain. These can now be unambiguously identified (modulo NNLO corrections), with the dynamical non-perturbative corrections to the structure function, shown in Figure 3. We define them from:

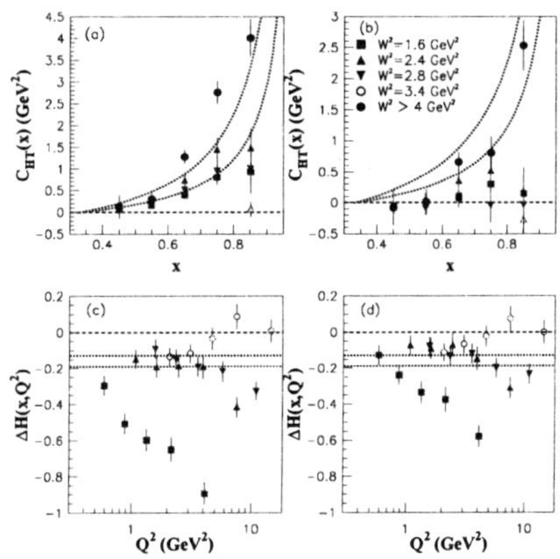

Figure 3: Coefficients C_{HT} (a,b), Eq.(3), and ΔH (c,d), Eq.(4), described in the text, plotted for various ranges of W^2. (b,d) and (a,c) are obtained with and without resummation, respectively. The dotted lines in (a,b) encompass previous extractions of the $\mathcal{O}(1/Q^2)$ terms. The dotted lines in (c,d) correspond to $\Delta H \propto -1/W^2$, with $2.3 < W^2 < 2.8$ GeV^2.

$$F_2^{exp}(x,Q^2) = F_2^{pQCD+TMC}(x,Q^2) + \frac{H(x,Q^2)}{Q^2}, \tag{3}$$

where F_2^{pQCD} is calculated either at NLO or resummed, and TMC are subsequently taken into account in both cases. As in previous extractions [20] we use:

$$H(x,Q^2) = F_2^{pQCD+TMC}(x,Q^2) C_{HT}(x), \tag{4}$$

which stems from the lack of knowledge of the anomalous dimensions for the twist-4 operators, and is however a reasonable assumption within the precision of the data (see also [21]). We extract C_{HT} from both the large W^2 data, and in the resonance region, separately. We perform a cut at large Q^2 ($Q^2 \leq 100$ GeV^2) since the precision with which the deviations of the data from pQCD would need to be known to render significative values of the power corrections is smaller than the error on the data in this region. As a result, our large W^2 extraction (full dots in Fig.3 a,b) is slightly larger than the values in [20,21] (short-dashed lines in Fig.3a,b), consistent with [12]. The low W^2 data yield a much smaller value for C_{HT}, as it is already visible from Figure 2 that shows a bend-over of the slope vs. Q^2 at low Q^2 (and W^2). The possible departure from the twist expansion is displayed in Fig. 3c,d, where we show the quantity

$\Delta H(x,Q^2)$, extracted from:

$$\frac{F_2^{exp}}{F_2^{pQCD+TMC}} = 1 + \frac{C_{HT}(x)}{Q^2} + \Delta H(x,Q^2). \qquad (5)$$

$\Delta H(x,Q^2)$ is negative for all bins except for the higher mass resonances, $W^2 \geq 3$ GeV^2, as expected if a cancellation among higher order inverse powers were to occur. The data at $2 \leq W^2 \leq 3$ GeV^2 are in agreement with a simple $\mathcal{O}(1/W^4)$ fit (dotted curves in Fig.3c,d). The data in the $N-\Delta$ transition region show a much steeper fall with Q^2 and they cannot be described by the inverse power expansion, a result similar to that found in our analysis of the moments of F_2 [13,12]. Resummation produces an overall reduction of the non-perturbative corrections (Fig.3 b,d) that does not modify our conclusion.

3 A model for duality.

We briefly present a dynamical model for the form of local duality and the Q^2 dependence that have been unravelled in the first part of our work [13,16]. The purpose of our previous study has been to define more clearly the issues to be addressed in a QCD interpretation of BG duality, namely the non trivial structure of scaling violations and of the power corrections extracted from the data.

At a scale $Q_o^2 < Q^2$ (and correspondingly at small invariant masses), partons are seen as organized in color neutral clusters with a Q^2-independent mass spectrum peaked at low mass, $M_C^2 \approx Q_o^2$, according to a mechanism similar to the "preconfinement" one envisaged for jet fragmentation in [2]. Within the preconfinement picture, the resonance spectrum in inclusive scattering is seen to follow in average the deep inelastic curve, modulo pQCD corrections [16]. We show in fact that the structure functions for both processes are determined by a convolution over the same Q^2-independent cluster mass spectrum, with the Q^2 dependent structure of the clusters. The residual Q^2 dependent violations of duality in the form of power corrections are a consequence, in our model, of phase space effects which modify the convolution integral, most strongly in the $N-\Delta$ transition region (figure 3 and [16]). Details of the model and a quantitative calculation of the parton distributions in hadrons are presented elsewhere [22].

I thank the National Institute for Theoretical Physics at the University of Adelaide and the Institute of Nuclear and Particle Physics at the University of Virginia, for support. I also thank Wally Melnitchouk, Andreas Schreiber, and Tony Thomas, for organizing a very interesting and stimulating workshop.

References

1. Yu.L. Dokshitzer, Proceedings of the *29th International Conference on High-Energy Physics (ICHEP 98)*, Edited by A. Astbury, D. Axen, J. Robinson. Singapore, World Scientific, 1999, p.305; hep-ph/9812252.
2. D. Amati and G. Veneziano, *Phys. Lett.* **83B**, 87 (1979).
3. Ya.I. Azimov, Yu.L. Dokshitzer, V.A. Khoze and S.I. Troyan, *Zeit. Phys.* C **27**, 65 (1985).
4. DELPHI Collaboration, P. Abreu *et al.*, *Phys. Lett.* B **459**, 397 (1999).
5. J. Breitweg *et al.*, *Phys. Lett.* B **487**, 53 (2000).
6. CDF Collaboration, A.N. Safonov for the collaboration, hep-ex/0007037.
7. J. Breitweg *et al.*, *Eur. Phys. J.* C **12**, 53 (1999).
8. E.D. Bloom and F.J. Gilman, *Phys. Rev.* D **4**, 2901 (1971); *Phys. Rev. Lett.* **25**, 1140 (1970).
9. A. De Rujula, H. Georgi and H.D. Politzer, *Ann. Phys.* **103**, 315 (1977).
10. I. Niculescu *et al.*, *Phys. Rev. Lett.* **85**, 1186 (2000).
11. X. Ji and P. Unrau, *Phys. Rev.* D **52**, 73 (1995).
12. I. Niculescu, C. Keppel, S. Liuti, and G. Niculescu, *Phys. Rev.* D **60**, 094001 (1999); S. Liuti, *Nucl. Phys. Proc. Suppl.* **74**, 380 (1999).
13. C. Armstrong, R. Ent, C.E. Keppel, S. Liuti, I. Niculescu and G. Niculescu, *Phys. Rev.* D **63**, 094008 (2001).
14. A. Bodek et al., *Phys. Rev.* D **20**, 1471 (1979); J. S. Poucher et al., *Phys. Rev. Lett.* **32**, 118 (1974); A. Bodek, Ph.D. Thesis, Massachusetts Institute of Technology, unpublished (1971).
15. L. W. Whitlow, *Phys. Lett.* B **282**, 475 (1992); *ibid* Ph.D. Thesis, Stanford University, unpublished (1990).
16. S. Liuti, R. Ent, C. Keppel and I. Niculescu, *submitted for publication.*
17. G. Curci, W. Furmanski and R. Petronzio, *Nucl. Phys.* B **175**, 27 (1980); H.T. Herrod and S. Wada, *Phys. Lett.* **96B**, 195 (1980).
18. R.G. Roberts, *Eur. Phys. J.* C **10**, 697 (1999).
19. G. Sterman and W. Vogelsang, Proceedings of "Physics at Run II: QCD and Weak Boson Physics Workshop: Final General Meeting, Batavia, Illinois, 4-6 Nov 1999; hep-ph/0002132.
20. U.K. Yang and A. Bodek, *Phys. Rev. Lett.* **82**, 2467 (1999).
21. S.I. Alekhin, *Phys. Rev.* D **63**, 094022 (2001).
22. S. Liuti, *in preparation.*

GLUONS, QUARKS, AND THE TRANSITION FROM NONPERTURBATIVE TO PERTURBATIVE QCD [a]

ANTHONY G. WILLIAMS

CSSM and Department of Physics and Mathematical Physics, University of Adelaide, Australia 5005

E-mail: awilliam@physics.adelaide.edu.au

Lattice-based investigations of two fundamental QCD quantities are described, namely the gluon and quark propagators in Landau gauge. We have studied the Landau gauge gluon propagator using a variety of lattices with spacings from $a = 0.17$ to 0.41 fm to explore finite volume and discretization effects. We also introduce the general method of "tree-level correction" to minimize the effect of lattice artefacts at large momenta. We demonstrate that it is possible to obtain scaling behavior over a very wide range of momenta and lattice spacings and to explore the infinite volume and continuum limits of the Landau-gauge gluon propagator. These results confirm the earlier conclusion that the Landau gauge gluon propagator is infrared finite. We study the Landau gauge quark propagator in quenched QCD using two forms of the $\mathcal{O}(a)$-improved propagator with the Sheikholeslami-Wohlert quark action with the nonperturbative value for the clover coefficient c_{sw} and mean-field improvement coefficients in our improved quark propagators. We again implement an appropriate form of tree-level correction. We find good agreement between our improved quark propagators. The extracted value of the infrared quark mass in the chiral limit is found to be 300 ± 30 MeV. We conclude that the momentum regime where the transition from nonperturbative to perturbative QCD occurs is $Q^2 \simeq 4$ GeV^2.

1 Introduction

Lattice gauge theory is currently the only known "first principles" approach to studying nonperturbative QCD. It is therefore important for lattice QCD to provide constraints and guidance for the construction of quark-based models [1] and to provide an indication of the momentum regime at which we can expect perturbative QCD to become applicable. The quark and gluon propagators are two of the most fundamental quantities in QCD. There has been considerable interest in the infrared behavior of the gluon propagator as a probe into the mechanism of confinement and by studying the scalar part of the quark propagator, the mass function, we can gain insight into the mechanisms of chiral symmetry breaking. Both are used as input for other quark-model calculations.

[a] The work reported here was performed in collaboration with P.O. Bowman, F.D.R. Bonnet, D.B. Leinweber, J.I. Skullerud and J. Zanotti

	Dimensions	β	a (fm)	Volume (fm^4)	Configurations
1w	$16^3 \times 32$	5.70	0.179	$2.87^3 \times 5.73$	100
1i	$16^3 \times 32$	4.38	0.166	$2.64^3 \times 5.28$	100
2	$10^3 \times 20$	3.92	0.353	$3.53^3 \times 7.06$	100
3	$8^3 \times 16$	3.75	0.413	$3.30^3 \times 6.60$	100
4	$16^3 \times 32$	3.92	0.353	$5.65^3 \times 11.30$	100
5	$12^3 \times 24$	4.10	0.270	$3.24^3 \times 6.48$	100
6	$32^3 \times 64$	6.00	0.099	$3.18^3 \times 6.34$	75

Table 1: Details of the lattices used to calculate the gluon propagator. Lattices 1w and 1i have the same dimensions and approximately the same lattice spacing, but were generated with the Wilson and improved actions respectively. Lattice 6 was generated with the Wilson action.

2 Gluon Propagator

We use an $\mathcal{O}(a^2)$ tree-level, tadpole-improved action [2] and for the tadpole (mean-field) improvement parameter we use the plaquette measure [3]. A full description and discussion of the gluon propagator results summarized here can be found elsewhere.[4,5,6]

Gauge fixing on the lattice is achieved by maximizing a functional, the extremum of which implies the gauge fixing condition. The usual Landau gauge fixing functional implies that $\sum_\mu \partial_\mu A_\mu = 0$ up to $\mathcal{O}(a^2)$. To ensure that gauge dependent quantities are also $\mathcal{O}(a^2)$ improved, we implement the analogous $\mathcal{O}(a^2)$ improved gauge fixing.[4] The dimensionless lattice gluon field $A_\mu(x)$ is calculated from the link variables in the usual way, which agrees with the continuum to $\mathcal{O}(a^2)$. We then calculate the scalar part of the propagator

$$D(x-y) = \sum_\mu \langle A_\mu(y) A_\mu(x) \rangle \, . \tag{1}$$

To isolate the nonperturbative behavior of the gluon propagator, we can divide the propagator by its lattice tree level form (i.e., that of lattice perturbation theory).[5] For the momentum space gluon propagator $D(q^2)$, we see that in the continuum $q^2 D(q^2)$ will approach a constant up to logarithmic corrections as $q^2 \to \infty$ because of asymptotic freedom. The continuum tree-level propagator is $1/q^2$. We also expect asymptotic freedom on the lattice despite finite lattice spacing artefacts. We *define* the lattice q_μ such that the lattice $D^{\text{tree}}(q) \equiv 1/q^2$, and use this momentum throughout. This is referred to as tree-level correction and we have seen that it significantly reduces discretization errors at large momenta. For the two actions considered here, this means that we work with the momentum variables defined as

$$q_\mu^W \equiv \frac{2}{a}\sin\frac{\hat{q}_\mu a}{2}, \qquad q_\mu^I \equiv \frac{2}{a}\sqrt{\sin^2\left(\frac{\hat{q}_\mu a}{2}\right)+\frac{1}{3}\sin^4\left(\frac{\hat{q}_\mu a}{2}\right)}, \tag{2}$$

for the Wilson and improved actions, respectively. All figures (quark and gluon propagators) have a cylinder cut imposed upon them, i.e. all momenta must lie close to the lattice diagonal. In Table 1 we show the various lattices that we have studied for the gluon propagator. In Fig. 1 we plot $q^2D(q^2)$ for a fine unimproved Wilson action and for our finest improved action. Despite having very different lattice spacings the agreement is excellent for the entire intermediate and high-momentum regime. The small discrepancy in the deep infrared due to finite volume effects is not apparent in this way of plotting the data. We plot $D(q^2)$ for five different lattice in Fig. 2 and see pleasing agreement for the results. Note that we are plotting bare quantities only and there is thus an overall wavefunction renormalization for the gluon propagator (i.e., $Z_3(\mu,a)$ for the renormalization point μ). The vertical scale is thus unimportant and only the variation with momentum is relevant. This way of presenting the data shows that there is a small residual finite volume dependence, where the infrared gluon propagator is *decreasing* with increasing lattice volume V. We have performed a fit as a function of $1/V$ and have seen that the large volume $\beta = 3.92$ lattice gives results which are very close to the infinite volume limit. In Fig. 3 we plot $D(q^2)$ in the intermediate and ultraviolet regime and compare with the three-loop perturbative QCD form. We see that the above $q \sim 2$ GeV the agreement is excellent, but that below this momentum scale nonperturbative effects are becoming apparent.

3 Quark Propagator

The Landau gauge quark propagator results summarized here have been presented and discussed in more detail elsewhere.[10] All $\mathcal{O}(a)$ errors in the fermion action can be removed by adding appropriate terms to the Lagrangian [7,8]. It is then usual to perform appropriate field transformations to improve the quark operators as well.[9]

In the continuum, the quark propagator has the following general form,

$$S(p) = \frac{1}{i\,\not{p}A^c(p)+B^c(p)} \equiv \frac{Z^c(p)}{i\,\not{p}+M^c(p)}. \tag{3}$$

We expect the lattice quark propagator to have a similar form, but with $\not{k}$

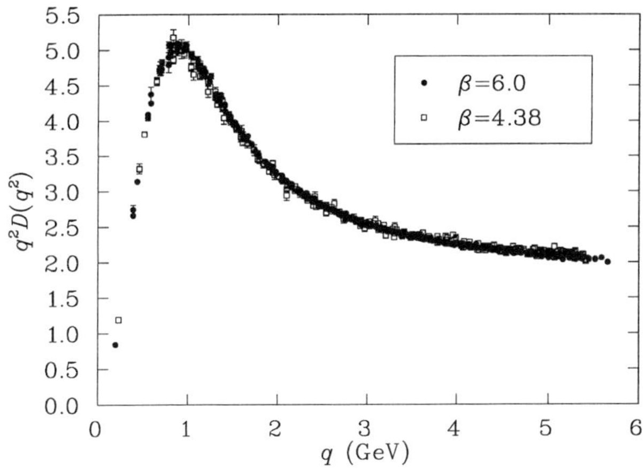

Figure 1: Comparison of the gluon propagator from the finest improved lattice (lattice 1i, $\beta = 4.38$) and the finest Wilson lattice (lattice 6, $\beta = 6.0$). Data has been cylinder cut and the appropriate tree-level corrections have been applied. The data from lattice 6 is half-cut whereas lattice 1i displays the full Brillouin zone. We have determined $Z_3(\text{improved})/Z_3(\text{Wilson}) = 1.08$ by matching the vertical scales of the data.

replacing $\not p$:

$$S(p) = \frac{Z(p)}{i\not k + M(p)} \tag{4}$$

where k is a new 'lattice momentum', $k_\mu = \frac{1}{a}\sin(\hat{p}_\mu a)$. We do not have sufficient space here to describe the hybrid tree-level correction that was used for the quark propagator results presented here, but a detailed description has recently been given.[10] We again use the cylinder cut to further reduce hypercubic discretization artefacts. As for the gluon propagator the results for $Z(p)$ are for the bare quantity only and contain an overall renormalization constant $Z_2(\mu, a)$. In Fig. 4 the vertical scales have been adjusted so that the two sets of results are renormalized and hence coincide at 2.1 GeV. In this figure we see the characteristic dip in the infrared for $Z(p)$, which occurs also in model Dyson-Schwinger equation studies [1] of dynamical chiral symmetry breaking. This dip has essentially disappeared by around 2 GeV. The improved action corresponding to S_R is the one we prefer and it gives the more expected ultraviolet behavior of the renormalized $Z(p)$, i.e., it tends towards a constant. In Fig. 5 we see the characteristic behavior of the quark mass function familiar from quark model studies [1] and that the transition to the perturbative regime is occurring at approximately 2 GeV. Note that there is no renormalization of

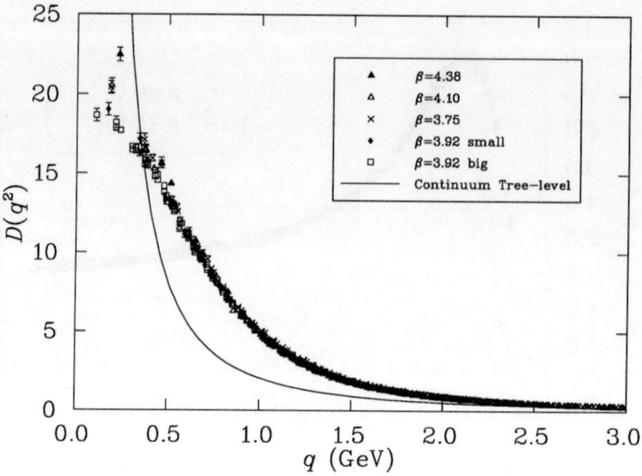

Figure 2: Comparison of the gluon propagator generated with an improved action on five different lattices. We find good agreement down to $q \simeq 500$ MeV. At the lowest accessible momenta the data points drop monotonically with increasing volume, but the lowest point (on the largest lattice) shows signs of having converged to its infinite volume value. For comparison with perturbation theory, a plot of the continuum, tree-level gluon propagator (i.e., $1/q^2$ appropriately scaled) has been included.

the quark mass function. At large momenta the mass function should become the running quark mass of perturbative QCD. Finally in Fig. 6 we present a simple quadratic extrapolation to the chiral limit for the available S_I data. The slight dip is not statistically significant and is almost certainly a residual lattice artefact. The infrared mass (i.e., at $p = 0$) in the chiral limit is approximately 330 ± 30 MeV, which is characteristic of the constituent quark mass scale.

4 Summary and Conclusions

The gluon propagator has been calculated on fine unimproved lattices and on a variety of improved lattices with an $\mathcal{O}(a^2)$ improved action in $\mathcal{O}(a^2)$ improved Landau gauge. The infrared behavior of this propagator strongly suggests the the Landau gauge gluon propagator is infrared finite. We have ruled out the $1/q^4$ behavior popular in some Dyson-Schwinger quark model studies [1] and indeed any infrared singularity appears to be very unlikely. The possible effects of lattice Gribov copies remains a very interesting question and we are currently carrying out similar studies across a variety of lattices in Laplacian gauge, which is a Landau-like smooth gauge fixing, but is free of Gribov copies.

We have used two different definitions of the $\mathcal{O}(a)$ improved quark propa-

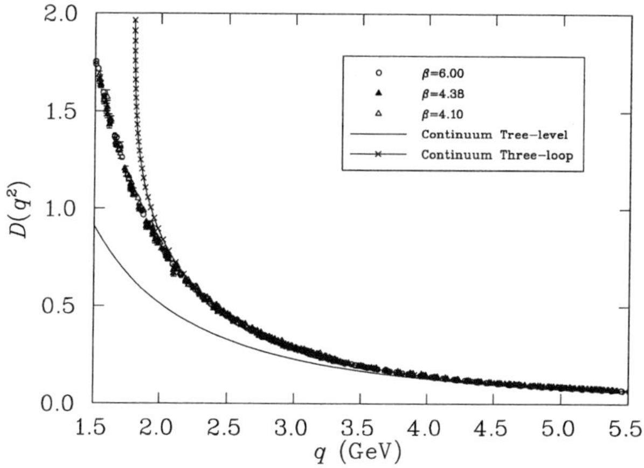

Figure 3: Comparison of the lattice gluon propagator with that obtained from perturbation theory, in the ultraviolet to intermediate regime. The continuum expressions are tree-level (i.e., $1/q^2$ appropriately scaled) and the three-loop perturbative QCD expression.

gator, corresponding to the quark propagators denoted S_I and S_R. We make use of asymptotic freedom to factor out the tree level behaviour, replacing it with the 'continuum' tree level behaviour $Z(p) = 1, M(p) = m$. This tree-level correction dramatically improves the data. We find that $M(0)$ approaches a value of 300 ± 30 MeV in the chiral limit, which is very much in keeping with the concept of a "constituent quark mass" and agrees with the infrared values of the quark mass commonly used in model studies.[1] We also find a significant dip in the value for $Z(p)$ at low momenta. This is again entirely consistent with what is found in model studies of dynamical chiral symmetry breaking [1]. An examination of Figs. 1, 3, 4, 5, and 6 provide a clear indication that perturbative QCD behavior is not becoming dominant in the gluon and quark propagators until we reach momenta of order $Q^2 \simeq 4$ GeV^2.

References

1. C.D. Roberts and A.G. Williams, *Prog. Part. Nucl. Phys.* **33**, 477 (1994).
2. *P. Weisz, Nucl. Phys.* B **212**, 1 (1983).
3. G.P. Lepage & P.B. Mackenzie, *Phys. Rev.* D **48**, 2250 (1993).
4. F.D.R. Bonnet, P.O. Bowman, D.B. Leinweber, A.G. Williams & D.G. Richards, *Aust. J. Phys.* **52**, 939 (1999); *Nucl. Phys.* B (Proc. Suppl.) **83**, 905 (2000).
5. D.B. Leinweber, J.I. Skullerud, A.G. Williams, and C. Parrinello, *Phys.*

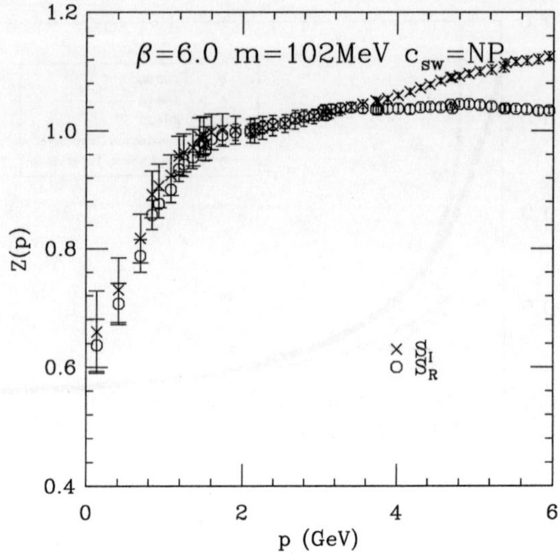

Figure 4: The tree-level corrected $Z(p)$ for c_{sw} = NP and $\kappa = 0.137$. The figure shows $Z(p)$ vs. momentum in physical units, and after rescaling ("renormalizing") so that $Z(2.1\text{GeV}) = 1$. The infrared agreement after rescaling is very good and we take the tree-level corrected $Z(p)$ from S_R with the nonperturbative c_{sw} as the best estimate for this quantity.

Rev. D **58**, 031501 (1998); *ibid.* **60**, 094507 (1999); Erratum – *ibid.* **61**, 079901 (2000); D.B. Leinweber, C. Parrinello, J.I. Skullerud, and A.G. Williams *Nucl. Phys.* B (*Proc. Suppl.*) **73**, 629 (1999); *ibid.*, 626 (1999).

6. F.D.R. Bonnet, P.O. Bowman, D.B. Leinweber, and A.G. Williams, *Phys. Rev.* D **62**, 051501 (2000); *ibid.* **64**, 034501 (2001).
7. M. Lüscher *et al.*, *Nucl. Phys.* B **478**, 365 (1996).
8. C. Dawson *et al.*, *Nucl. Phys.* B (*Proc. Suppl.*) **63**, 877 (1998).
9. G. Heatlie *et al.*, *Nucl. Phys.* B **352**, 266 (1991).
10. J.I. Skullerud and A.G. Williams, *Nucl. Phys.* B (*Proc. Suppl.*) **83**, 209 (2000); *Phys. Rev.* D **63**, 054508 (2001); J.I. Skullerud, D.B. Leinweber, and A.G. Williams, to appear in *Phys. Rev.* D, hep-lat/0102013.

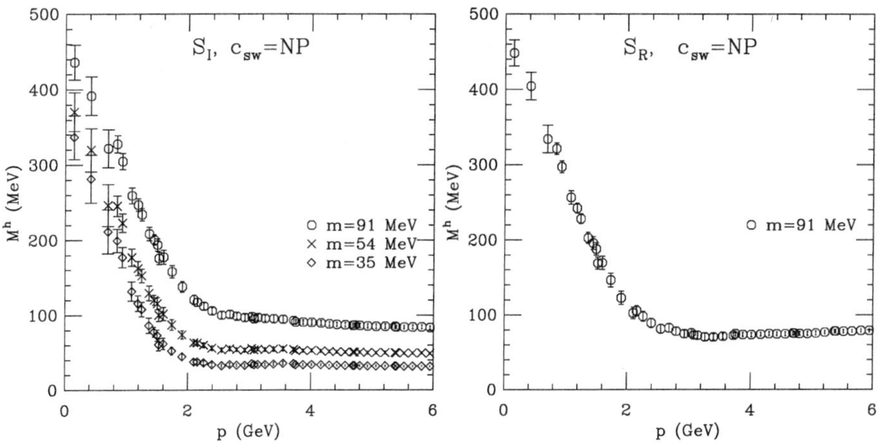

Figure 5: The tree-level corrected $M(p)$, for c_{SW} = NP, using S_I (left) and S_R (right). We find good agreement between the two data sets, both in the infrared and ultraviolet. The residual disagreement at intermediate momenta is a pointer to lattice artifacts that we have not brought under full control, even with nonperturbative improvement and tree-level correction.

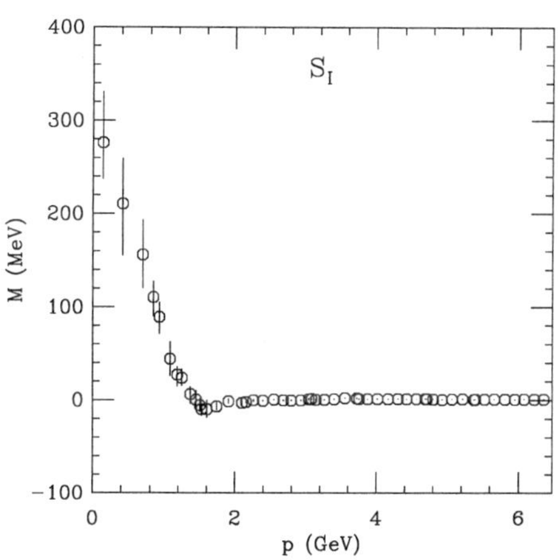

Figure 6: The tree-level corrected mass function from S_I with c_{SW} = NP, with the bare mass extrapolated to zero using a quadratic fit. The small dip at $p \sim 1.6$ GeV is not statistically significant and may be due to residual lattice artifacts. The non-zero values for $M(p)$ in the chiral limit are entirely due to dynamical chiral symmetry breaking and provide a direct measure of this effect.

ESTIMATING LOW ENERGY MODEL PARAMETERS FROM DEEP INELASTIC SCATTERING

L. P. HOYT and A. I. SIGNAL
Institute of Fundamental Sciences
Massey University
Private Bag 11 222, Palmerston North
New Zealand
E-mail: A.I.Signal@massey.ac.nz

Model calculations of parton distributions are one of the ways to connect our knowledge of physics at low energy scales with the high energy scales where α_{QCD} is small. There now exist a number of these calculations giving qualitative agreement with experimental data. We present a best-fit analysis for the valence u and d quark distributions calculated in the MIT bag model and in a chirally symmetric extension of the bag model. This analysis yields best-fit model parameters in agreement with other low energy fits, and provides new constraints on the parameters governing the pion cloud of the nucleon.

1 Introduction

Over the past thirty years the data from deep inelastic experiments has steadily increased in quality and precision. However, this data has not greatly affected our qualitative understanding of the non-perturbative physics at lower energy scales, despite the fact that the shapes of the structure functions and parton distributions contain important information about the low energy structure of the proton. There appear to be two reasons for this. Firstly it is difficult to use the QCD evolution equations to go from experimental scales ($Q^2 > 1$ GeV^2) down to low scales: often the evolution is unstable in this direction, plus there is no systematic way of dealing with corrections beyond next-to-leading order. Secondly it is difficult to interpret the parton distributions in terms of the parameters (e.g. confinement radius) and processes (e.g. chiral symmetry breaking) that are of interest at low energy scales.

In view of these difficulties, a more fruitful approach may be to begin in the low energy sector and attempt to work towards the experimental scales. That is, starting from a model of the nucleon, where we can vary one or more of the low energy parameters of interest, we calculate the parton distributions, which can then be compared with those obtained from experiment. Provided these calculations are reasonably robust, we can then observe the effects on the calculated distributions of varying the model parameters and attempt to use best fit methods to estimate the values of the model parameters that give good agreement with the observed distributions.

In this talk we present our initial efforts in this direction. We use the MIT bag model[1] to calculate the parton distributions for the u and d valence quarks. These are then compared with the MRST99[2] and CTEQ5[3] parametrizations of the experimental data. We observe the effects of varying the three parameters of the model plus the calculation, and attempt to minimize the χ^2 difference between the experimental parametrizations and the bag model calculations. We also extend the model to include a pion cloud, which can contribute to the quark distributions via the Sullivan [4] process. Again we compare the calculations with the parametrizations of the experimental data, and attempt a best fit analysis.

2 MIT Bag Model Parton Distributions

For our purposes, it is sufficient to use the spherical cavity approximation of the MIT bag model[5]. The most important parameter is the bag radius R. Fits to low energy properties such as the hadron spectrum and magnetic moments give good results for $R \approx 1$ fm. When a pion cloud is added to the model to restore chiral symmetry, as in the Cloudy Bag model [6], fits to the same low energy properties give R in the range 0.8 – 1.0 fm.

We adopt here the approach of the Adelaide group for calculating bag model parton distributions

$$q_f(x) = p^+ \sum_n \delta(p^+(1-x) - p_n^+) \left|\langle n|\psi_{+,f}(0)|p\rangle\right|^2 . \tag{1}$$

The detailed calculation is given in reference [7], but we outline some of the important points here. In order to ensure that the calculated distributions have the correct support (i.e. $0 \leq x \leq 1$) the initial state $|p\rangle$ and the intermediate state $|n\rangle$ must be momentum eigenstates. To make a momentum eigenstate from the static bag model the Peierls-Yoccoz projection [8] is used, which is a non-relativistic approximation, valid for the magnitude of the momentum smaller than the particle mass. This corresponds to $x < \sim 0.7$ for parton distribution calculations.

The possible intermediate states $|n\rangle$ arising from the action of the operator $\psi_f(0)$ on the initial state $|p\rangle$ are either 2-quark states, which dominate the valence distributions, or 3-quark plus 1-antiquark states. For the 2-quark intermediate states the contributions to the parton distributions are given by

$$q^{\uparrow\downarrow}_{(2),f}(x) = \frac{M}{(2\pi)^3} \sum_m \langle\mu|P_{f,m}|\mu\rangle \int_{p_{\min}}^{\infty} dp_n \frac{|\phi_2(p_n)|^2}{|\phi_3(0)|^2} |\tilde{\Psi}^{\uparrow\downarrow}_{+,f,m}(p_n)|^2 \tag{2}$$

where $\tilde{\Psi}$ is the Fourier transform of the MIT bag ground state wavefunction and

$$p_{\min} = \left| \frac{M^2(1-x)^2 - M_n^2}{2M(1-x)} \right| . \tag{3}$$

The mass of the 2-quark intermediate state M_n is not determined in the model, and is a parameter that we must set. The contributions from the 3-quark plus 1-antiquark intermediate states can be calculated in a similar manner. These shape of these contributions is well approximated by $(1-x)^7$, so we use this function for the 3-quark plus 1-antiquark contributions to the valence distributions, with the strength determined by the requirement that the valence number sum rules are satisfied.

The one-gluon exchange interaction which splits the nucleon and Δ by 300 MeV is also responsible for the different x-dependencies of the u and d valence distributions [9]. The 2-quark intermediate state will be in either a spin singlet or triplet, whose masses are split by about 200 MeV by the spin-spin dependence of the interaction. From the $SU(6)$ wavefunction of the proton we have

$$u_2(x) = \frac{3}{2} q_s(x) + \frac{1}{2} q_t(x), \quad d_2(x) = q_t(x) \tag{4}$$

where the subscripts denote the spin configuration of the quark pair.

Our calculated distributions are at a scale $Q^2 = \mu^2$ appropriate to the bag model, where $\mu \sim k_\perp$ is not determined in the model and is here regarded as another parameter. In order to carry out the comparison with experimental distributions, we must evolve our calculated distributions. We use NLO evolution evaluated using the algorithm of Kosower[10] which is extremely efficient for valence distributions. In figure 1 we show the effects of varying each of the three parameters of our model calculations. Increasing the bag radius R gives a valence distribution with a sharper peak, and decreases the distribution at large x, which agrees with the uncertainty principle. Increasing M_n pushes the distribution to smaller x. This occurs because the peak of $q_2(x)$ occurs at approximately $x = 1 - M_n/M$. Finally increasing the bag scale μ^2 pushes the distribution to larger x as expected from the behaviour of QCD evolution.

We attempt to obtain a 'best fit' to the experimental data on structure functions by minimising the χ^2 statistic for our valence distributions compared to the MRST99 parametrizations of the valence distributions at $Q^2 = 1.25$ GeV^2. We use the difference between the MRST99 and CTEQ5 parametrizations as indicative of the experimental uncertainties in the distributions. The difference is a few per cent in the low and medium-x regions, but becomes quite

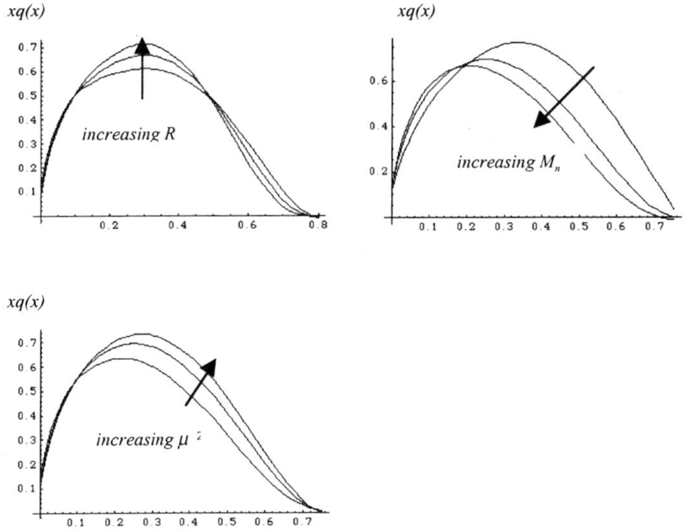

Figure 1: The effects on the bag model quark distribution of varying the model parameters R, M_n and μ^2.

large for $x > 0.6$, as a consequence of the lack of data in this region. As we expect the approximations used in the bag model calculations to break down in the large-x region, we therefore limit the range of comparison to $x \leq 0.6$. We also impose a lower limit of $x \geq 0.05$ as the bag model calculation also breaks down for small x. This is because small x corresponds to large distances on the light-cone, however the maximum possible spatial separation in the bag model calculation is $4R$, corresponding to $x \sim 0.1$ at the bag scale μ^2. We use a Levenberg-Marquandt method to find the minimum of χ^2 in the (R, M_n, μ^2) space. Our best fit distributions are shown in figure 2. The fit is not partic-

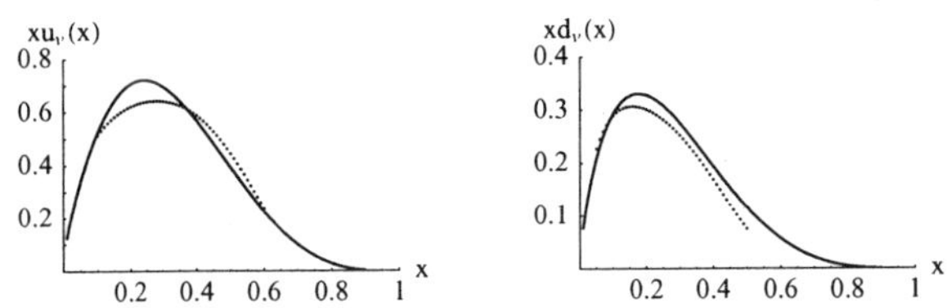

Figure 2: The best fit valence u and d distributions calculated using the MIT bag model (dotted lines) and the MRST99 valence distributions.

ularly good ($\chi^2 = 5.4$/d.o.f), with the peaks of the calculated distributions being in different positions to the data, as well as being somewhat lower and broader than the experimental distributions. The best fit value of the bag radius is $R = 0.733 \pm 0.011$ fm, which is somewhat lower than low energy estimates. The best fit value of the intermediate mass is $M_n = 832 \pm 23$ MeV, which is slighly larger than previous calculations have used [7,11], while the best fit value of the bag scale is $\mu^2 = 0.197 \pm 0.016$ GeV^2, in good agreement with expectations. The statistical uncertainties are at the 90% confidence limit.

3 Pion Cloud Model Distributions

We can improve our calculated distributions by including the effects of the pion cloud of the nucleon. The Meson Cloud Model (MCM)[12] includes the contributions from the virtual γ scattering from either hadron of a meson-plus-baryon component in the Fock space wavefunction of the physical nucleon. This model has been used successfully to describe the violation of the Gottfried sum rule[13,14]. The contributions from the meson cloud to the nucleon parton distributions can be written as a convolution:

$$\delta q(x) = \sum_{BM} \left\{ \int_x^1 \frac{dy}{y} f_{BM/N}(y) q^B(\frac{x}{y}) + \int_x^1 \frac{dy}{y} f_{MB/N}(y) q^M(\frac{x}{y}) \right\}, \tag{5}$$

where q^B and q^M are the parton distributions of the baryon and meson respectively. The fluctuation function $f(y)$ satisfies

$$f_{MB/N}(y) = f_{BM/N}(1-y) \tag{6}$$

and will depend on the form factor at the NBM vertex. The key assumption of the model is that the meson-plus-baryon state is long-lived compared to DIS timescales, so that the convolution is valid. The MCM naturally suppresses states with large masses, so that at the few percent level (compatible with the uncertainties in the experimental data) we only need to consider the smallest mass states, which come from $N \to N\pi$ and $N \to \Delta\pi$ interactions.

Using the $SU(6)$ wavefunctions the contributions to the valence distributions can be found in terms of the baryon distributions u^p, d^p, u^{Δ^0} and d^{Δ^+}), which we calculate using the bag model,and the pion valence distribution, v^π, for which we use the parametrization of the Dortmund group[15].

The other input we need for the MCM calculation are the form factors at the NBM vertex. We use an exponential form for these (a dipole form gives very similar results)

$$G_{BM}(y, k_\perp^2) = \exp\left[\frac{m_N^2 - m_{BM}^2(y, k_\perp^2)}{2\Lambda_{BM}^2}\right], \tag{7}$$

which introduces two new parameters, $\Lambda_{N\pi}$ and $\Lambda_{\Delta\pi}$, into the calculations of the valence distributions. There are two suggested sets of values for these parameters in the literature. From an analysis of high energy pp events Holtmann, Speth and Szurek (HSS) proposed[16] $\Lambda_{N\pi} = 1.10$ GeV and $\Lambda_{\Delta\pi} = 0.98$ GeV, while from an analysis of the Gottfried sum rule violation Melnitchouk, Speth and Thomas (MST) suggested[13] $\Lambda_{N\pi} = 0.8$ GeV and $\Lambda_{\Delta\pi} = 1.0$ GeV. We shall use both sets of MCM parameters, attempting to find a the best-fit in each case, as well as allowing both $\Lambda_{N\pi}$ and $\Lambda_{\Delta\pi}$ to vary freeely. As increasing either $\Lambda_{B\pi}$ increases the number of pions in the cloud, the effect of increasing one (or both) Λ is to increase the contribution from the pion valence distribution, which is peaked at greater x than the proton distributions, and hence to move the proton distributions to the right.

We have performed a best-fit analysis in the (R, M_n, μ^2) parameter space to find best-fit valence distributions with both the HSS and MST sets of $\Lambda_{B\pi}$ parameters. The results of these analyses are shown in figures 3 and 4. The

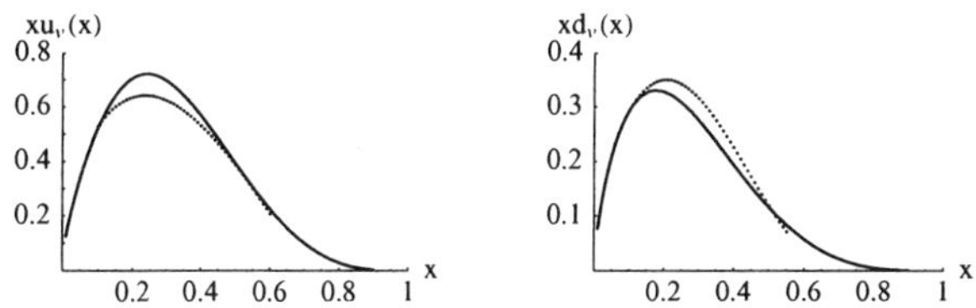

Figure 3: The best fit valence u and d distributions calculated using the MIT bag model and the HSS pion cloud parameters (dotted lines) and the MRST99 valence distributions.

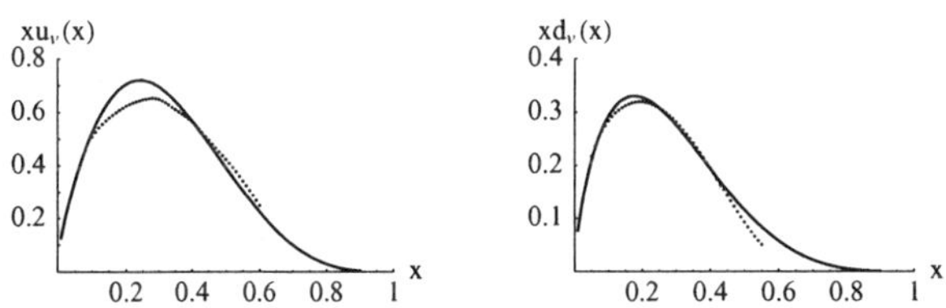

Figure 4: The best fit valence u and d distributions calculated using the MIT bag model and the MST pion cloud parameters (dotted lines) and the MRST99 valence distributions.

MCM has improved the fit in both cases, though the minimum χ^2 for both was 4.1/d.o.f, which are not very good fits. The positions of the distribution peaks are in better agreement with the data, however the widths and heights of these peaks are not in particularly good agreement. The best-fit parameters in the HSS (MST) cases were

$$
\begin{aligned}
R &= 0.771 \pm 0.016\text{fm} && (0.777 \pm 0.014\text{fm}) \\
M_n &= 778 \pm 23\text{MeV} && (763 \pm 23\text{MeV}) \\
\mu^2 &= 0.213 \pm 0.020\text{GeV}^2 && (0.192 \pm 0.016\text{GeV}^2)
\end{aligned} \tag{8}
$$

Compared to the no pion cloud case earlier, the MCM has had the effect of increasing the bag radius R and decreasing the intermediate state mass M_n, while the bag scale is unchanged.

Finally we can allow $\Lambda_{N\pi}$ and $\Lambda_{\Delta\pi}$ to be free parameters, and attempt to find a best fit in the five-dimensional parameter space. These investigations are still at a preliminary stage, however we can report that it is possible to improve the quality of the fit by allowing the $\Lambda_{B\pi}$ to vary. In figure 5 we show one such fit, which has $\chi^2 = 2.3$/d.o.f. We observe that the peaks of

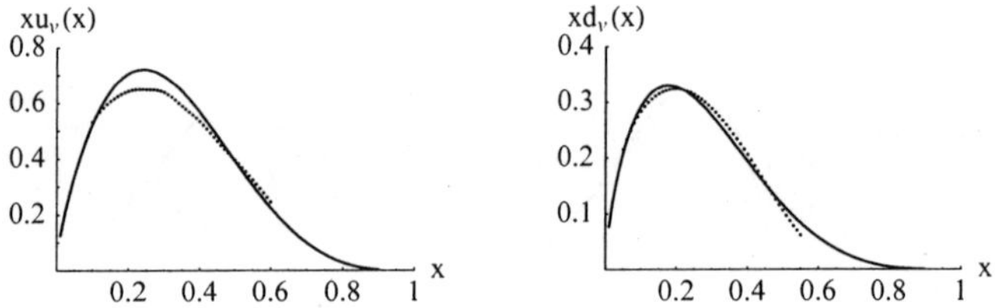

Figure 5: The best fit valence u and d distributions calculated using the MIT bag model and allowing the pion cloud parameters to vary (dotted lines) and the MRST99 valence distributions.

the distributions agree very well with the data apart from the height of $u_v(x)$. Unfortunately this fit has $\Lambda_{\Delta\pi} = 1.31$GeV, which is rather large, and would give a Gottfried sum rule violation of the wrong sign.

4 Summary

We have attempted to relate the experimental data on valence quark distributions to the parameters of the MIT bag model. Using best-fit methods we found a fair fit to the data. This fit can be improved by including in the model calculations the effects of the pion cloud, which is required from axial symmetry considerations. These improved fits support a bag radius of around 0.75 – 0.80 fm, but not any smaller as this would broaden the distributions too much. The best fit values of the intermediate state mass M_n and the bag scale μ^2 are in agreement with our expectaions based on the bag model. Allowing the parameters of the meson cloud model to vary can also improve the fit to the data. These investigations are at a preliminary stage, but appear to support a fairly soft $NN\pi$ form factor. In further work we shall continue investigating the form factor dependence of the valence distributions, which may be able to provide new constraints on the pion cloud parameters. In addition, we shall take into account the pion cloud contribution to the $N-\Delta$ mass difference, which will reduce the strength of the spin-spin splitting in the model. Also we need to address the non-relativistic deficiencies of the Peierls-Yoccoz projection, which causes the calculated distributions to fall off too rapidly at large-x. Finally we hope to include sea distributions and spin-dependent distributions in our fits in the near future.

Acknowledgements

AIS would like to thank his previous collaborators Tony Thomas and Andreas Schreiber with whom he had many enlightening discussions. This work is supported by the Marsden Fund of the Royal Society of New Zealand.

1. A. Chodos, R. L. Jaffe, K. Johnson, C.B. Thorn and V. Weisskopf, *Phys. Rev.* D **9**, 3471 (1974); A. Chodos, R. L. Jaffe, K. Johnson and C.B. Thorn, *Phys. Rev.* D **10**, 2599 (1974).
2. A. D. Martin, R. G. Roberts, W. J. Stirling and R. S. Thorne, Eur. Phys. J. C **14**, 133 (2000).
3. CTEQ, H. L. Lai *et al*, Eur. Phys. J. C **12**, 375 (2000).
4. J. D. Sullivan, *Phys. Rev.* D **5**, 1732 (1972).
5. T. DeGrand, R. L. Jaffe, K. Johnson and J. Kiskis, *Phys. Rev.* D **12**, 2060 (1975).
6. A. W. Thomas, Adv. Nucl. Phys. **13**, 1 (1984).
7. A. W. Schreiber, A. I. Signal and A. W. Thomas, *Phys. Rev.* D **44**, 2653 (1991).
8. R. E. Peierls and J. Yoccoz, Proc. Phys. Soc. A **70**, 381 (1957).
9. F. E. Close and A. W. Thomas, *Phys. Lett.* B **212**, 481 (1988).
10. D. A. Kosower, *Nucl. Phys.* B **506**, 439 (1997).
11. C. Boros and A. W. Thomas, *Phys. Rev.* D **60**, 074017 (1999).
12. J. Speth and A. W. Thomas, Adv. Nucl. Phys. **24**, 83 (1998); S. Kumano, Phys. Rep. **303**, 103 (1998).
13. W. Melnitchouk, J. Speth, and A. W. Thomas, *Phys. Rev.* D **59**, 014033 (1998).
14. M. Alberg and E. M. Henley, *Nucl. Phys.* A **663**, 301 (2000).
15. M. Glück, E. Reya, and I. Schienbein, Eur. Phys. J. C **10**, 313 (1999).
16. H. Holtmann, A. Szczurek and J. Speth, *Nucl. Phys.* A **569**, 631 (1996).

4. Form Factors

MEASUREMENT OF G_{E_P}/G_{M_P} TO $Q^2 = 5.6$ GEV^2 VIA RECOIL POLARIZATION AT JEFFERSON LAB

OLIVIER GAYOU
FOR THE HALL A COLLABORATION

College of William and Mary, Williamsburg, Virginia 23187
Université Blaise Pascal/IN2P3, F-63177 Aubière, France
E-mail: gayou@jlab.org

The measurement of the elastic form factors is a key ingredient to any complete understanding of the internal structure of the nucleons, and ultimately of the strong force. Precise data are essential to impose stringent tests on any QCD-based theory. The electromagnetic interaction provides a unique tool to investigate these form factors. In elastic electron scattering off a proton, the electron interacts with the nucleon exchanging a virtual photon. The electron-photon interaction is fully understood from QED, hence making the hadron vertex the only unknown of the reaction. The current at this vertex can be written as in Eq. (1)

$$J^\mu = \overline{U}(p') \left\{ \gamma^\mu F_1(Q^2) + \frac{i\sigma^{\mu\nu} q_\nu}{2M_p} F_2(Q^2) \right\} U(p) \tag{1}$$

where $q_\nu = p'_\nu - p_\nu$, $U(p)$ and $\overline{U}(p')$ are the incoming and outgoing proton spinors respectively, and Q^2 is the 4-momentum transfer. $F_1(Q^2)$ and $F_2(Q^2)$ are the form factors, that contain the information about the internal structure of the proton. $F_1(Q^2)$ is the non-spin flip Dirac form factor, and $F_2(Q^2)$ the spin-flip Pauli form factor. These can be rewritten in the Sachs formalism, in terms of electric and magnetic form factors, as in Eq. (2)

$$\begin{aligned} G_E(Q^2) &= F_1(Q^2) - \kappa\tau F_2(Q^2) \\ G_M(Q^2) &= F_1(Q^2) + \kappa F_2(Q^2) \end{aligned} \tag{2}$$

with $\tau = \frac{Q^2}{4M_p^2}$ and κ is the anomalous magnetic moment of the proton. The electric and magnetic form factors are images of the charge and magnetic current distributions respectively. In elastic scattering, where the one-photon exchange dominates[1], they depend only on the four-momentum transfer squared Q^2.

These form factors have been measured by Rosenbluth separation over the last 50 years. The unpolarized Rosenbluth ep cross section is[2]

$$\frac{d\sigma}{d\Omega} = \frac{\alpha^2 E'_e \cos^2\frac{\theta_e}{2}}{4E_e^3 \sin^4\frac{\theta_e}{2}} \left[G_{E_p}^2 + \frac{\tau}{\epsilon} G_{M_p}^2 \right] \left(\frac{1}{1+\tau} \right) \tag{3}$$

where $\epsilon = \left[1 + 2(1+\tau)\tan^2\frac{\theta_e}{2}\right]^{-1}$ is the virtual photon longitudinal polarization. The form factors can be separately extracted by measuring this cross section at a given Q^2 over a range of ϵ values obtained by varying the beam energy and the electron scattering angle. These data[3] were fitted with a dipole shape, $G_D = \left(1 + \frac{Q^2}{0.71}\right)^{-2}$, Fourier transform of an exponential radial distribution. The magnetic form factor of the proton G_{M_p} was accurately measured to 2-3% level up to $Q^2 = 10$ GeV^2, as shown in Fig. 1b. However, with increasing Q^2, the magnetic term in the cross section is enhanced by the factor τ, so that it becomes more difficult to separate the electric term. Fig. 1a shows $\frac{G_{E_p}}{G_D}$ as a function of Q^2: there is an obvious discrepancy in the data from different experiments, and a lack of accuracy, at $Q^2 > 1$ GeV^2.

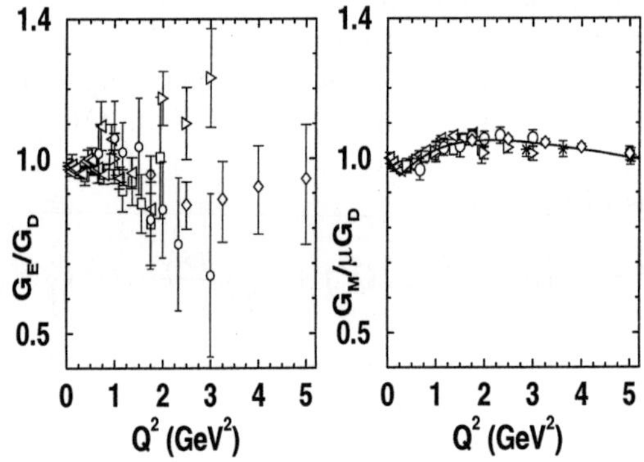

Figure 1. Electric (a, left) and magnetic (b, right) form factors measured by Rosenbluth separation, compared with the dipole shape.

An alternative way to extract the small electric term is to measure the interference term $G_{E_p}G_{M_p}$ via recoil polarization. This was given first in 1974 by Akhiezer and Rekalo[4], and discussed in detail in 1981 by Arnold, Carlson and Gross[5]. In this method, a longitudinally polarized electron beam hits a proton target, and the polarization is partially transferred to the recoil proton. The only two non-zero components of this polarization transfer in the one-photon exchange approximation are:

$$I_0 P_t = -2\sqrt{\tau(1+\tau)}G_{E_p}G_{M_p}\tan\frac{\theta_e}{2} \tag{4}$$

$$I_0 P_\ell = \frac{1}{M_p}\left(E_e + E_{e'}\right)\sqrt{\tau\left(1+\tau\right)}G_{M_p}^2 \tan^2\frac{\theta_e}{2} \tag{5}$$

where $I_0 \propto G_{E_p}^2 + \frac{\tau}{\epsilon}G_{M_p}^2$ is the unpolarized cross section. Measuring simultaneously these two components and taking their ratio directly leads to the ratio of the form factors

$$\frac{G_{E_p}}{G_{M_p}} = -\frac{P_t}{P_\ell}\frac{\left(E_e + E_{e'}\right)}{2M_p}\tan\frac{\theta_e}{2} \tag{6}$$

This technique was first used at Bates[6], and later at Jefferson Lab[7] in 1998, giving the first spectacular evidence that the charge and magnetic current distributions in the proton are different, as shown in Fig. 2. The purpose of the experiment E99-007 at Jefferson Lab, presented in this article, was to extend this measurement to $Q^2 = 5.6$ GeV^2.

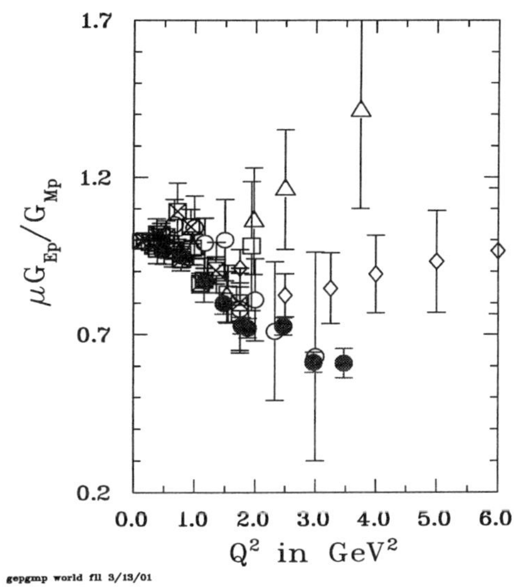

Figure 2. Ratio $\mu\frac{G_{E_p}}{G_{M_p}}$ as a function of Q^2 for Rosenbluth separation data (open symbols) and recoil polarization data (filled circles).

A 4.6 GeV polarized electron beam produced by the CEBAF accelerator at Jefferson Lab hit the Hall A 15 cm-long cryogenic hydrogen target. The polarization of the beam was measured to be 70 to 75% by two independent polarimeters in the hall, based on the Compton and Møller processes respectively, and the current was around 50 μA. The proton was detected in one of the two High Resolution Spectrometers (HRS)[7,8]. Background from target walls and pion electroproduction within the acceptance of the HRS was eliminated by requiring the corresponding electron to be detected. Solid angle matching was required to maximize the number of elastic protons detected. At $Q^2 = 5.6$ GeV^2, to detect all the protons in the 7 msr acceptance of the HRS, it was necessary to have an electron detector with an acceptance of 42 msr. For this purpose, a 2.55×1.35 m^2 lead-glass calorimeter was assembled, in an array of 17 columns and 9 rows of 15×15 cm^2 blocks. At each kinematics, the distance to the target was adjusted according to the Jacobian to match the proton solid angle. The electron hitting the lead-glass develops an electromagnetic shower, emitting Čerenkov light. This light is detected in photomultipliers, which send a signal to ADCs and TDCs for energy and timing reconstruction. Simple software cuts on timing and angular correlations with the proton allow a very good selection of elastic events, as shown in Fig. 3.

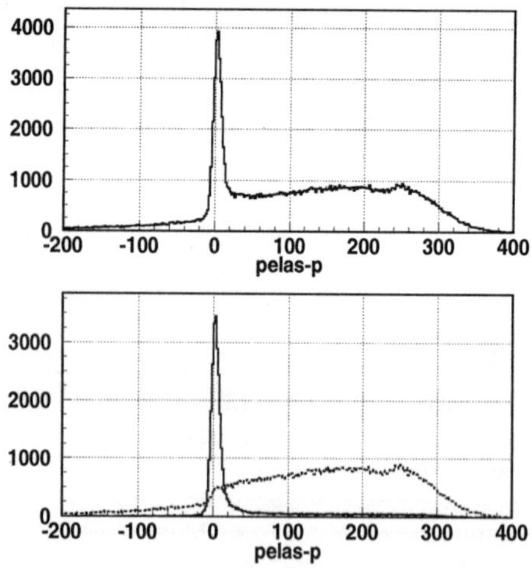

Figure 3. Selection of elastic events: difference between the elastic proton momentum calculated from its angle, and the measured momentum. This quantity should peak at 0 for elastic events. This plot illustrates the background in the acceptance of the HRS (top), and the selection made by cuts in the calorimeter (bottom): accepted events are under the fuill line peaked at 0, and rejected events are under the flat dotted line

The polarization of the elastic protons was measured in the Focal Plane Polarimeter, located in the detector hut at the top of the left HRS. The proton scatters in a 100 cm analyser block of polyethylene, and asymmetries in the azimuthal angular distribution of this scattering are related to the polarization components perpendicular to the momentum at the focal plane. The polarization components at the target, which are the quantities of interest, are then obtained by rotating the spin through the spectrometer using a differential algebra-based code, COSY[9], to transport the spin event by event.

The form factor ratio was measured in the spring of 2000 at $Q^2 = 3.0$ GeV^2, which successfully tested the use of a calorimeter. Then in November-December of 2000, another overlap point with Ref.[7] at $Q^2 = 3.5$ GeV^2, and 3 new data points at $Q^2 = 4.0$, 4.8 and 5.6 GeV^2 were measured. Preliminary results are presented in Fig. 4.

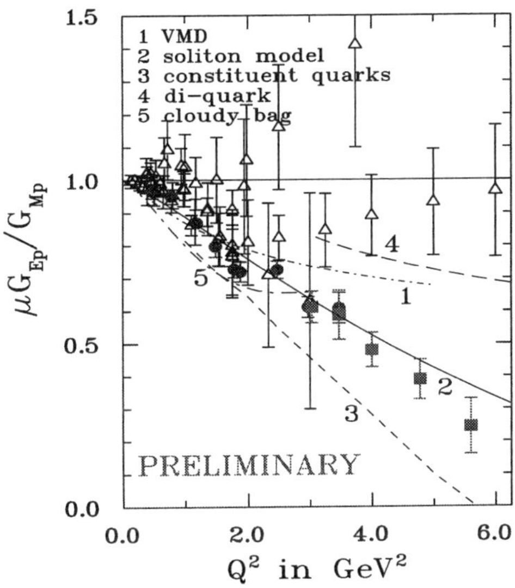

Figure 4. Ratio $\mu \frac{G_{E_p}}{G_{M_p}}$ as a function of Q^2 for Rosenbluth separation data (open symbols) and recoil polarization data from 1998 (filled circles) to $Q^2 = 3.5$ GeV^2 and from this experiment to $Q^2 = 5.6$ GeV^2 (filled squares).

The errors shown here are statistical only. The major sources of systematic uncertainties come from the knowledge of spin precession, and are based on the resolution of the target quantities, and the uncertainty on the geometric shape and position of the magnetic elements of the spectrometer. They are currently in the process of being calculated, and are expected to be smaller than the statistical errors at all kinematics. The main result of this experiment is that the linear decrease of the ratio of the electric to magnetic form factors of the proton observed in Ref.[7] continues to $Q^2 = 5.6$ GeV^2. In a non-relativistic approach, this result is interpreted as the charge distribution extending further out than the magnetic current distribution.

The calculations of proton form factors in the Q^2-range discussed here are all based on phenomenology. Shown in Fig. 4 are the vector meson dominance (VMD) model[10], a calculation based on the solitonic nature of the nucleon[11], the relativistic constituent-quark model[12], the di-quark model[13] and the cloudy bag model[14].

In Fig. 5, the data are compared with the perturbative QCD (pQCD) prediction[15] that the asymptotic behaviour of the form factors is $F_1 \propto \frac{1}{Q^4}$ and $F_2 \propto \frac{1}{Q^6}$, so that $Q^2\frac{F_2}{F_1} \rightarrow constant$. The data clearly indicates that this asymptotic regime is not reached yet.

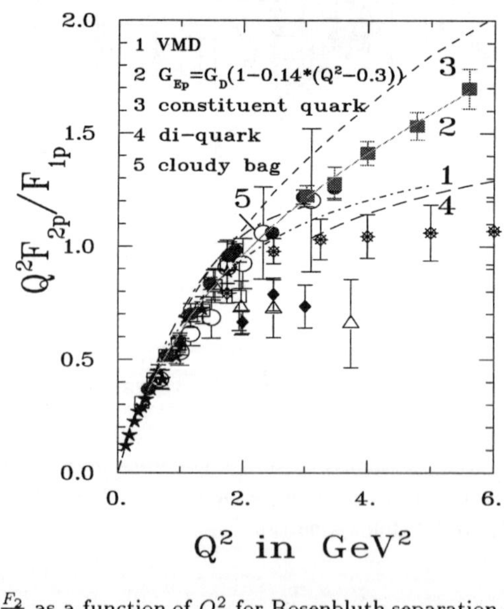

Figure 5. Ratio $Q^2\frac{F_2}{F_1}$ as a function of Q^2 for Rosenbluth separation data (open symbols) and recoil polarization data from 1998 (filled circles) to $Q^2 = 3.5$ GeV^2 and from this experiment to $Q^2 = 5.6$ GeV^2 (filled squares). Perturbative QCD predicts a constant for this ratio as an asymptotic limit.

A recent work form J. Ralston *et al.*[16] raised the question of the behaviour of the ratio $\sqrt{Q^2}\frac{F_2}{F_1}$. As shown in Fig. 6, this ratio reaches a constant value starting at about $Q^2 = 2.0$ GeV^2. This behaviour is clearly a spectacular result that present and future theoretical descriptions of the nucleon form factors should include; in particular they should answer the question of the compatibility with the pQCD prediction.

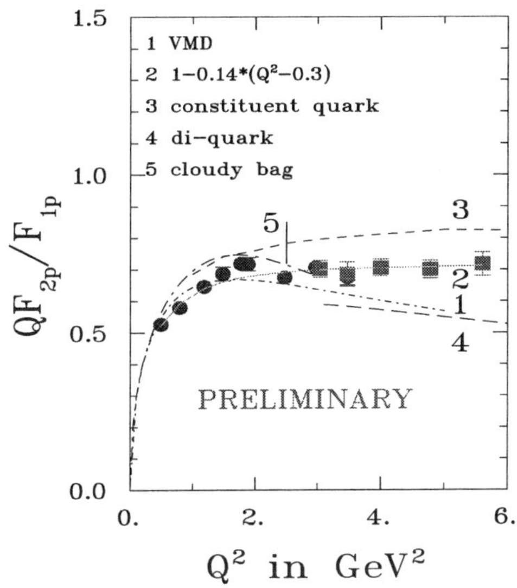

Figure 6. Ratio $\sqrt{Q^2}\frac{F_2}{F_1}$ as a function of Q^2.

In conclusion, we have presented the first precise measurement of the proton form factor ratio G_{E_p}/G_{M_p} to $Q^2 = 5.6$ GeV^2. The result show that the linear decrease observed in Ref.[7] continues to higher momentum transfer, prompting the question whether the electric form factor will cross 0 and become negative at moderately higher Q^2. A future experiment at yet higher Q^2 is in preparation at Jefferson Lab.

Acknowledgments

The collaboration thanks the Hall A technical staff and the Jefferson Lab Accelerator Division for their outstanding support during the experiment. The

Southern Universities Research Association (SURA) operates the Thomas Jefferson National Accelerator Facility for the United States Department of Energy under contract DE-AC05-84ER40150. This work was also supported by the U.S. National Science Foundation, the Italian Istituto Nazionale di Fisica Nucleare (INFN), the French Commissariat à l'Energie Atomique and Centre National de la Recherche Scientifique (CNRS), and the Natural Sciences and Engineering Council of Canada.

References

1. E. E. Chambers and R. Hofstadter, *Phys. Rev.* **103**, 1454 (1956).
2. M. N. Rosenbluth, *Phys. Rev.* **79**, 615 (1950).
3. J. Litt *et al.*, *Phys. Lett.* B **31**, 40 (1970);
   Ch. Berger *et al.*, *Phys. Lett.* B **35**, 87 (1971);
   L. E. Price *et al.*, *Phys. Rev.* D **4**, 45 (1971);
   W. Bartel *et al.*, *Nucl. Phys.* B **58**, 429 (1975);
   L. Andivahis *et al.*, *Phys. Rev.* D **50**, 5491 (1994);
   R. C. Walker *et al.*, *Phys. Rev.* D **49**, 5671 (1994).
4. A. I. Akhiezer and M. P. Rekalo, *Sov. J. Part. Nucl.* **3**, 277 (1974).
5. R. Arnold, C. Carlson and F. Gross, *Phys. Rev. C* **23**, 363 (1981).
6. B. Milbrath *et al.*, *Phys. Rev. Lett.* **80**, 452 (1998);
   B. Milbrath *et al.*, *Phys. Rev. Lett.* **82**, 2221(E) (1999).
7. M. K. Jones *et al.*, *Phys. Rev. Lett.* **84**, 1398 (2000).
8. O. Gayou *et al.*, submitted to *Phys. Rev. C Brief Report* (2001).
9. M. Berz, COSY INFINITY version 7, Michigan State University NSCL Technical Report MSUCL-977 (1995).
10. G. Höhler *et al.*, *Nucl. Phys. B* **114**, 505 (1976);
    F. Iacello, A. D. Jackson, and A. Landé, *Phys. Lett.* **43B**, 191 (1973);
    M. F. Gari and W. Krümpelman, *Z. Phys. A* **322**, 689 (1985);
    P. Mergell, U. G. Meissner and D. Drechsel, *Nucl. Phys. A* **596**, 367 (1996).
11. G. Holzwarth, *Z. Phys. A* **356**, 339 (1996).
12. P. L. Chung and F. Coester, *Phys. Rev.* D **44**, 229 (1991);
    I. G. Aznauryan, *Phys. Lett.* B **316**, 391 (1993).
13. P. Kroll, M. Schürmann and W. Schweiger, *Z. Phys. A* **338**, 339 (1991).
14. D. H. Lu, A. W. Thomas and A. G. Williams, *Phys. Rev. C* **57**, 2628 (1998).
15. S. J. Brodsky and G. R. Farrar, *Phys. Rev.* D **11**, 1309 (1975).
16. J. Ralston *et al.*, in Proc. of 7th International Conference on Intersection of Particle and Nuclear Physics, Quebec City (2000), p.302.

PHYSICAL HADRON PROPERTIES FROM LATTICE DATA AT LARGE QUARK MASSES

A. W. THOMAS

CSSM, Adelaide University, Adelaide SA 5005, AUSTRALIA
E-mail: athomas@physics.adelaide.edu.au

We review recent progress in the understanding of the functional dependence of hadron masses and electromagnetic properties on quark mass. This is a fascinating area in which there has been a very productive interplay between hadron models, chiral perturbation theory and lattice QCD. The results lead one to expect to be able to deduce quite accurate hadron properties from the next generation of supercomputers, with speeds of tens of Tera-flops, rather than hundreds.

1 Introduction

Current lattice simulations with dynamical quarks (i.e., full QCD) are extremely demanding in terms of computer time. The essential point is that the time taken is slower than m_q^{-3} – e.g., see Ref. [1]. The state-of-the-art for hadron masses is that m_q typically lies above 60 MeV, although there is a preliminary result from CP-PACS at about 40 MeV [2]. To put it more bluntly, the quark masses for which simulations currently exist are at masses a factor of 8-20 too high. This means an increase of computing power to hundreds of Tera-flops would be needed to calculate realistic hadron properties. Even with the current remarkable rate of increase this will take a decade or more.

Clearly, if one wants to compare the properties of physical hadrons with data now, rather than waiting another decade, it is necessary to extrapolate as a function of quark mass. On the other hand, one should also realize that the lattice data obtained so far represents a wealth of information on the properties of hadrons within QCD itself over a range of quark masses. Just as the study of QCD as a function of N_c has taught us a great deal, so the behaviour as a function of m_q can give us great insight into hadronic physics and guide our model building – as well as the extrapolation to realistic masses.

Because of the limited space available to outline a great deal of recent work on this problem, we first summarise the conclusions which emerge from our investigations over the past three years. We then choose just a couple of illustrations of the reasoning behind these conclusions.

Summary:

- In the region of quark masses $\bar{m} > 60$ MeV or so (m_π greater than typically 400-500 MeV) hadron properties are smooth, slowly varying functions of something like a constituent quark mass, $M \sim M_0 + c\bar{m}$ (with $c \sim 1$).

- Indeed, $M_N \sim 3M, M_{\rho,\omega} \sim 2M$ and magnetic moments behave like $1/M$.

- As $\bar{m}$ decreases below 60 MeV or so, chiral symmetry leads to rapid, non-analytic variation, with $\delta M_N \sim \bar{m}^{3/2}, \delta\mu_H \sim \bar{m}^{1/2}$ and $\delta < r^2 >_{\text{ch}} \sim \ln \bar{m}$.

- Chiral quark models like the cloudy bag provide a natural explanation of this transition. The scale is basically set by the inverse size of the pion source – the inverse of the bag radius in the bag model.

- When the pion Compton wavelength is smaller than the size of the composite source chiral loops are strongly suppressed. On the other hand, as soon as the pion Compton wavelength is larger than the source one begins to see rapid, non-analytic chiral corrections.

These are remarkable results that will have profound consequences for our further exploration of hadron structure within QCD as well as the analysis of the vast amount of data now being taken concerning unstable resonances. In terms of immediate results for the structure of the nucleon, we note that the careful incorporation of the correct chiral behaviour of QCD into the extrapolation of its properties calculated on the lattice has produced:

- A method of extrapolating hadron masses (so far applied to the full QCD data for the N, Δ [3] and ρ [4]), which incorporates the correct non-analytic behaviour required by chiral symmetry in QCD.

- The best values of the p and n magnetic moments from QCD [5].

- The best value of the sigma commutator [6].

- Improved values for the magnetic moments of the hyperons [7].

- Improved values for the charge radii of the baryon octet [8].

- Resolution of a longstanding discepancy of typically 50% between the extrapolated moments of the non-singlet distribution $u - d$ and the experimentally measured moments [9,10].

- An improved lattice estimate of the strangeness magnetic moment of the proton G_M^s [11].

2 Extrapolation of Hadron Masses in Full QCD

Key to a correct extrapolation of hadron properties to the physical quark masses, which are almost zero, are the chiral properties of QCD. In particular, spontaneous chiral symmetry breaking requires the existence of Goldstone bosons whose masses vanish in the limit of zero quark mass (the chiral limit). As a corollary to this, there must be contributions to hadron properties from Goldstone boson loops. These loops have the unique property that they give rise to terms in an expansion of most hadronic properties as a function of quark mass which are not analytic. As a simple example, consider the nucleon mass. The most important chiral corrections to M_N come from the processes $N \to N\pi \to N$ (σ_{NN}) and $N \to \Delta\pi \to N$ ($\sigma_{N\Delta}$). (We will come to what it means to say these are the most important shortly.) We write $M_N = M_N^{\rm bare} + \sigma_{NN} + \sigma_{N\Delta}$. In the heavy baryon limit one has

$$\sigma_{NN} = -\frac{3g_A^2}{16\pi^2 f_\pi^2} \int_0^\infty dk \frac{k^4 u^2(k)}{k^2 + m_\pi^2}, \tag{1}$$

where $u(k)$ is a natural high momentum cut-off which is the Fourier transform of the source of the pion field (e.g. in the cloudy bag model (CBM) it is $3j_1(kR)/kR$, with R the bag radius [12]). From the point of view of PCAC it is natural to identify $u(k)$ with the axial form-factor of the nucleon, a dipole with mass parameter 1.02 ± 0.08GeV [13].

Quite independent of the form chosen for the ultra-violet cut-off, one finds that σ_{NN} is a non-analytic function of the quark mass. The non-analytic piece of σ_{NN} is independent of the form factor and gives

$$\sigma_{NN}^{LNA} = -\frac{3g_A^2}{32\pi f_\pi^2} m_\pi^3 \sim \bar{m}^{\frac{3}{2}}. \tag{2}$$

This has a branch point, as a function of $\bar{m}$, at $\bar{m} = 0$. Such terms can only arise from Goldstone boson loops.

As a practical question it is important to establish just how significant this non-analytic behaviour is in practice. If the pion mass is given in GeV, $\sigma_{NN}^{LNA} = -5.6 m_\pi^3$ and at the physical pion mass it is just -17 MeV. However, at only three times the physical pion mass, $m_\pi = 420$MeV, it is -460MeV – half the mass of the nucleon. If one's aim is to extract physical nucleon properties from lattice QCD calculations this is extremely important. The most sophisticated lattice calculations with dynamical fermions are only just becoming feasible at such low masses and to connect to the physical world one must extrapolate from $m_\pi \sim 500$MeV to $m_\pi = 140$MeV. Clearly one must have control of the chiral behaviour.

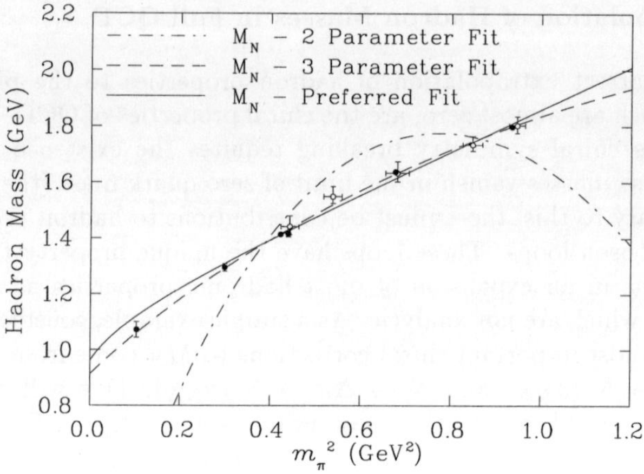

Figure 1. A comparison between phenomenological fitting functions for the mass of the nucleon – from Ref. [3]. The two parameter fit corresponds to using Eq.(3) with γ set equal to the value known from χPT. The three parameter fit corresponds to letting γ vary as an unconstrained fit parameter. The solid line is the two parameter fit based on the functional form of Eq.(4).

Figure 1 shows recent lattice calculations of M_N as a function of m_π^2 from CP-PACS [2] and UKQCD [14]. The dashed line indicates a fit which naively respects the presence of a LNA term,

$$M_N = \alpha + \beta m_\pi^2 + \gamma m_\pi^3, \tag{3}$$

with α, β and γ fitted to the data. While this gives a very good fit to the data, the chiral coefficient γ is only -0.761, compared with the value -5.60 required by chiral symmetry. If one insists that γ be consistent with QCD the best fit one can obtain with this form is the dash-dot curve. This is clearly unacceptable.

A better procedure suggested by Leinweber et al. [3], which also involves just three parameters, is to evaluate σ_{NN} and $\sigma_{N\Delta}$ with the same ultra-violet form factor, with mass parameter Λ, and to fit M_N as

$$M_N = \alpha + \beta m_\pi^2 + \sigma_{NN}(m_\pi, \Lambda) + \sigma_{N\Delta}(m_\pi, \Lambda). \tag{4}$$

Using a sharp cut-off ($u(k) = \theta(\Lambda - k)$) these authors were able to obtain analytic expressions for σ_{NN} and $\sigma_{N\Delta}$ which reveal the correct LNA behaviour – and next to leading (NLNA) in the $\Delta\pi$ case, $\sigma_{N\Delta}^{NLNA} \sim m_\pi^4 \ln m_\pi$. These expressions also reveal a branch point at $m_\pi = M_\Delta - M_N$, which is important if one is extrapolating from large values of m_π to the physical value. The solid curve in Fig. 1 is a two parameter fit to the lattice data using Eq.(4), but fixing Λ at a value suggested by CBM simulations to be equivalent to the preferred 1 GeV dipole. A small increase in Λ is necessary to fit the lowest mass data point, at $m_\pi^2 \sim 0.1$ GeV^2, but clearly one can describe the data very well while preserving the exact LNA and NLNA behaviour of QCD.

A similar analysis has recently been carried out for the ρ meson, where the LNA behaviour arises from the process $\rho \to \pi\omega$ and the NLNA behaviour from $\rho \to \pi\pi$ [4]. Figure 2 summarises the analysis of the CP-PACS and UKQCD data for the nucleon, the delta and the ρ meson. Clearly a careful chiral extrapolation can change the physical hadron mass by more than 100 MeV, compared with a naive linear extrapolation (shown for the Δ by a straight line). This figure also illustrates the ratio of 2/3 for the slope of the vector meson to the (average of the) nucleon and delta slopes at large quark masses – as expected in a constituent quark model.

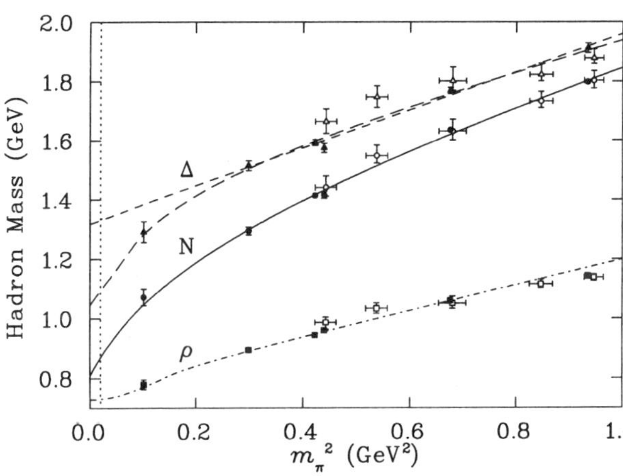

Figure 2. Three parameter fits, as described in the text, to the N, Δ and ρ masses as a function of pion mass, including the correct chiral behaviour. The comparison with the straight line fit for the Δ shows that a naive linear extrapolation can lead to errors of hundreds of MeV at the physical pion mass. The data is from CP-PACS (solid) and UKQCD (open) with 5% rescaling (within systematic errors) to give the best agreement between the two data sets.

2.1 The Sigma Commutator

The analysis of the lattice data for M_N, incorporating the correct non-analytic behaviour, can yield interesting new information concerning the sigma commutator of the nucleon:

$$\sigma_N = \frac{1}{3}\langle N|[Q_{i5},[Q_{i5},H_{QCD}]]|N\rangle = \langle N|\bar{m}(\bar{u}u+\bar{d}d)|N\rangle. \tag{5}$$

This is a direct measure of chiral SU(2) symmetry breaking in QCD, and the widely accepted experimental value is 45 ± 8MeV [6]. (Although there are recent suggestions that it might be as much as 20 MeV larger [15].) Using the Feynman-Hellmann theorem one can also write

$$\sigma_N = \bar{m}\frac{\partial M_N}{\partial \bar{m}} = m_\pi^2 \frac{\partial M_N}{\partial m_\pi^2}. \tag{6}$$

Historically, lattice calculations have evaluated $< N|(\bar{u}u+\bar{d}d)|N >$ at large quark mass and extrapolated this scale dependent quantity to the "physical" quark mass, which had to be determined in a separate calculation. The latest result with dynamical fermions, $\sigma_N = 18 \pm 5$ MeV [16], illustrates how difficult this procedure is. On the other hand, if one has a fit to M_N as a function of m_π which is consistent with chiral symmetry, one can evaluate σ_N directly using Eq.(6). Using Eq.(4) with a sharp cut-off yields $\sigma_N \sim 55$ MeV, while a dipole form gives $\sigma_N \sim 45$ MeV [6]. The residual model dependence can only be removed by more accurate lattice data at low m_π^2. Nevertheless, the result $\sigma_N \in (45,55)$ MeV is in very good agreement with the data. In contrast, the simple cubic fit, with γ inconsistent with chiral constraints, gives ~ 30 MeV. Until the experimental situation regarding σ_N improves, it is not possible to draw definite conclusions regarding the strangeness content of the nucleon. However, the fact that two-flavour QCD reproduces the current prefered value should certainly stimulate some thought and a lot of work.

3 Electromagnetic Properties

It is a general consequence of quantum mechanics that the long-range charge structure of the proton comes from its π^+ cloud ($p \to n\pi^+$), while for the neutron it comes from its π^- cloud ($n \to p\pi^-$). However, it is not often realized that the LNA contribution to the nucleon charge radius goes like $\ln m_\pi$ and diverges as $\bar{m} \to 0$ [17]. This cannot be reproduced by a constituent quark model [18].

While there are only limited (and indeed quite old) lattice data for hadron charge radii, recent experimental progress in the determination of hyperon charge radii has led us to examine the extrapolation procedure for obtaining charge data from the lattice simulations [8]. Figure 3 shows the extrapolation of the lattice data [19] for the charge radius of the proton. Clearly the agreement with experiment is much better once the log required by chiral symmetry is correctly included, than if, for example, one simply made a linear extrapolation in the quark mass (or m_π^2). Full details of the results for all the octet baryons may be found in Ref. [8].

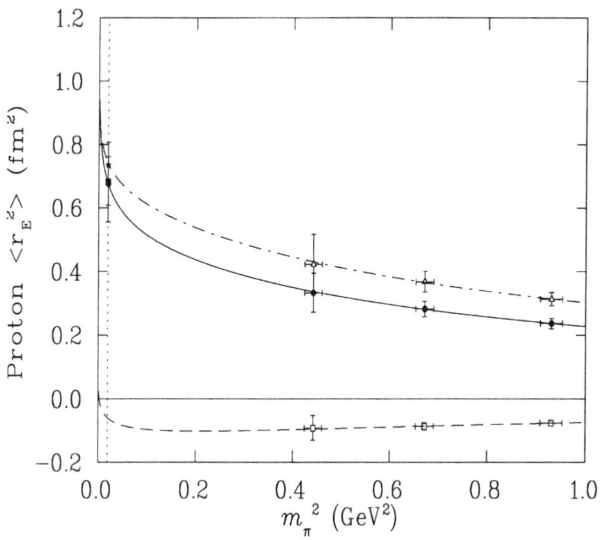

Figure 3. Fits to lattice results for the squared electric charge radius of the proton – from Ref. [8]. Fits to the contributions from individual quark flavours are also shown: the u-quark sector results are indicated by open triangles and the d-quark sector results by open squares. Physical values predicted by the fits are indicated at the physical pion mass, where the full circle denotes the result predicted from the first extrapolation procedure and the full square denotes the baryon radius reconstructed from the individual quark flavor extrapolations. (N.B. The latter values are actually so close as to be indistinguishable on the graph.) The experimental value is denoted by an asterisk.

The situation for baryon magnetic moments is also very interesting. The LNA contribution in this case arises from the diagram where the photon couples to the pion loop. As this involves two pion propagators the expansion of the proton and neutron moments is $\mu^{p(n)} = \mu_0^{p(n)} \mp \alpha m_\pi + \mathcal{O}(m_\pi^2)$. Here $\mu_0^{p(n)}$ is the value in the chiral limit and the linear term in m_π is proportional to $\bar{m}^{\frac{1}{2}}$, a branch point at $\bar{m} = 0$. The coefficient of the LNA term is $\alpha =$

$4.4\mu_N \mathrm{GeV}^{-1}$. At the physical pion mass this LNA contribution is $0.6\mu_N$, which is almost a third of the neutron magnetic moment. *No constituent quark model can or should get better agreement with data than this* [18].

Just as for M_N, the chiral behaviour of $\mu^{p(n)}$ is vital to a correct extrapolation of lattice data. One can obtain a very satisfactory fit to some rather old data, which happens to be the best available, using the simple Padé [5]:

$$\mu^{p(n)} = \frac{\mu_0^{p(n)}}{1 \pm \frac{\alpha}{\mu_0^{p(n)}} m_\pi + \beta m_\pi^2} \tag{7}$$

The data can only determine two parameters and Eq.(7) has just two free parameters while guaranteeing the correct LNA behaviour as $m_\pi \to 0$ **and** the correct behaviour of heavy quark effective theory (HQET) at large m_π^2. The extrapolated values of μ^p and μ^n at the physical pion mass, $2.85 \pm 0.22\mu_N$ and $-1.90 \pm 0.15\mu_N$, respectively, are currently the best estimates from non-perturbative QCD. For more details of this fit we refer to Ref. [5], while for the application of similar ideas to other members of the nucleon octet we refer to Ref. [7] and for the strangeness magnetic moment of the nucleon we refer to Ref. [11].

Incidentally, from the point of view of the naive quark model it is interesting to examine the ratio of the absolute values of the proton and neutron magnetic moments as a function of m_π^2. The agreement of the constituent quark result, namely 3/2, with the experimental value to within a few percent is usually taken as a major success. However, it has been recently pointed out that this close agreement is little more than an accident of the quark mass [18].

4 Conclusion

We refer to the introductory section for a summary of the main achievements so far from these studies of hadron properties versus quark mass. Clearly there is still a great deal to be done. One can test the chiral extrapolation techniques which we have developed against new, full QCD data, as well as existing quenched and partially quenched data which is available at much lower quark mass. The really exciting aspect of this work, provided it satisfies all the tests thrown at it, is the following. The next generation of supercomputers will permit full QCD calculations at quark masses 2–3 times the physical values. Our work suggests that this should enable us to make quite accurate extrapolations of hadronic properties. In this way it should be possible to make meaningful comparisons with experimental data in the next few years, rather than waiting a decade for machines with speeds of 100's of Tera-flops.

Acknowledgments

It is a pleasure to acknowledge the contributions of many collaborators in the work described here, especially D. Leinweber, W. Detmold, W. Melnitchouk, Stewart Wright and Ross Young. This work was supported by the Australian Research Council and Adelaide University.

References

1. T. Lippert, S. Gusken and K. Schilling, *Nucl. Phys. Proc. Suppl.* **83**, 182 (2000).
2. S. Aoki *et al.* [CP-PACS-Collaboration], *Phys. Rev.* D **60**, 114508 (1999) [hep-lat/9902018].
3. D. B. Leinweber, A. W. Thomas, K. Tsushima and S. V. Wright, *Phys. Rev.* D **61**, 074502 (2000) [hep-lat/9906027].
4. D. B. Leinweber et al., hep-lat/0104013.
5. D. B. Leinweber, D. H. Lu and A. W. Thomas, *Phys. Rev.* D **60**, 034014 (1999) [hep-lat/9810005].
6. D. B. Leinweber, A. W. Thomas and S. V. Wright, *Phys. Lett.* B **482**, 109 (2000) [hep-lat/0001007].
7. E. J. Hackett-Jones, D. B. Leinweber and A. W. Thomas, *Phys. Lett.* B **489**, 143 (2000) [hep-lat/0004006].
8. E. J. Hackett-Jones, D. B. Leinweber and A. W. Thomas, *Phys. Lett.* B **494**, 89 (2000) [hep-lat/0008018].
9. W. Detmold et al., distributions," hep-lat/0103006.
10. W. Detmold, in these preceedings.
11. D. B. Leinweber and A. W. Thomas, *Phys. Rev.* D **62**, 074505 (2000).
12. S. Theberge, A. W. Thomas and G. A. Miller, *Phys. Rev.* **D22**, 2838 (1980); A. W. Thomas, *Adv. Nucl. Phys.* **13**, 1 (1984).
13. A. W. Thomas and W. Weise, "The Structure of the Nucleon" (Wilcy-VCH, Berlin,2001).
14. C. R. Allton *et al.* [UKQCD], *Phys. Rev.* **D60**, 034507 (1999).
15. M. Knecht, *PiN Newslett.* **15**, 108 (1999) [hep-ph/9912443].
16. **SESAM** Collaboration, S. Gusken *et al.*, *Phys. Rev.* D **59**, 054504 (1999).
17. D. B. Leinweber and T. D. Cohen, *Phys. Rev.* D **47**, 2147 (1993).
18. D. B. Leinweber, A. W. Thomas and R. D. Young, *Phys. Rev. Lett.* **86**, 5011 (2001) [hep-ph/0101211].
19. D. B. Leinweber et al., *Phys. Rev.* D **43**, 1659 (1991).

MESON CLOUD CONSIDERATIONS IN THE STRANGE MAGNETIC MOMENT OF THE NUCLEON FROM LATTICE QCD

DEREK B. LEINWEBER and ANTHONY W. THOMAS

Special Research Center for the Subatomic Structure of Matter (CSSM) and Department of Physics and Mathematical Physics, University of Adelaide 5005

E-mail: dleinweb@physics.adelaide.edu.au & athomas@physics.adelaide.edu.au

Charge symmetry between the light u and d quarks in octet baryons provides relations among various quark-sector contributions to baryon magnetic moments which may be used to gain information on the strange quark contribution to the nucleon magnetic moment, $G_M^s(0)$. In analyzing these relations, it is important to consider the nonanalytic contributions to baryon magnetic moments arising from the meson cloud. In this report we consider both the π- and K-meson contributions to baryon magnetic moments. We find that the contribution from $\Xi \to \Sigma\overline{K}$ is large and plays a pivotal role in the final determination of $G_M^s(0)$. With the inclusion of the $\overline{K}$ cloud, we find $G_M^s(0)$ is definitely negative with $G_M^s(0) = -0.4\,^{+0.2}_{-0.4}\ \mu_N$.

1 Introduction

In calculating the electromagnetic form factors of hadrons in a field theoretic formalism, one encounters two topologically distinct ways of performing the electromagnetic current insertion in the three-point correlation function.[1] Fig. 1 displays skeleton diagrams for these two possible insertions (with Euclidean time increasing to the right). These diagrams may be dressed with an arbitrary number of gluons and quark loops. The left-hand diagram illustrates the connected insertion of the current to one of the "valence" [a] quarks of the baryon. In the right-hand diagram the external field produces a $q\,\overline{q}$ pair which in turn interacts with the valence quarks of the baryon via gluons. This diagram allows the strange quark to contribute to nucleon matrix elements.

Direct calculations of the right-hand diagram for magnetic form factors have been attempted in lattice QCD simulations.[3,4] The calculations are computationally challenging and the fact that they must be carried out at finite momentum transfer makes it difficult to achieve a comfortable signal to noise ratio in the hadron correlation functions.

The approach to $G_M^s(0)$ we pursue here [1] is based on the approximation of charge symmetry in the nucleon moments; i.e. equal u and d current quark

[a] Note that the term "valence" used here differs from that commonly used in the framework of deep-inelastic scattering. Here "valence" simply describes the quark whose quark flow line runs continuously from $0 \to x_2$. These lines can flow backwards as well as forwards in time and therefore have a sea contribution associated with them.[2]

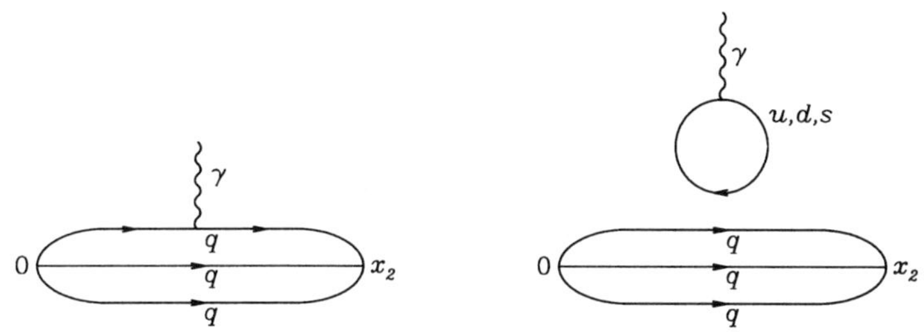

Figure 1: Diagrams illustrating the two topologically distinct insertions of the current within the framework of lattice QCD. Within full QCD, these skeleton diagrams for the connected (left) and disconnected (right) current insertions may be dressed by an arbitrary number of gluons and quark loops.

masses and negligible electromagnetic corrections. This approximation is expected to be valid at the 1% level [5] and is certainly sufficient for the analysis of $G_M^s(0)$. In this case, the three-point correlation functions for octet baryons lead to the following equalities for electromagnetic current matrix elements:

$$\begin{aligned} p &= e_u\, u_p + e_d\, d_p + O_N\,, & n &= e_d\, u_p + e_u\, d_p + O_N\,, \\ \Sigma^+ &= e_u\, u_{\Sigma^+} + e_s\, s_\Sigma + O_\Sigma\,, & \Sigma^- &= e_d\, u_{\Sigma^+} + e_s\, s_\Sigma + O_\Sigma\,, \\ \Xi^0 &= e_s\, s_\Xi + e_u\, u_\Xi + O_\Xi\,, & \Xi^- &= e_s\, s_\Xi + e_d\, u_\Xi + O_\Xi\,. \end{aligned} \tag{1}$$

Here, O denotes the contributions from the quark-loop sector – shown on the right-hand side of Fig. 1. The baryon label represents the magnetic moment. Subscripts allow for environment sensitivity in the quark moments.[1,6,7]

Focusing now on the nucleon, we note that for magnetic properties, O_N contains sea-quark-loop contributions from primarily u, d, and s quarks. By definition

$$O_N = \frac{2}{3}\,{}^{\ell}G_M^u - \frac{1}{3}\,{}^{\ell}G_M^d - \frac{1}{3}\,{}^{\ell}G_M^s\,, \tag{2}$$

$$= \frac{{}^{\ell}G_M^s}{3}\left(\frac{1 - {}^{\ell}R_d^s}{{}^{\ell}R_d^s}\right)\,, \quad \text{where} \quad {}^{\ell}R_d^s \equiv \frac{{}^{\ell}G_M^s}{{}^{\ell}G_M^d}\,, \tag{3}$$

and the leading superscript, ℓ, reminds the reader that the contributions are loop contributions. Note that, charge symmetry provides ${}^{\ell}G_M^u = {}^{\ell}G_M^d$.

The strange-quark loop contributions to the nucleon magnetic moment, G_M^s may be isolated from (1) and (3) in the following two phenomenologically useful forms,

$$G_M^s = \left(\frac{{}^\ell R_d^s}{1 - {}^\ell R_d^s}\right)\left[2p + n - \frac{u_p}{u_{\Sigma^+}}\left(\Sigma^+ - \Sigma^-\right)\right], \tag{4}$$

and

$$G_M^s = \left(\frac{{}^\ell R_d^s}{1 - {}^\ell R_d^s}\right)\left[p + 2n - \frac{u_n}{u_{\Xi^0}}\left(\Xi^0 - \Xi^-\right)\right]. \tag{5}$$

Incorporating the experimentally measured baryon moments provides

$$G_M^s = \left(\frac{{}^\ell R_d^s}{1 - {}^\ell R_d^s}\right)\left[3.673 - \frac{u_p}{u_{\Sigma^+}}(3.618)\right], \tag{6}$$

and

$$G_M^s = \left(\frac{{}^\ell R_d^s}{1 - {}^\ell R_d^s}\right)\left[-1.033 - \frac{u_n}{u_{\Xi^0}}(-0.599)\right], \tag{7}$$

where all moments [b] are expressed in nuclear magnetons (μ_N). We stress that *these expressions for G_M^s are exact consequences of QCD, under the assumption of charge symmetry.*

The SAMPLE collaboration has reported [9] new values for the strange-quark contribution to the nucleon magnetic moment, $G_M^s(0)$, in which the original proton data are combined with new data from parity violation experiments on the deuteron. Now $G_M^s(0)$ is determined independently of the poorly understood nucleon anapole moment. Presently [9]

$$G_M^s(Q^2 = 0.1 \text{ GeV}^2) = +0.14 \pm 0.29 \pm 0.53\ \mu_N, \tag{8}$$

with statistical and systematic errors reported consecutively. Equation (7) provides a particularly favorable case for the determination of G_M^s with minimal dependence on the valence-quark ratio. It also holds the promise to unambiguously determine the sign of $G_M^s(0)$.

Equating (6) and (7) provides a linear relationship between u_p/u_{Σ^+} and u_n/u_{Ξ^0} which must be satisfied within QCD. Fig. 2 displays this relationship by the dashed and solid line, the latter corresponding to values for which $G_M^s(0) > 0$ when ${}^\ell R_d^s$ is in the anticipated range $0 < {}^\ell R_d^s < 1$. Since the line does not pass through the point (1, 1) corresponding to the simple quark model assumption of universality, the experimentally measured baryon moments are signaling that there must be an environment effect exceeding 12% in both ratios

[b] Note that the measured magnetic moments are all known sufficiently accurately [8] that the experimental errors play no role in our subsequent analysis.

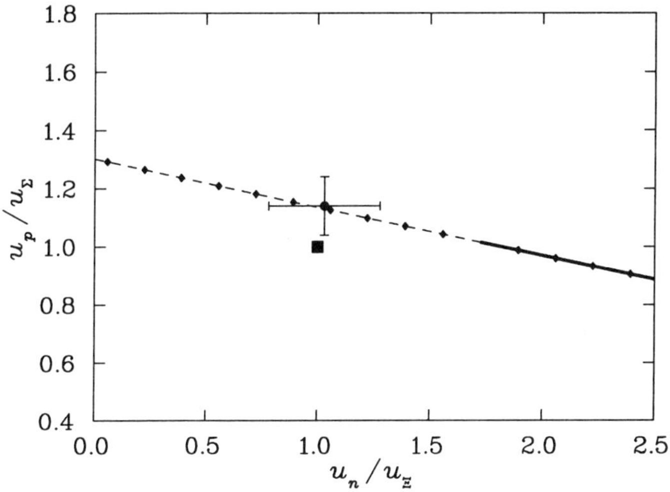

Figure 2: The consistency relation between u_p/u_{Σ^+} and u_n/u_{Ξ^0} which must be satisfied within QCD. The part of the straight line which is dashed corresponds to $G_M^s(0) < 0$, while the solid part of the line has $G_M^s(0) > 0$. The standard quark model assumption of intrinsic quark moments independent of their environment is indicated by the filled square at $(1, 1)$. The diamonds on the constraint line mark 0.1 μ_N changes in $G_M^s(0)$ when ${}^{\ell}R_d^s = 0.5$. The lattice QCD prediction (after an appropriate chiral extrapolation, discussed in following sections) is illustrated by the filled circle with standard errors indicated.

or approaching 20% or more in at least one of the ratios. Moreover, a positive value for $G_M^s(0)$ requires an environment sensitivity exceeding 70% in the u_n/u_{Ξ^0} ratio. Hence the experimental indication that $G_M^s(0) > 0$ challenges the intrinsic magnetic moment concept which is fundamental to the constituent quark model.

2 Chiral Corrections

One of the major challenges at present in connecting lattice calculations of hadronic properties with the physical world is that computational limitations restrict the accessible quark masses to values an order of magnitude larger than the physical values. At such large masses one is far from the region where chiral perturbation theory is applicable. Yet one knows that for current quark masses near zero there is important non-analytic structure (as a function of the quark mass) which must be treated correctly if we are to compare with physical hadron properties.

Recent analyses of the strange magnetic form factor of the nucleon [7,10] have been made possible by a breakthrough in the treatment of these chiral corrections for the nucleon magnetic moments.[11] The simple Padé approximant

$$\mu = \frac{\mu_0}{1 - \frac{\chi}{\mu_0} m_\pi + c\, m_\pi^2}\,, \tag{9}$$

not only builds in the usual magnetic moment of a Dirac particle at moderately large m_π^2 but has the correct leading non-analytic (LNA) behavior of chiral perturbation theory

$$\mu = \mu_0 + \chi\, m_\pi\,, \tag{10}$$

with χ a model-independent constant.

Our focus here is to extrapolate existing lattice QCD estimates of valence quark contributions to baryon moments (u_p, u_n, u_{Σ^+}, u_{Ξ^0}) to the physical mass regime. The leading non-analytic pieces of these various contributions to the baryon magnetic moments are proportional to

$$\beta\, \frac{m_N}{8\,\pi\, f_\pi^2}\, m_\pi \equiv \chi\, m_\pi\,, \tag{11}$$

where the pion decay constant $f_\pi = 93$ MeV. Isolating a particular quark flavor contribution only requires setting the electric charge of all other quark flavors to zero. Separation of the contributions of the LNA terms into valence and sea-quark sectors has been resolved in earlier work.[7]

Previous investigations[7,10] of the strangeness contribution to nucleon magnetic moments focused on the LNA contributions having their origin in the pion cloud alone. The pion cloud provides significant non-linear behavior and gives rise to an enormous environmental sensitivity of the quark sector contributions to baryon magnetic moments. For example, u_n/u_Ξ was found to be 1.51 ± 0.37 providing $G_M^s(0) = -0.16 \pm 0.18$. Since these first studies, there is growing evidence that the K-meson cloud may play an important role in the magnetic properties of the Ξ baryon. It is difficult to describe the magnetic moment of the Ξ^0 baryon[12] using the Padé of Eq. (9). Moreover, the physical value[13] of u_Ξ obtained from Eq. (1) is $-0.599(14)\ \mu_N$ whereas consideration of the pion cloud alone[7] provides $-0.37(3)\ \mu_N$. Here we extend our approach to include contributions from K mesons.

When incorporating higher mass meson-baryon intermediate states, it is important to consider the energy difference between the baryon and its meson cloud disassociation. As the source of the meson cloud is finite in size, high mass intermediate states will ultimately be suppressed by the form factor describing the size of the baryon. Moreover, only low mass intermediate states will provide interesting curvature that cannot be accommodated by the fit parameters μ_0 and c in Eq. (9). For this reason we focus on contributions from $\overline{K}$ ($s\bar{u}$ or $s\bar{d}$) mesons where the strange quark in the hyperon can flow into the meson at the meson-baryon vertex. For these intermediate states, the effective mass of the meson is reduced by the mass splitting between the baryons involved at the vertex. This contrasts the case for K ($u\bar{s}$ or $d\bar{s}$) mesons where both the mass of the meson and the mass splitting between baryons with $\Delta s = 1$ is required to generate the intermediate state.

Chiral perturbation theory provides the following relationship between the pion and kaon masses as the mass of the non-strange quark varies:

$$m_K^2 = m_K^{(0)2} + \frac{1}{2} m_\pi^2 \,, \tag{12}$$

where, $m_K^{(0)}$ is the mass of the kaon in the SU(2) chiral limit. To place the LNA contribution associated with the kaon cloud in the denominator of the Padé, as done for the pion, we note that $(m_K - m_K^{(0)})$ vanishes in the SU(2) chiral limit, so the baryon moment may be described as

$$\mu = \frac{\mu_0}{1 - \frac{\chi_\pi}{\mu_0} m_\pi - \frac{\chi_K}{\mu_0}\left(m_K - m_K^{(0)}\right) + c\, m_\pi^2} \,, \tag{13}$$

preserving the celebrated chiral and heavy-quark features of Eq. (9). Further refinements of the Padé will be reported in a forthcoming publication.[14] Finally the effective mass reduction of the K meson, associated with the baryon mass splittings discussed above, is incorporated by redefining $m_K^{(0)}$ in Eqs. (12) and (13) to be the kaon mass in the chiral limit less the mass splitting. For example in the case $\Xi \to \Sigma K$

$$m_K^{(0)} = \sqrt{\left(m_K^{\rm phys}\right)^2 - \frac{1}{2}\left(m_\pi^{\rm phys}\right)^2} - \left(m_\Xi^{\rm phys} - m_\Sigma^{\rm phys}\right) . \tag{14}$$

The coefficients β and χ for the relevant quark sectors and baryons are summarized in Table 1. For the calculation of $G_M^s(0)$, the LNA chiral behavior of u_n and u_{Ξ^0} is crucial. The coupling of the kaon cloud in Ξ^0 is large and equal to that for the pion cloud to the neutron. Moreover, the sea-quark contribution is smaller for the kaon cloud in Ξ^0, such that the kaon cloud coupling for the u-valence quark in Ξ^0 exceeds that for the pion cloud in the neutron. The latter observation makes the incorporation of the kaon cloud crucial, even though the curvature associated with the heavier meson is reduced relative to that for the pion.

Table 1: Coefficients β and χ of (9) and (11) for the leading nonanalytic (LNA) contribution of the u-quark sectors to the magnetic moment of various baryons. The valence contribution is obtained by subtracting the sea-quark loop contribution from the u-flavor sector contribution. We express the LNA coefficients χ in terms of the usual SU(6) constants F and D, as well as giving numerical values for χ (which is related to β through Eq.(11)). The one-loop corrected values [15] for F and D are used where $F = 0.40$ and $D = 0.61$. The u-quark charge has been normalized to 1.

Baryon	Channel	β, χ	u-flavor sector	u-sea-quark loop
n	$N \to N\pi$	β_π	$(F+D)^2$	$(9F^2 - 6FD + 5D^2)/3$
		χ_π	4.41	2.65
Ξ^0	$\Xi \to \Xi\pi$	β_π	$-(F-D)^2$	$(F-D)^2$
		χ_π	−0.19	0.19
	$\Xi \to \Sigma K$	β_K	$(F+D)^2$	$(F+D)^2/2$
		χ_K	4.41	2.21
	$\Xi \to \Lambda K$	β_K	0	$(3F-D)^2/6$
		χ_K	0	0.251
p	$N \to N\pi$	β_π	$-(F+D)^2$	$(9F^2 - 6FD + 5D^2)/3$
		χ_π	−4.41	2.65
Σ^+	$\Sigma \to \Sigma\pi$	β_π	$-2F^2$	$2F^2$
		χ_π	−1.38	1.38
	$\Sigma \to \Lambda\pi$	β_π	$-2D^2/3$	$2D^2/3$
		χ_π	−1.07	1.07
	$\Sigma \to NK$	β_K	0	$(D-F)^2$
		χ_K	0	0.19

3 Results

Fig. 3 compares the extrapolation of the valence u-quark magnetic moment contributions[16] to the neutron and Ξ^0. The onset of significant curvature from the pion cloud near the chiral limit is evident. We do not show the analogous curves for the proton and Σ^+ as they differ little from our earlier results.[7] The results from these extrapolations taken at the physical pion mass are plotted in Fig. 2 as the filled circle with standard errors. The fact that this point lies in precise agreement with the charge-symmetry relation reinforces the importance of including the kaon cloud in the extrapolation process. u_Ξ now takes the value $-0.54(5)\ \mu_N$ in agreement with $-0.599(14)\ \mu_N$ from experiment.

Having determined the valence quark magnetic moment ratios of Eqs. (6)

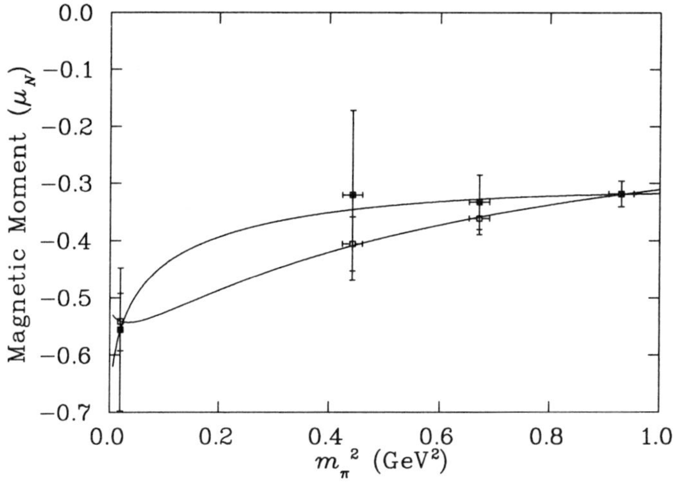

Figure 3: Extrapolation of the valence u-quark magnetic moment contributions to the neutron (solid symbols) and Ξ^0 (open symbols).

and (7) we can predict the value of $G_M^s(0)$ as a function of the ratio of strange to light quark contributions to the sea-quark loop, ${}^{\ell}R_d^s$. Fig. 4 reports results from Eq. (7) based on u_n/u_Ξ. The predictions are perfectly consistent, with a more precise value from Eq. (7) than Eq. (6). The unambiguous result is that $G_M^s(0)$ is less than zero when ${}^{\ell}R_d^s$ lies in the expected range $0 < {}^{\ell}R_d^s < 1$. Taking ${}^{\ell}R_d^s = 0.5 \pm 0.1$ consistent with published estimates[3] provides

$$G_M^s(0) = -0.4\,^{+0.2}_{-0.4}\ \mu_N\,. \tag{15}$$

Acknowledgments

This work was supported by the Australian Research Council.

References

1. D.B. Leinweber, *Phys. Rev.* D **53**, 5115 (1996) [hep-ph/9512319].
2. T. D. Cohen and D. B. Leinweber. *Comments Nucl. Part. Phys.* **21**, 137 (1993) [hep-ph/9212225]; A.W. Thomas, *Austral. J. Phys.* **44**, 173 (1991).
3. S. J. Dong, K. F. Liu and A. G. Williams, *Phys. Rev.* D **58**, 074504 (1998) [hep-ph/9712483].
4. W. Wilcox, *Nucl. Phys. Proc. Suppl.* **94**, 319 (2001) [hep-lat/0010060].

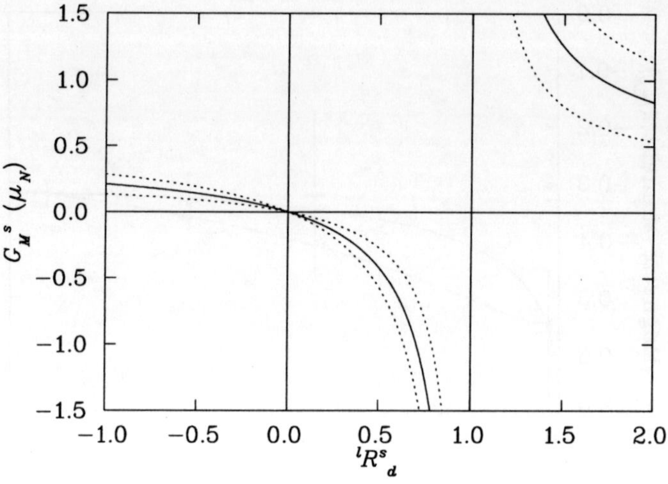

Figure 4: The strange quark contribution to the nucleon magnetic moment $G_M^s(0)$ obtained from (7) plotted as a function of ${}^lR_d^s$ reflecting the relative contribution of strange to light sea-quark loop contributions. The solid curve illustrates the central values and the dashed curves denote the standard error.

5. G.A. Miller, B.M. Nefkens and I. Slaus, *Phys. Rept.* **194** 1 (1990).
6. D. B. Leinweber, *Phys. Rev.* D **45**, 252 (1992); *ibid.*, *Phys. Rev.* D **47**, 5096 (1993) [hep-ph/9302266].
7. D. B. Leinweber and A. W. Thomas, *Phys. Rev.* D **62**, 074505 (2000) [hep-lat/9912052].
8. Particle Data Group, *Eur. Phys. J.* C, **15**, 1 (2000).
9. SAMPLE Collaboration (A. Blake, *et al.*), "2000 Bates Report," pp. 2–11, (2001).
10. D. B. Leinweber and A. W. Thomas, *Nucl. Phys.* A **684**, 35 (2001) [nucl-th/0007058].
11. D.B. Leinweber, D.H. Lu and A.W. Thomas, *Phys. Rev.* D **60**, 034014 (1999) [hep-lat/9810005].
12. E. J. Hackett-Jones, D. B. Leinweber and A. W. Thomas, *Phys. Lett.* B **489**, 143 (2000) [hep-lat/0004006].
13. C. W. Wong, [hep-lat/0103021].
14. D. B. Leinweber and A. W. Thomas, in preparation.
15. E. Jenkins, M. Luke, A. V. Manohar, and M. J. Savage, *Phys. Lett.* B **302**, 482 (1993), *ibid.* B **388** (1996) E866, [hep-ph/9212226].
16. D. B. Leinweber, *et al.*, *Phys. Rev.* D **43**, 1659 (1991).

ELECTROMAGNETIC INTERACTIONS IN LIGHT FRONT DYNAMICS

J.-F. MATHIOT
Laboratoire de Physique Corpusculaire, Université Blaise-Pascal,
24 avenue des Landais, F-63177 Aubière Cedex, France
E-mail: mathiot@in2p3.fr

We discuss the calculation of electromagnetic observables in the context of Light-Front Dynamics. We show why one should be very careful to clearly identify the non-physical contributions to any approximate calculation which arise from the explicit breaking of covariance in Light Front Dynamics.

1 Light Front Dynamics and rotational invariance

1.1 Covariant formulation of Light-Front Dynamics

Among the various ways to deal with relativity in the description of bound states, we will focus on Light Front Dynamics (LFD). In the standard formulation of LFD [1], the wave function of a system is defined on a plane characterized by the equation $t + z/c = 0$. The usual Schrödinger, equal time, formalism is easily recovered in LFD by letting c going to infinity.

The formulation of relativistic systems in LFD has many advantages. Maybe the most important one is the absence of vacuum fluctuations, while all particles are on the mass-shell but off energy-shell. Accordingly, a meaningful decomposition of the state vector describing the system under study in terms of Fock components of definite number of particles is possible. Of course, the number of Fock components to be considered in any practical calculation depends on the dynamics of the system, and on the kinematical regime one is interested in.

The most serious drawback of this formulation is, however, that the position of the light-front $t + z = 0$ (with $c = 1$) is not rotational invariant. Since the rotations in the zx and zy planes change the position of the light-front, the associated generators depend on the dynamics and cannot be reduced to kinematical transformations [2]. In practice, this means one needs to know the complete dynamics in order to write down the general structure of a bound state of definite angular momentum. It also means that any electromagnetic operator should have the same (dynamical) transformation properties for matching the bound state wave function one's. This is essential to guarantee that any physical amplitude (or cross-section) does not depend on the particular choice of the light-front we start with.

We therefore need an explicit procedure to exhibit in a convenient way these dynamical transformations. This is achieved in the Covariant formulation of LFD (CLFD) [3]. Our starting point is the invariant definition of the light-front by $\omega.x = 0$ where ω is an (unspecified) light-like four vector ($\omega^2 = 0$). The standard formulation of LFD can be easily recovered with the particular value of $\omega = (1, 0, 0, -1)$.

This definition of the light-front is explicitly invariant by any four-dimensional rotation, or any three-dimensional rotation and Lorentz boost. Consequently, these transformations become ω-dependent, but do not necessitate the knowledge of the dynamics of the system to construct them explicitly. The dynamics enters now into the ω dependence of the wave function and of the electromagnetic operator, in such a way that any physical amplitude should not depend on the particular position of the light-front, i.e. on ω, unless approximations have been made. In that case, which is almost always true in practice, the explicit covariance of the approach enables us to exhibit the ω dependence of the amplitude and to extract the physical part from the non-physical ω-dependent one, as we shall explain below for simple examples.

1.2 Two-body wave function

The state vector $\phi(p)$ describing any bound state system is decomposed in Fock components according to:

$$\begin{aligned}
\phi(p) = {} & (2\pi)^{3/2}\int \phi_1(k_1, p, \omega\tau) a^\dagger(\vec{k}_1)|0\rangle \delta^{(4)}(k_1 - p - \omega\tau) 2(\omega\cdot p) d\tau \frac{d^3k_1}{(2\pi)^{3/2}\sqrt{2\varepsilon_{k_1}}} \\
& + (2\pi)^{3/2}\int \phi_2(k_1, k_2, p, \omega\tau) a^\dagger(\vec{k}_1) a^\dagger(\vec{k}_2)|0\rangle \\
& \times \delta^{(4)}(k_1 + k_2 - p - \omega\tau) 2(\omega\cdot p) d\tau \frac{d^3k_1}{(2\pi)^{3/2}\sqrt{2\varepsilon_{k_1}}} \frac{d^3k_2}{(2\pi)^{3/2}\sqrt{2\varepsilon_{k_2}}} \\
& + (2\pi)^{3/2}\int \phi_3(k_1, k_2, k_3, p, \omega\tau) a^\dagger(\vec{k}_1) a^\dagger(\vec{k}_2) a^\dagger(\vec{k}_3)|0\rangle \\
& \times \delta^{(4)}(k_1 + k_2 + k_2 - p - \omega\tau) 2(\omega\cdot p) d\tau \\
& \times \frac{d^3k_1}{(2\pi)^{3/2}\sqrt{2\varepsilon_{k_1}}} \frac{d^3k_2}{(2\pi)^{3/2}\sqrt{2\varepsilon_{k_2}}} \frac{d^3k_3}{(2\pi)^{3/2}\sqrt{2\varepsilon_{k_3}}} + \cdots ,
\end{aligned} \tag{1}$$

where $\varepsilon_{k_i} = \sqrt{\vec{k}_i^{\,2} + m_i^2}$ and m_i is the mass of the particle i of momentum k_i.

Note that in this decomposition the momentum conservation law involves the position ω of the light front plane. For the two-body component for instance:

$$p = k_1 + k_2 - \omega\tau \ . \tag{2}$$

The parameter τ, conjugate to an energy, is uniquely determined from the on-mass-shell conditions for the elementary particles. It measures how far the particles are off-energy shell. This conservation law implies also that the Center of Mass (CM) system defined by $\vec{k_1}+\vec{k_2}=\vec{0}$ does not imply that $\vec{p}=\vec{0}$!

Let us first write down the general structure of a scalar bound state made of two scalar particles. Its wave function should depend on two invariant variables. These can be of course $s=(k_1+k_2)^2$ and $t=(k_1-k_2)^2$. More conveniently, it is appropriate [3] to use the variables $\vec{k}$ and $\vec{n}$ defined as:

$$\begin{aligned} \vec{k} &= \mathcal{L}^{-1}(\vec{\mathcal{P}})\vec{k_1} \ , \\ \vec{n} &= \mathcal{L}^{-1}(\vec{\mathcal{P}})\vec{\omega} \ . \end{aligned} \tag{3}$$

where $\mathcal{L}$ is the Lorentz boost and $\vec{\mathcal{P}}=\vec{k_1}+\vec{k_2}$. It is interesting to note that $\vec{k}$ coincides, in the CM system, with the relative momentum between particles 1 and 2. It can be shown [3] that through a Lorentz boost, or a rotation, the vectors $\vec{k}$ and $\vec{n}$ are rotated only. This implies that one can choose as the two invariants, $\vec{k}^2$ and $\vec{k}\cdot\vec{n}$. We can thus parametrize the two-body scalar wave function by :

$$\varphi(\vec{k}^2, \vec{k}.\vec{n}) \tag{4}$$

In the non relativistic limit, defined by letting c go to infinity, we have $\vec{n} \equiv \vec{n}/c \to 0$ when $c \to \infty$, so that the relativistic wave function (4) tends to the Schrödinger one $\varphi(\vec{k}^2)$. Note that the usual light-cone coordinates can easily be defined in terms of these variables:

$$x = \frac{\omega.k_1}{\omega\cdot p} = \frac{1}{2}\left(1 - \frac{\vec{n}.\vec{k}}{\varepsilon_k}\right) \ , \ \vec{k_T}^{\,2} = \vec{k}^2 - (\vec{n}\cdot\vec{k})^2 \ . \tag{5}$$

This two-body wave function is solution of the following Light-Front integral equation [3]. It reads, for equal constituent masses m:

$$\left[4(\vec{k}^{\,2}+m^2) - M^2\right]\varphi(\vec{k}^2, \vec{k}\cdot\vec{n}) = -\frac{m^2}{2\pi^3}\int \varphi(\vec{k'}^{\,2}, \vec{k'}\cdot\vec{n})K(\vec{k}^{\,\prime}, \vec{k}, \vec{n}, M^2)\frac{d^3k'}{\varepsilon_{k'}} \ , \tag{6}$$

where the kernel of the interaction has been denoted by K and M is the boundstate mass.

The structure of the pion wave function, i.e. a scalar bound system made of two spin 1/2 particles, can now easily be found. The general decomposition of the wave function, compatible with the quantum numbers of the pion is:

$$\psi = \frac{1}{\sqrt{2}}\bar{u}(k_2)\left[A_1 + A_2\frac{\not\omega}{\omega\cdot p}\right]\gamma_5 v(k_1)\ . \tag{7}$$

For simplicity, we do not take into account isospin. The pion wave function has thus two independent components on the light-front. All other structures involving for instance the individual momenta k_1 or k_2 can be expressed in terms of the ones above using the Dirac equation for the on-mass shell constituent quarks. The representation of this wave function in terms of the variables $\vec{k}$ and $\vec{n}$ is thus:

$$\psi = \frac{1}{\sqrt{2}}w_2^t\left(g_1 + \frac{i\vec{\sigma}\cdot[\vec{n}\times\vec{k}]}{k}g_2\right)w_1\ , \tag{8}$$

where w is the two-component nucleon spinor normalized to $w^\dagger w = 1$. The component proportional to g_1 is the only one which survives in the non-relativistic limit, while the component g_2 is the component which dominates in the pion form factor at very large momentum transfer [3].

The second wave function of particular interest is the J=1 bound state wave function of two spin 1/2 particles: this is the case for the deuteron, but also for charmonium states. Its general structure has six independent components:

$$\Psi^\lambda_{\sigma_2\sigma_1}(\vec{k},\vec{n}) = w^\dagger_{\sigma_2}\psi^\lambda(\vec{k},\vec{n})\sigma_y w^\dagger_{\sigma_1}\ , \tag{9}$$

with:

$$\begin{aligned}\vec{\psi}(\vec{k},\vec{n}) = {} & f_1\frac{1}{\sqrt{2}}\vec{\sigma} + f_2\frac{1}{2}\left[\frac{3\vec{k}(\vec{k}\cdot\vec{\sigma})}{\vec{k}^2} - \vec{\sigma}\right] + f_3\frac{1}{2}\left[3\vec{n}(\vec{n}\cdot\vec{\sigma}) - \vec{\sigma}\right] \\ & + f_4\frac{1}{2k}\left[3\vec{k}(\vec{n}\cdot\vec{\sigma}) + 3\vec{n}(\vec{k}\cdot\vec{\sigma}) - 2(\vec{k}\cdot\vec{n})\vec{\sigma}\right] \\ & + f_5\sqrt{\frac{3}{2}}\frac{i}{k}\left(\vec{k}\times\vec{n}\right) + f_6\frac{\sqrt{3}}{2k}\left[(\vec{k}\times\vec{n})\times\vec{\sigma}\right]\ ,\end{aligned} \tag{10}$$

In the non relativistic limit, the f_1 component tends to the S-state wave function, while the f_2 component tends to the D-state, and all other components go zero. These relativistic components can easily be evaluated in perturbation theory, starting from the non-relativistic wave function in the r.h.s. of eq.(6), the relativistic kernel and calculating the relativistic components with the l.h.s. of eq.(6) [4].

Note that among the relativistic components, the f_5 component dominates. This behavior has been traced back to the contribution of pair terms (Z diagrams) in a $1/m$ expansion, when a pseudo-scalar pion-nucleon coupling is considered [5].

2 The Electromagnetic Current on the Light Front

For simplicity, we consider a two-body system to start with, but all our results are obviously applicable to many-body systems. Let us consider a typical electron scattering experiment. In the Feynman approach, the amplitude of this process is factorized in the product of the electromagnetic vertex of the projectile, the propagator $1/q^2$ of the virtual photon and the electromagnetic vertex of the bound system (hadronic vertex). This latter is then expressed through the physical form factors. The complete set of (light-cone time ordered) diagrams should, of course, reproduce the same observable cross-section. However, any separate time ordered amplitude does not in general factorize in this formalism. One way to circumvent this problem is to choose a particular orientation of the momentum transfer so that $\omega{\cdot}q = 0$, where q is the photon momentum (or $q^+ = 0$ in standard LFD). This is always possible if $Q^2 \leq 0$. In that case, one can show that the two processes (at the electromagnetic and hadronic vertices) always factorize like in the Feynman approach [3]. One can thus concentrate only on the hadronic vertex.

2.1 Non physical form factors

The question of extracting the physical form factors from the elementary electromagnetic amplitude in LFD is rather general and appears for a system with any spin. The sum of all the on-energy-shell amplitudes does not depend on the four-vector ω, whereas any given amplitude can be ω-dependent. The amplitude corresponding to a separate diagram can thus depend on ω.

The matrix element of the exact current, $J_\rho = \langle p'|J_\rho(0)|p\rangle$, can be represented as the sum of all possible contributions:

$$J_\rho = J_\rho^{(1)} + J_\rho^{(2)} + \cdots + J_\rho^{(n)} + \cdots , \tag{11}$$

where the superscripts is a short notation for indicating the various levels of approximation. In the case of the deuteron form factors for instance, the terms in (11) correspond to the impulse approximation, meson exchange current contribution, or the contribution of an increasing number of Fock components. Any contribution in (11) consists of the sum of two parts: a ω-independent one denoted by $F_\rho^{(n)}$, and a ω-dependent one denoted by $B_\rho^{(n)}(\omega)$:

$$J_\rho^{(n)} = F_\rho^{(n)} + B_\rho^{(n)}(\omega) \; . \tag{12}$$

Since the matrix element of the exact current does not depend on ω, the ω-dependent parts should cancel each other:

$$B_\rho^{(1)}(\omega) + B_\rho^{(2)}(\omega) + \cdots + B_\rho^{(n)}(\omega) + \cdots = 0 \; , \tag{13}$$

and, hence, the matrix element is determined by the sum of the ω-independent parts only:

$$J_\rho = F_\rho^{(1)} + F_\rho^{(2)} + \cdots + F_\rho^{(n)} + \cdots \; . \tag{14}$$

It follows that *the problem of restoring the independence of any physical amplitude on* ω *is reduced to the separation, and omission, of the "pure"* ω*-dependent terms.*

We emphasize that this definition of the form factors, after separating out ω-dependent contributions, still leads to approximate form factors. However, our procedure guarantees that the form factors do not receive any spurious, non-physical contributions. Since these spurious, ω-dependent, contributions can be as large as the physical ones, in particular at high momentum transfer, our procedure is the most adequate when one wants to draw any conclusion on a particular model, or to compare various relativistic approaches among themselves. Moreover, it leads to an unambiguous determination of the form factors, independently of the orientation of the light front or of the choice of the component of the current which is used to calculate the amplitude.

2.2 Physical observables

The case of scalar bound states is trivial. The general decomposition of the current matrix element is given here by:

$$J_\rho = (p+p')_\rho F(Q^2) + \omega_\rho B_1(Q^2) \; . \tag{15}$$

The invariant functions F and B_1 could depend in principle on $\omega{\cdot}p$ and $\omega{\cdot}p'$. However, the four-vector ω is defined up to an arbitrary number, and, hence, the theory is invariant with respect to the replacement $\omega \to \alpha\omega$, where α is a number. The form factors F and B_1 can therefore depend only on the ratio $\omega{\cdot}p'/\omega{\cdot}p$. But since $\omega{\cdot}q = 0$, we have $\omega{\cdot}p'/\omega{\cdot}p = 1$, and the functions F and B_1 (and similar functions for the case of spin 1/2 and 1 systems) depend on Q^2 only.

It is then obvious to extract the physical form factor $F(Q^2)$: one has just to contract J_ρ with ω^ρ:

$$F(Q^2) = \frac{J\cdot\omega}{2\omega\cdot p} \ . \tag{16}$$

This is the procedure used in standard LFD where $J\cdot\omega \to J^+$.

The situation is however not as simple as this for J=1 or J=1/2 systems like the deuteron or the nucleon form factors. The electromagnetic form factors of a system with spin 1 (the deuteron, for instance) are defined by the following decomposition of the electromagnetic vertex:

$$\langle\lambda'|J_\rho|\lambda\rangle = e_\mu^{*\lambda'}(p')T_\rho^{\mu\nu}e_\nu^\lambda(p) \ . \tag{17}$$

Here $e_\mu^\lambda(p)$ is the deuteron polarization vector, p and p' are the initial and final deuteron momenta, λ and λ' are the corresponding helicities. The physical tensor $T_\rho^{\mu\nu}$ is now decomposed into a physical part $J_\rho^{\mu\nu}$ which gives the physical form factors of the deuteron (charge, magnetic and quadrupole), and the spurious contributions $B_\rho^{\mu\nu}$. These latter have the general form:

$$\begin{aligned} B_\rho^{\mu\nu} = & \frac{M^2}{2(\omega\cdot p)}\omega_\rho\left[B_1 g^{\mu\nu} + B_2\frac{q^\mu q^\nu}{M^2} + B_3 M^2\frac{\omega^\mu\omega^\nu}{(\omega\cdot p)^2} + B_4\frac{q^\mu\omega^\nu - q^\nu\omega^\mu}{2\omega\cdot p}\right] \\ & + B_5 P_\rho M^2\frac{\omega^\mu\omega^\nu}{(\omega\cdot p)^2} + B_6 P_\rho\frac{q^\mu\omega^\nu - q^\nu\omega^\mu}{2\omega\cdot p} + B_7 M^2\frac{g_\rho^\mu\omega^\nu + g_\rho^\nu\omega^\mu}{\omega\cdot p} \\ & + B_8 q_\rho\frac{q^\mu\omega^\nu + q^\nu\omega^\mu}{2\omega\cdot p} \ . \end{aligned} \tag{18}$$

One can thus calculate the tensor $J_\rho^{\mu\nu}$ from the complete current $T_\rho^{\mu\nu}$, estimated in any model, by standard algebraic manipulations [6]. Note that contrarily to the pion form factor, one needs to extract the B's in order to get unambiguous determination of the physical form factors. The standard procedure which consists in calculating $\langle\lambda'|J^+|\lambda\rangle$ for various helicity states in terms of the physical form factors only, assuming all B's to be zero, is not well defined in the sense that the physical form factors thus depend on the particular choice made for the four helicity states in order to calculate the three form factors [6].

The situation is very similar for the nucleon form factor [7].

3 Applications

3.1 Perturbative calculations

Since one can do an "exact" calculation, order by order in perturbation theory in terms of a given small parameter (the coupling constant for instance), the non-physical form factors B are zero at each order of the expansion. We checked this explicitly for the anomalous magnetic moment of the electron [8]. In that case, both calculations, in standard LFD [1] and in CLFD [3] give identical results.

3.2 Non-perturbative calculations in peculiar limiting situations

In given kinematical conditions, an approximate non-perturbative calculation can approach the exact solution, so that the non-physical contributions also go to zero. We have investigated three such cases:

- When the relevant momenta get small, the contribution of the B terms, of relativistic origin, should disappear. We checked this explicitly in the case of the radiative correction to the leptonic decay width of the charmonium states [9].
- The $B \to D$ semileptonic decay in the heavy quark limit [10].
- The pion form factor in the asymptotic high Q^2 regime [3].

3.3 Fully non-perturbative systems

In that case, one can not avoid to extract the physical from the non-physical form factors since for instance one usually has to truncate the Fock state expansion. In that case, the covariant formulation of LFD assumes its full advantage. This is the case for the calculation of the nucleon form factors at moderate Q^2, a subject of intense experimental activity at TJNAF [11].

4 Conclusions

The covariant formulation of LFD is a very powerful tool to control the approximation made in practical calculations in LFD. This is of utmost importance in order to understand electromagnetic interactions in any channel. The separation of physical form factors from non-physical form factors is mandatory for moderate Q^2 observables, and for bound state systems. This is of particular interest for TJNAF experiments.

The next step is now the calculation of the bound state wave function with its many relativistic components.

Acknowledgments

I would like to thank my collaborators, F. Bissey, J. Carbonell, Th. Cousin, B. Desplanques, J.-J. Dugne and V. Karmanov for numerous discussions, comments and advice on the subject presented in this talk.

References

1. S.J. Brodsky, H.C. Pauli and S.S. Pinsky, *Phys. Rep.* **301**, 299 (1998).
2. P.A.M. Dirac, *Rev. Mod. Phys.* **21**, 392 (1949).
3. J. Carbonell, B. Desplanques, V.A. Karmanov and J.-F. Mathiot, *Phys. Rep.* **300**, 215 (1998).
4. J. Carbonell and V.A. Karmanov, *Nucl. Phys.* A **581**, 625 (1995).
5. B. Desplanques, V.A. Karmanov and J.-F Mathiot, *Nucl. Phys.* A **589**, 697 (1995).
6. V.A. Karmanov and A.V. Smirnov, *Nucl. Phys.* A **546**, 691 (1992); V.A. Karmanov and A.V. Smirnov, *Nucl. Phys.* A **575**, 520 (1994).
7. V.A. Karmanov and J.-F. Mathiot, *Nucl. Phys.* A **602**, 388 (1996).
8. J.-J. Dugne, V.A. Karmanov and J.-F. Mathiot, *The charge and mass perturbative renormalization in explicitly covariant LFD*, hep-ph/0101156.
9. F. Bissey, J.-J. Dugne and J.-F. Mathiot, *Dynamical relativistic corrections to the leptonic decay width of heavy quarkonia*, in preparation.
10. F. Bissey and J.-F. Mathiot, *Eur. Phys. Jour.* **16**, 131 (2000).
11. O. Gayou, contribution to this workshop

SOFT QCD MODELING OF MESON ELECTROMAGNETIC FORM FACTORS

PETER C. TANDY

Center for Nuclear Research, Department of Physics, Kent State University, Kent, OH 44242, USA

E-mail: tandy@cnr2.kent.edu

We summarize recent progress in soft QCD modeling based on the set of Dyson–Schwinger equations truncated to ladder-rainbow level. We pay particular attention to electromagnetic elastic and transition form factors of the pion. This covariant approach accommodates quark confinement and implements the QCD one-loop renormalization group behavior. The dressed quark propagators are compared to the most recent lattice-QCD data.

1 Introduction

High energy electroweak probes utilize perturbative QCD to quantify the structure of hadrons in terms of parton structure functions. Their detailed behavior samples various aspects of the quark-gluon dynamics of the hadron bound state. This information is intrinsically non-perturbative and unravelling it from sets of structure functions will require a connection be made with non-perturbative QCD calculations and models. Although some lattice QCD studies have begun to produce moments of structure functions [1], the opportunities are very limited at present. The most extensive hadronic models were designed to study the mass spectrum and decays and usually contain elements that limit their use for high energy lepton scattering. Examples include: non-relativistic kinematics, a lack of manifest Lorentz covariance, no quark sea, no dynamical gluons, no renormalization group behavior for change of scale, and no confinement of quarks. Studies of deep inelastic scattering within the Nambu–Jona-Lasinio model [2] have helped clarify some of the issues confronting work within a quark field theory format.

Here we summarize recent progress in soft QCD modeling based on the set of Dyson–Schwinger equations [DSEs] of the theory [3,4]; we pay particular attention to electromagnetic couplings to the pion. This covariant approach accommodates quark confinement and implements the QCD one-loop renormalization group behavior. Deep inelastic scattering phenomena have not been treated with the present model; a simplified related approach that retains all the essential features has recently produced excellent results for the pion valence quark distribution amplitude [5].

Meson electromagnetic form factors in impulse approximation are described by two diagrams, in one the photon couples to the quark, in the other to the antiquark. A form factor can be associated with each diagram. For example, the photon coupling to the antiquark of a flavor $a\bar{b}$ pseudoscalar meson produces [6]

$$2\,P_\nu\,F_{a\bar{b}\bar{b}}(Q^2) = N_c \int^\Lambda \frac{d^4q}{(2\pi)^4}\,\mathrm{Tr}\big[S^a(q)\,\Gamma^{a\bar{b}}_{\rm ps}(q,q_+;P_-)\,S^b(q_+) \times i\Gamma^b_\nu(q_+,q_-)\,S^b(q_-)\,\bar{\Gamma}^{a\bar{b}}_{\rm ps}(q_-,q;-P_+)\big]\,, \qquad (1)$$

where $q = k + \frac{1}{2}P$, $q_\pm = k - \frac{1}{2}P \pm \frac{1}{2}Q$, $P_\pm = P \pm \frac{1}{2}Q$. We work in the isospin symmetry limit, so for the pion we have $F_\pi(Q^2) = F_{u\bar{u}u}(Q^2)$. The charged and neutral kaon form factors are given by $F_{K^+} = \frac{2}{3}F_{u\bar{s}u} + \frac{1}{3}F_{u\bar{s}\bar{s}}$ and $F_{K^0} = -\frac{1}{3}F_{d\bar{s}d} + \frac{1}{3}F_{d\bar{s}\bar{s}}$ respectively. To implement such a study, one needs a consistent QCD modeling of the photon-quark dressed vertex Γ_ν, the dressed quark propagator $S(q)$, and the meson Bethe-Salpeter (BS) amplitude Γ_{ps}. The first two are related by the vector Ward-Takahashi identity[WTI]; the second two are related by the axial-vector WTI.

The DSE for the renormalized quark propagator in Euclidean space is

$$S(p)^{-1} = i\,Z_2\,\not p + Z_4\,m(\mu) + Z_1 \int^\Lambda \frac{d^4q}{(2\pi)^4}\,g^2 D_{\mu\nu}(k)\frac{\lambda^a}{2}\gamma_\mu S(q)\Gamma^a_\nu(q,p)\,, \qquad (2)$$

where $D_{\mu\nu}(k)$ is the dressed-gluon propagator, $\Gamma^a_\nu(q;p)$ is the dressed-quark-gluon vertex, and $k = p - q$. The solution has the form $S(p)^{-1} = i\not p A(p^2) + B(p^2)$ and is renormalized at spacelike μ^2 according to $A(\mu^2) = 1$, and $B(\mu^2) = m(\mu)$, where $m(\mu)$ is the current quark mass. The notation $\int^\Lambda$ represents a translationally invariant regularization at scale Λ. The same regularization occurs consistently at all stages as indicated in Eq. (1). One takes $\Lambda \to \infty$ as the final step.

The $q\bar{q}\gamma$ vertex $\Gamma_\mu(p_+,p_-)$ corresponding to total momentum $Q = p_+ - p_-$ satisfies the Bethe-Salpeter equation [BSE]

$$\Gamma_\mu(p_+,p_-) = Z_2\,\gamma_\mu + \int^\Lambda \frac{d^4q}{(2\pi)^4}\,K(p,q;Q)\,S(q_+)\,\Gamma_\mu(q_+,q_-)\,S(q_-)\,, \qquad (3)$$

where we use the notation $p_+ = p + \eta Q$ and $p_- = p - (1-\eta)Q$ for the outgoing and incoming quark momenta respectively at the vertex. Here η, which specifies how the total momentum is shared between quark and antiquark (thus defining a relative momentum p), is arbitrary and physical observables should not depend on it. The kernel K is the renormalized, amputated $q\bar{q}$ scattering kernel that is irreducible with respect to a pair of $q\bar{q}$ lines. Other electroweak processes [7] require, e.g., the dressed $q\bar{q}W$ vertex which is defined in an analogous way.

When the homogeneous version of Eq. (3) has vector solutions at (discrete) timelike total momenta $Q^2 = -m^2$, these are the vector meson bound states. The vector vertex $\Gamma_\mu(p_+, p_-)$ will have simple poles at those locations. This correspondence holds for all other transformation characters (pseudoscalar, axial vector, etc), labelled by quantum numbers J^{PC}. The homogeneous solutions $\Gamma_M(p_+, p_-)$ are the bound state BS amplitudes and they are normalized in the canonical way. Explicit representations of the $\Gamma_M(p_+, p_-)$ require an expansion in a complete set of covariants constructed from gamma matrices and momenta. For example, the general representation for pseudoscalar bound states is [8]

$$\Gamma_{ps}(k_+, k_-; P) = \gamma_5\left[iE + \not{P}\, F + \not{k}\, G + \sigma_{\mu\nu}\, k_\mu P_\nu\, H\right], \tag{4}$$

where the invariant amplitudes E, F, G and H are Lorentz scalar functions $f(k^2; k \cdot P; \eta)$. Note that these functions depend on the momentum partitioning parameter η because the total and relative momenta have been employed in the covariants. However, physical observables are independent of this parameter; this is verified numerically within the present approach as long as the set of employed covariants is complete [6].

2 Ladder-Rainbow Model

To solve the BSE, we use a ladder truncation, with an effective quark-antiquark interaction that reduces to the perturbative running coupling at large momenta [8,9]. In conjunction with the rainbow truncation for the quark DSE, the ladder truncation of the BSE preserves both the vector WTI for the $q\bar{q}\gamma$ vertex and the axial-vector WTI. The latter ensures the existence of almost massless pseudoscalar mesons which are the Goldstone bosons connected with dynamical chiral symmetry breaking [8]. In combination with the impulse approximation, this ladder-rainbow truncation ensures electromagnetic current conservation [6].

The ladder truncation of the BSE, Eq. (3), is $K(p, q; P) \to -\mathcal{G}(k^2)\, D^{\rm free}_{\mu\nu}(k)\, \frac{\lambda^i}{2}\gamma_\mu \otimes \frac{\lambda^i}{2}\gamma_\nu$ where $k = p - q$, and $D^{\rm free}_{\mu\nu}$ is the free gluon propagator in Landau gauge. The corresponding rainbow truncation of the quark DSE, Eq. (2), is given by $\Gamma^i_\nu(q, p) \to \gamma_\nu \lambda^i/2$ together with $g^2 D_{\mu\nu}(k) \to \mathcal{G}(k^2) D^{\rm free}_{\mu\nu}(k)$. This truncation was found to be particularly suitable for the flavor octet pseudoscalar and vector mesons since the next-order contributions in a quark-gluon skeleton graph expansion, have a significant amount of cancellation between repulsive and attractive corrections [10].

Table 1. Results [6,9,11,12,13] for light mesons in GeV compared to experiment [14].

	expt. (estimates)	calc. (fitted)
$m^{u=d}_{\mu=1\text{GeV}}$	5 - 10 MeV	5.5 MeV
$m^{s}_{\mu=1\text{GeV}}$	100 - 300 MeV	125 MeV
- $\langle\bar{q}q\rangle^0_\mu$	$(0.236\ \text{GeV})^3$	$(0.241\ \text{GeV})^3$
m_π	0.1385 GeV	0.138
f_π	0.0924 GeV	0.093
m_K	0.496 GeV	0.497
f_K	0.113 GeV	0.109
m_ρ	0.770 GeV	0.742
f_ρ	0.216 GeV	0.207
$m_{K^\star}$	0.892 GeV	0.936
$f_{K^\star}$	0.225 GeV	0.241
m_ϕ	1.020 GeV	1.072
f_ϕ	0.236 GeV	0.259
r^2_π	0.44 fm^2	0.45
r^2_{K+}	0.34 fm^2	0.38
$r^2_{K^0}$	-0.054 fm^2	-0.086
$r^2_{\pi\gamma\gamma}$	0.42 fm^2	0.39
$g_{\pi\gamma\gamma}$	0.50	0.50
$g_{\rho\pi\gamma}$	0.57	0.54
$g_{\rho\pi\pi}$	6.02	4.85
$g_{\phi KK}$	4.64	4.63
$g_{K^{\star+}\pi^+K^0}$	4.53	4.6

The model is completely specified once a form is chosen for the "effective coupling" $\mathcal{G}(k^2)$. We employ the Ansatz [9]

$$\frac{\mathcal{G}(k^2)}{k^2} = \frac{4\pi^2\, D\, k^2}{\omega^6}\, \mathrm{e}^{-k^2/\omega^2} + \frac{4\pi^2\, \gamma_m\, \mathcal{F}(k^2)}{\frac{1}{2}\ln\left[\tau + \left(1 + k^2/\Lambda^2_{\text{QCD}}\right)^2\right]}\,, \tag{5}$$

with $\gamma_m = 12/(33 - 2N_f)$ and $\mathcal{F}(s) = (1 - \exp\frac{-s}{4m_t^2})/s$. The ultraviolet behavior is chosen to be that of the QCD running coupling $\alpha(k^2)$; the ladder-rainbow truncation then generates the correct perturbative QCD structure of the DSE-BSE system of equations. The first term implements the strong infrared enhancement in the region $0 < k^2 < 1\,\text{GeV}^2$ phenomenologically required to produce a realistic value for the chiral condensate. In recent years, there has been some progress in understanding the interplay between ghosts and gluons in Landau-gauge QCD, indicating that ghosts could play an important role for this enhancement [4]. However, this is not yet ready for incorporation into phenomenological studies of hadron properties. We use $m_t = 0.5\,\text{GeV}$, $\tau = e^2 - 1$, $N_f = 4$, $\Lambda_{\text{QCD}} = 0.234\,\text{GeV}$, and a renormalization scale $\mu = 19\,\text{GeV}$ which is well into the perturbative domain [8,9]. The remaining parameters, $\omega = 0.4\,\text{GeV}$ and $D = 0.93\,\text{GeV}^2$ along with the quark masses, are fitted to give a good description of $\langle \bar{q}q \rangle$, $m_{\pi/K}$ and f_π. The value of f_K, the ground state vector masses and decay constants, and the other quantities in Table 1 are then produced without any free parameters [6,9,11]. Note that the complete set of covariants of the BS amplitudes, see Eq. (4), are needed to satisfy the axial-vector WTI [8]. The pseudovector amplitudes F and G contribute about 30% of the masses and decay constants of the pseudoscalars; they dominate the asymptotic behavior of the pion charge form factor [15].

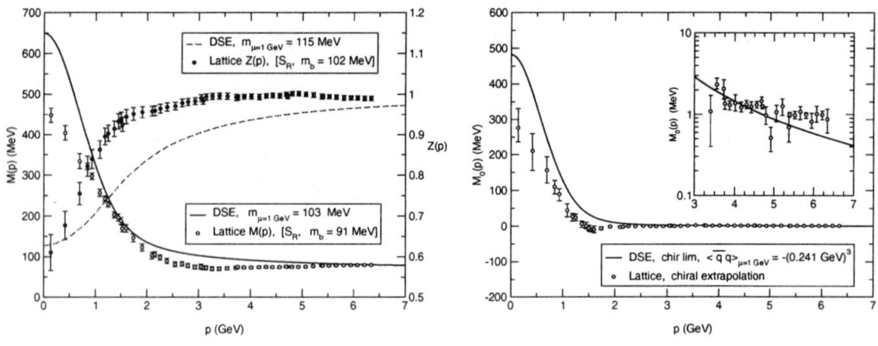

Figure 1. *Left Panel*: DSE solution [9] for quark propagator amplitudes compared to recent lattice data [16]. *Right Panel*: The chiral limit DSE mass function compared to the linear extrapolation of lattice data to the chiral limit [16].

In Fig. 1 we compare the DSE model [9] propagator amplitudes defined by $S(p) = Z(p^2)[i\not{p} + M(p^2)]^{-1}$ with the most recent results from lattice QCD [16]. In contrast to a previous comparison [17], the data here has more lattice artifacts removed and the DSE calculations use the indicated current mass values to match the lattice mass function at 6 GeV. There is agreement in the qualitative infrared structure of the mass function particularly in the way the infrared enhancement sets in. Since the lattice simulation produces the regulated but un-renormalized propagator, the scale of the field renormalization function Z is arbitrary and only the shape is a meaningful comparison. For this reason, we have scaled the lattice data for Z so that $Z(5\text{ GeV}) = 1$. The ladder-rainbow DSE model typically produces a Z that saturates much slower than does the lattice Z; this may signal a deficiency of the bare gluon-quark vertex. The lattice work [16] also produced a linear extrapolation to the chiral limit mass function $M_0(p)$ and we compare this to the DSE result with the insert emphasizing high momentum. The known one-loop renormalization group UV behavior of chiral QCD, which is preserved by this DSE model, is

$$M_0(p^2) \simeq \frac{2\pi^2\gamma_m}{3} \frac{-\langle\bar{q}q\rangle^0}{p^2\left(\frac{1}{2}\ln\left[\frac{p^2}{\Lambda^2_{\rm QCD}}\right]\right)^{1-\gamma_m}}, \tag{6}$$

with $\langle\bar{q}q\rangle^0$ being the renormalization-point-independent chiral condensate [8]. Although the lattice data does not distinguish clearly between this fall-off and a flat behavior, the scale of the data is consistent with a condensate that is less than 10% higher than that of the DSE model. That is, we make a tentative assignment of $\langle\bar{q}q\rangle_{\mu=1\text{ GeV}} = -(248\text{ GeV})^3$ for this lattice data.

3 Meson Elastic and Transition Form Factors

Our results [6] for $F_\pi(Q^2)$ are shown in Fig. 2; the charge radii are given in Table 1. Up to about $Q^2 = 3\,\text{GeV}^2$, our results for $F_\pi(Q^2)$ and $F_K(Q^2)$ can be fit-

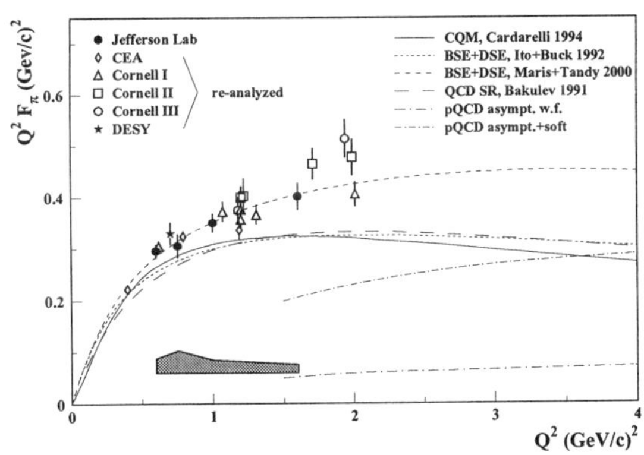

Figure 2. Our results for the pion charge form factor [6] compared to new data from JLab [18] including a re-analysis of older data. Also shown are various earlier model results. Figure taken from the PhD thesis of J. Volmer [18].

ted quite well by a monopole [6]. Asymptotically, the behavior is $Q^2F(Q^2) \to c$ up to logarithmic corrections. However, numerical limitations prevent us from accurately determining these constants. Around $Q^2 = 3\,\text{GeV}^2$, our result for Q^2F_π is well above the pQCD result [19] $16\pi f_\pi^2\alpha_s(Q^2) \sim 0.2\,\text{GeV}^2$, and clearly not yet asymptotic. The calculated time-like electromagnetic form factor displays vector meson bound state poles from which one can extract the strong coupling constants $g_{\rho\pi\pi}$ and $g_{\phi KK}$ respectively [12]. The results are given in Table 1 and are close to the experimental data and to the results from direct evaluation [20].

The impulse approximation for the $\gamma^\star\pi\gamma$ vertex with $\gamma^\star$ momentum Q is

$$\Lambda_{\mu\nu}(P,Q) = i\frac{\alpha}{\pi f_\pi}\,\epsilon_{\mu\nu\alpha\beta}\,P_\alpha Q_\beta\,g_{\pi\gamma\gamma}\,F_{\gamma^\star\pi\gamma}(Q^2) \tag{7}$$
$$= \frac{N_c}{3}\int\frac{d^4q}{(2\pi)^4}\text{Tr}\left[S(q)\,i\Gamma_\nu(q,q')\,S(q')\,i\Gamma_\mu(q',q'')\,S(q'')\,\Gamma_\pi(q'',q;P)\right]\,.$$

where the momenta follow from momentum conservation. In the chiral limit, the value at $Q^2 = 0$, corresponding to the decay $\pi^0 \to \gamma\gamma$, is given by the axial anomaly and its value $g^0_{\pi\gamma\gamma} = \frac{1}{2}$ is a direct consequence of only gauge invariance and chiral symmetry; this value is reproduced by our calculations [15] and corresponds well with the experimental width of 7.7 eV. In Fig. 3 (left) we show our results [13] with realistic quark masses, normalized to the experimental $g_{\pi\gamma\gamma}$. For such moderate Q^2, both the data and our DSE model are close to the

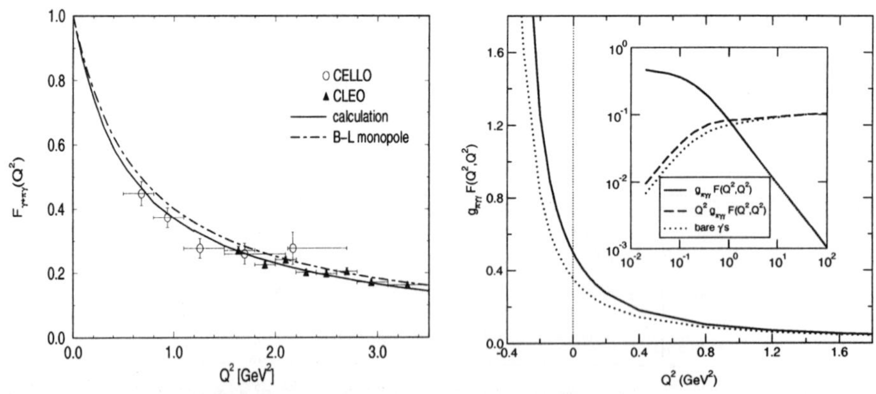

Figure 3. *Left Panel*: The $\gamma^\star\pi\gamma$ form factor [13], with data from CLEO and CELLO [21]. *Right Panel*: The form factor for $\gamma^\star\pi\gamma^\star$ for equal virtuality with the insert showing the asymptotic behavior of $Q^2F(Q^2)$.

monopole shape fitted to the Brodsky–Lepage asymptotic limit [22] $F \sim 8\pi^2 f_\pi^2$ from pQCD in the factorized approximation. As pointed out recently [23], it is advantageous to consider the process where both photons are virtual; the light-cone operator product expansion yields [22,24]

$$g_{\pi\gamma\gamma}\, F_{\pi\gamma\gamma}(Q_1^2, Q_2^2) \to 4\pi^2 f_\pi^2 \,\{ \frac{J(\omega)}{2(Q_1^2+Q_2^2)} + \mathcal{O}(\frac{\alpha_s}{\pi}, \frac{1}{Q^4}) \} \,. \tag{8}$$

Here the asymmetry is $\omega = (Q_1^2 - Q_2^2)/(Q_1^2 + Q_2^2)$, and the leading coefficient

$$J(\omega) = \frac{4}{3} \int_0^1 dx \, \frac{\phi_\pi(x)}{1 - \omega^2 (2x-1)^2} \,, \tag{9}$$

samples the pion bound state through the distribution amplitude $\phi_\pi(x)$. The normalization of the latter produces the model-independent result $J(0) = 4/3$ for the case of equal virtuality. The model-dependent $J(1)$ for the asymmetric case has received the estimates: 2 from use of $\phi_\pi^{\rm asym}(x)$ [22], 1.8 from a non-perturbative chiral quark model [23], 1.6 from a fit to CLEO data [23], and 4/3 from analysis of the impulse approximation [25,26]. A numerical confirmation of the asymptotic behavior within the present DSE model requires a level of detail in the representation of the analytic structure of dressed propagators and BS amplitudes that is not currently available. However, for equal virtuality there is no restriction, and our asymptotic result shown in Fig. 3 (right) produces $J(0) = 1.2$ which is within 10% of the 4/3 result of Eq. (9) from the light-cone operator product expansion.

Acknowledgments

I am grateful to the University of Adelaide's Special Research Centre for the Subatomic Structure of Matter for their hospitality and financial support. This work draws heavily upon the results of collaborations with Pieter Maris and Dennis Jarecke. I would also like to thank Craig Roberts, Tony Thomas, and Tony Williams for useful discussions. This work was partly funded by the National Science Foundation under grants Nos. PHY-0071361 and PHY-9722429, and benefited from the resources of the National Energy Research Scientific Computing Center and the Ohio Supercomputer Center.

References

1. M. Göckeler, R. Horsley, E. M. Ilgenfritz, H. Perlt, P. Rakow, G. Schierholz and A. Schiller, *Phys. Rev.* D **53**, 2317 (1996); G. Schierholz, these proceedings.

2. H. Weigel, E. Ruiz Arriola and L. Gamberg, *Nucl. Phys.* **560**, 383 (1999); E. Ruiz Arriola, these proceedings.
3. C. D. Roberts and S. M. Schmidt, *Prog. Part. Nucl. Phys.* **45**, S1:1 (2000).
4. R. Alkofer and L. von Smekal, *Phys. Rep.* (2001), in press, [hep-ph/0007355].
5. M. B. Hecht, C. D. Roberts and S. M. Schmidt, *Phys. Rev.* C **63**, 025213 (2001).
6. P. Maris and P. C. Tandy, *Phys. Rev.* C **62**, 055204 (2000)
7. C. Ji and P. Maris, in press, nucl-th/0102057.
8. P. Maris and C. D. Roberts, *Phys. Rev.* C **56**, 3369 (1997).
9. P. Maris and P. C. Tandy, *Phys. Rev.* C **60**, 055214 (1999).
10. A. Bender, C. D. Roberts and L. von Smekal, *Phys. Lett.* B **380**, 7 (1996).
11. P. Maris and P. C. Tandy, *Phys. Rev.* C **61**, 045202 (2000).
12. P. Maris and P.C. Tandy, KSUCNR-109-00, in progress, proceedings of the Confinement Research Program at the Erwin Schroedinger Institute for Mathematical Physics, Vienna, May-July, 2000.
13. P. Maris, *Nucl. Phys. Proc. Suppl.* **90**, 127 (2000).
14. Particle Data Group, C. Caso *et al.*, *Eur. Phys. J.* C **3**, 1 (1998).
15. P. Maris and C. D. Roberts, *Phys. Rev.* C **58**, 3659 (1998).
16. J. I. Skullerud and A. G. Williams, *Phys. Rev.* D **63**, 054508 (2001); J. I. Skullerud, D. B. Leinweber and A. G. Williams, hep-lat/0102013.
17. P. Maris, nucl-th/0009064, to appear in conference proceedings; C. D. Roberts, nucl-th/0007054, to appear in conference proceedings.
18. J. Volmer, PhD thesis, Amsterdam 2000; J. Volmer *et al.* [The Jefferson Lab F(pi) Collaboration], *Phys. Rev. Lett.* **86**, 1713 (2001)
19. G. R. Farrar and D. R. Jackson, *Phys. Rev. Lett.* **43**, 246 (1979).
20. D. W. Jarecke, P. Maris and P. C. Tandy, KSUCNR-101-01, in progress, (2001).
21. H. J. Behrend *et al.* [CELLO Collaboration], *Z. Phys.* C **49**, 401 (1991); J. Gronberg *et al.* [CLEO Collaboration], *Phys. Rev.* D **57**, 33 (1998).
22. G. P. Lepage and S. J. Brodsky, *Phys. Rev.* D **22**, 2157 (1980).
23. I. V. Anikin, A. E. Dorokhov and L. Tomio, *Phys. Lett.* B **475**, 361 (2000).
24. M. K. Chase, *Nucl. Phys.* B **167**, 125 (1980).
25. D. Kekez and D. Klabucar, *Phys. Lett.* B **457**, 359 (1999).
26. C. D. Roberts, *Fizika* B **8**, 285 (1999); P. C. Tandy, *Fizika* B **8**, 295 (1999); D. Klabucar and D. Kekez, *Fizika* B **8**, 303 (1999).

NUCLEON FORM FACTORS IN THE COVARIANT DIQUARK–QUARK MODEL

R. ALKOFER

Institute for Theoretical Physics, Tübingen University, Auf der Morgenstelle 14, D-72076 Tübingen, Germany, E-mail: Reinhard.Alkofer@uni-tuebingen.de

M. OETTEL

CSSM, Adelaide University, Adelaide, SA, 5005, Australia E-mail: moettel@physics.adelaide.edu.au

The nucleon is described as a bound state of a quark and an extended diquark. Hereby the notion "diquark" refers to the modelling of separable correlations in the two–quark Green's functions. Binding of quarks and diquarks takes place via an exchange interaction. Fully Poincaré covariant nucleon amplitudes are calculated for free constituent propagators as well as for dressed propagators which parameterise confinement. The corresponding results for space–like form factors differ quantitatively but not qualitatively for various *ansätze* for the propagators. These results do not allow one to draw definite conclusions on the permissibility of different dressing functions. Results for kaon photoproduction, on the other hand, exclude a whole class of constituent propagators.

1 Motivation

The central aim of the studies reported here is to develop a phenomenological QCD–based model for baryon structure. Recent experimental results emphasise the complicated nature of baryons. An example is provided by the ratio of the electric to the magnetic form factor of the proton measured at Jefferson Lab[1]: This ratio surprisingly decreases with increasing photon virtuality. We will see later on how this can be understood in our model.

Theoretical issues such as confinement, dynamical breaking of chiral symmetry and the formation of relativistic bound states can be understood and related to the properties of the non-perturbative propagators of QCD. In the Dyson–Schwinger approach we have obtained remarkable progress on this interrelation and other kindred questions during the last few years, see the recent reviews[2,3]. Amongst many important results we want to highlight the following: Using the general structure of the ghost Dyson–Schwinger equation in Landau gauge, Slavnov–Taylor identities and multiplicative renormalisibility one only needs the assumption that QCD Green's functions can be expanded in asymptotic series in the infrared to demonstrate confinement of transverse gluons[4,5]. Especially, the infrared behaviour of the gluon and ghost propagators are uniquely related. This prediction can be tested in future lattice calculations, such as those reported in ref.[6].

Corresponding studies of quark confinement are under way, and lattice results for the quark propagator[7] will serve to guide them. Since results are not available yet, and because studies of baryon properties require that one avoid unphysical thresholds a pragmatic way to proceed is via the parameterisation of quark propagators, i.e., by multiplying dressing functions to the quark propagators:

$$S^{(k)}(p) = \frac{i\not{p} - m_q}{p^2 + m_q^2} f_k \left(p^2/m_q^2\right). \tag{1}$$

The trivial dressing function $f_0 \equiv 1$ corresponds to the bare propagator. "Confining" ansätze include[8]

$$f_1(x) = \frac{1}{2}\left\{\frac{x+1}{x+1-i/d} + \frac{x+1}{x+1+i/d}\right\}, \tag{2}$$

$$f_2(x) = 1 - \exp\left[-d\left(1+x\right)\right], \tag{3}$$

$$f_3(x, x^*) = \tanh\left[d\left(1+x\right)\left(1+x^*\right)\right]. \tag{4}$$

The dressing functions f_k behave qualitatively differently for time-like momenta: $f_1(x)$ and $f_3(x)$ change sign (as in the case of a tree–level propagator) and the function $f_2(x)$, that models the exponential type propagator, increases drastically. Away from the real axis, the behaviour of the above model propagators can be read off easily from the definitions (2)–(4). By construction the propagators $S^{(1)}$ have complex conjugated poles at $p^2 = m^2(-1 \pm i/d)$ where m represents a parameter that would be interpreted as a mass if and only if the poles were on the real axis. The entire propagator $S^{(2)}$ oscillates for a nonzero imaginary part of p^2. The propagator $S^{(3)}$ is built such that its dressing function f_3 asymptotically approaches unity in all directions of the complex p^2 plane. For asymptotically large space-like momenta the three model propagators $S^{(k)}$, $(k = 1, 2, 3)$ match up with the bare propagator $S^{(0)}$.

Diquarks appear in many phenomenological models. The quantum numbers of one–gluon–exchange suggest that not only colour singlet $\bar{q}q$ pairs, i.e., mesons, are bound but also colour triplet qq pairs. Closer analysis shows though that in model Bethe–Salpeter equations beyond the ladder approximation[9,10,11] the absence of poles in the quark propagator implies the absence of poles in the colour triplet qq correlations as well. The latter correlations are the ones which we will call diquarks in the following. It is interesting to note that diquarks are seen in lattice calculations, see ref.[12]

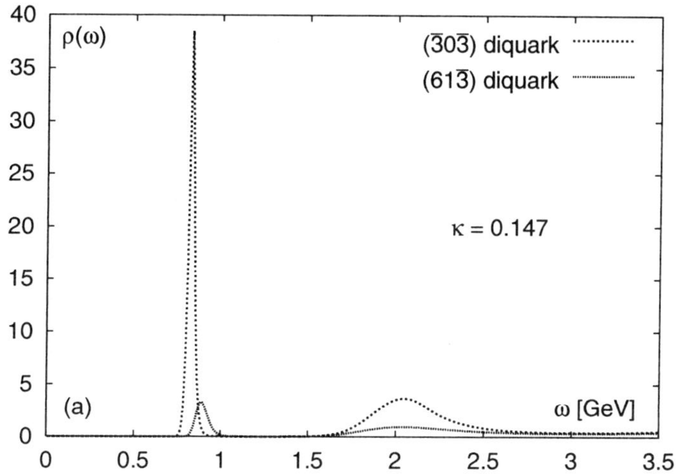

Figure 1. Spectral functions for diquarks from lattice data. $\bar{3}0\bar{3}$ denotes scalar diquarks, $61\bar{3}$ axialvector ones. (Adopted from ref.[12].)

and references therein. One might even be puzzled by the fact that the corresponding spectral functions for these confined objects are very similar to meson ones, as can be seen in fig. 1.

2 The relativistic Faddeev problem

Equipped with models for the quark propagator we investigate the relativistic three–quark problem. We neglect any three–particle irreducible interaction graphs between the quarks, which defines the well–known Faddeev problem. For the two–quark correlations we use diquarks in the scalar and axialvector channel as discussed above, i.e., we use a separable *ansatz* for the quark–quark t–matrix of the form

$$t^{\mathrm{sep}}_{\alpha\gamma,\beta\delta}(p,q,P) = \chi_{\gamma\alpha}(p)\, D(P)\, \bar{\chi}_{\beta\delta}(q) \; + \; \chi^{\mu}_{\gamma\alpha}(p)\, D^{\mu\nu}(P) \bar{\chi}^{\nu}_{\beta\delta}(q)\,. \qquad (5)$$

Here, P is the total momentum of the incoming and the outgoing quark-quark pair, p and q are the relative momenta between the quarks. $\chi_{\alpha\beta}(p)$ and $\chi^{\mu}_{\alpha\beta}(p)$ are vertex functions of quarks with a scalar and an axialvector diquark, respectively. We parameterise the finite size of the diquark vertices by a dipole form. We take the associated width parameter, that directly influences the proton electric radius, to be of the order 300–400 MeV. The inclusion of axialvector

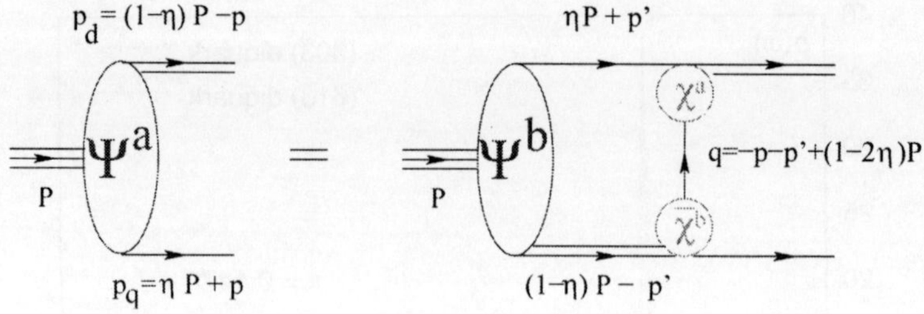

Figure 2. The Bethe–Salpeter equation for the spinorial baryon-quark-diquark wave function Ψ.

diquarks is the minimal requirement to describe decuplet baryons and, see below, is crucial for describing the nucleon electromagnetic form factors correctly. The diquark propagators $D^{[\mu\nu]}$ are taken to be free propagators of a spin–0 [1] particle multiplied by the dressing functions defined in eqs. (2) - (4).

Having imposed the separable *ansatz* (5) the Faddeev equations reduce to a coupled system of Bethe–Salpeter equations describing baryons as bound states of quarks and diquarks which interact by quark exchange. This interaction is by virtue of the colour degree of freedom attractive. For the nucleon these equations read

$$\int \frac{d^4p'}{(2\pi)^4} K(p,p',P) \begin{pmatrix} \Psi^5 \\ \Psi^{\mu'} \end{pmatrix} (p',P) = 0. \tag{6}$$

The interaction part of the kernel K,

$$K(p,p',P) = (2\pi)^4\, \delta(p-p')\, S^{-1}(p_q) \begin{pmatrix} D^{-1} & 0 \\ 0 & (D^{-1})^{\mu'\mu} \end{pmatrix} (p_d) + \\ \frac{1}{2} \begin{pmatrix} \chi\, S^T(q)\, \bar{\chi} & -\sqrt{3}\chi^{\mu'}\, S^T(q)\, \bar{\chi} \\ -\sqrt{3}\chi\, S^T(q)\, \bar{\chi}^{\mu} & -\chi^{\mu'}\, S^T(q)\, \bar{\chi}^{\mu} \end{pmatrix}, \tag{7}$$

is given by the quark exchange, for the definition of the momenta involved see Fig. 2. We have solved these equations without further reduction and thus obtained fully Poincaré covariant spinorial wave functions $\Psi^{[\mu]}$.

3 Electromagnetic form factors

In a study employing free quark and diquark propagators[14] we calculated the nucleon electromagnetic form factors. Gauge invariance and correct charge and norm were guaranteed by coupling the photon to all possible places in

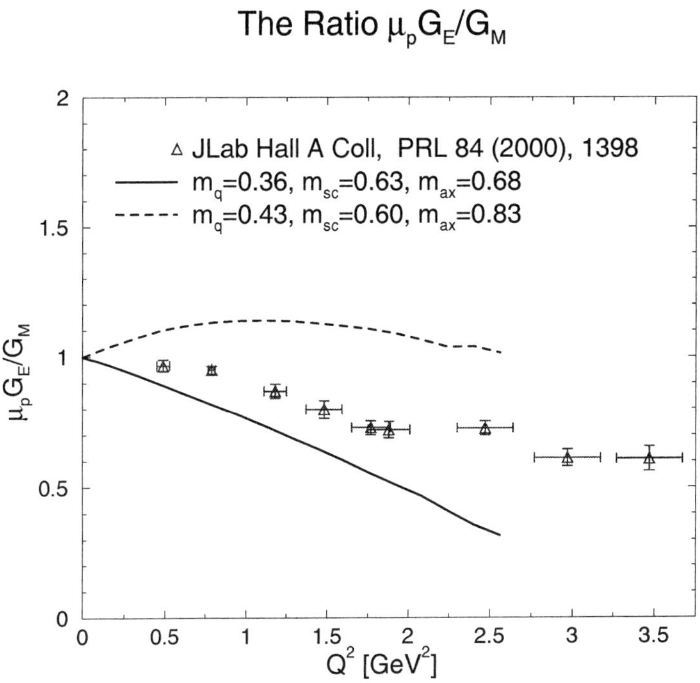

Figure 3. Ratio of electric to magnetic form factor of the proton as calculated in our model with two parameter sets. m_q, m_{sc} and m_{ax} are the masses of quark, scalar and axialvector diquarks, respectively (in GeV). The data are taken from ref.[1].

the kernel of the Bethe–Salpeter equation given in eq. (7) [15]. The results for the electric form factors (up to momentum transfers of 2.5 GeV^2) are in good agreement with the experimental data, nevertheless it turned out to be impossible to obtain a simultaneous, correct description of the nucleon magnetic moments and the mass of the Δ isobar. Due to the free particle thresholds $m_q > 411$ MeV had to be chosen to obtain a bound Δ and these constituent quark masses yielded proton magnetic moments $\mu_p \approx 1.9$. For lower masses $m_q = 360$ MeV we found $\mu_p \approx 2.5$, thus illustrating the necessity to avoid the free–particle poles for the quarks. We found that 20–25 % axialvector correlations (measured by the ratio of the norm contributions stemming from Ψ and Ψ^μ) are needed to describe the ratio of neutron to proton magnetic moments. This conclusion is confirmed when looking at the ratio of electric to magnetic form factors of the proton, see fig. 3: The parameter set with 30 %

Table 1. Nucleon magnetic moments and e.m. radii with different confinement parameterisations. The nucleon mass is used as input. The parameters are further constrained by describing approximately the spectrum of octet hyperons (c.f. ref.[13]) as given here.

	Expon.	c.c.poles	Non-anal.	Exp.
Λ [GeV]	1.13	1.13	1.12	1.12
Σ [GeV]	1.30	1.22	1.21	1.19
Ξ [GeV]	1.37	1.37	1.33	1.32
μ_p	2.83	2.70	2.33	2.79
μ_n	−2.37	−2.08	−1.82	−1.91
$(r_{el})_p$ [fm]	0.81	0.78	0.74	0.84
$(r_m)_p$ [fm]	1.06	0.83	0.77	0.84

axialvector correlations explains the data semi-quantitatively, while the one with very small axialvector correlations does not (for more details see ref.[14]).

It is required by Ward identities to use dressed vertex functions when employing the effective parameterisations of confinement (2) - (4), for details see ref.[8]. The resulting nucleon magnetic moments are given in table 1: Their values have increased as compared to ones calculated with free propagators, and they are as close to the experimental values as can be reasonably expected. Please note that our model does not contain effects like the pion cloud, which certainly does give a contribution to these quantities.

We would like to mention that strong and weak form factors have been calculated within this approach using free propagators[14] and confined ones[8] (see also ref. [16] for similar calculations). Given all the calculations for static quantities and space-like form factors one concludes that the use of confined propagators leads to better results. However, they do not allow one to distinguish between these different parameterisations. Therefore we have to test the model propagators at sufficiently large time–like momenta.

4 Kaon photoproduction

Kaon photoproduction is a comparatively simple process in our model because flavour algebra and parity dictates that in impulse approximation only diagrams with scalar diquarks contribute, for a detailed discussions see ref.[8]. Comparison with data reveals that the exponential propagator, see eq. (3),

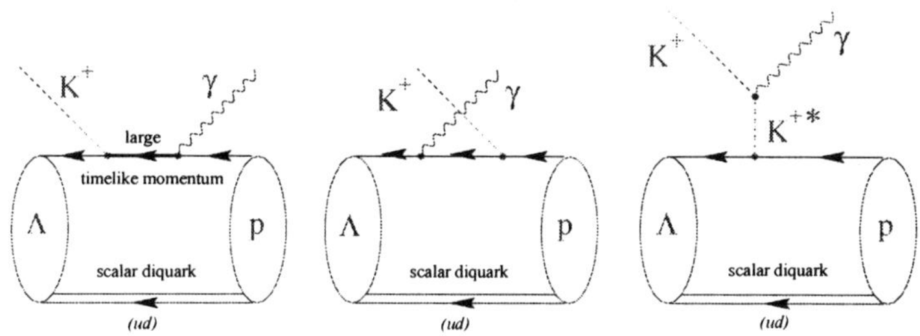

Figure 4. The three diagrams contributing mainly to kaon photoproduction in our model. The right one tests the quark propagator at large time–like momenta. The left one is a purely hadronic contribution.

provides cross sections which are in disagreement with data by orders of magnitude. Hereby the left-most diagram in fig. 4 overwhelms all other contributions. For the other two parametrisations the purely hadronic contribution, the right-most diagram in fig. 4, dominates. Given the theoretical problems with non–analytic propagators[8], the one with complex–conjugated poles (see eq. (2)), provides therefore the best fit to experiment.

5 Summary and Outlook

We have described a model for baryons respecting full Poincaré invariance. This is possible by assuming separability in the quark–quark t–matrix. This provides us with an effective definition for extended diquarks. Baryons are then described by the solutions of a Bethe–Salpeter equation. The binding mechanism is quark exchange: Colour antisymmetry means that it leads to an attractive interaction.

We have modelled quark and diquark confinement via a parameterisation of quark and diquark propagators. This has improved the results for static quantities and space–like form factors of the nucleon. However, these observables have not allowed to discriminate between different propagator types. This has been possible by considering kaon photoproduction[8]. According to our analysis the propagators with complex conjugate poles are best suited for future investigations. Further insight into permissible dressing mechanisms will come from studies of nucleon structure functions which are under way[17].

Acknowledgements

We thank Wally Melnitchouk, Andreas Schreiber, Peter Tandy, and Tony Thomas for organising this highly interesting workshop.

We are grateful to Steven Ahlig, Christian Fischer, Hugo Reinhardt, Craig Roberts, Sebastian Schmidt, Lorenz von Smekal and Herbert Weigel for their collaboration on related issues in this model and helpful discussions. R.A. thanks Sebastian Schmidt for support.

The work reported here has been supported by a Feodor–Lynen fellowship of the Alexander-von-Humboldt foundation for M.O., by the DFG under contract nos. We 1254/4-2 and Schm 1342/3-1, by COSY under contract no. 4137660 and by the Australian Research Council.

References

1. M. K. Jones *et al.*, *Phys. Rev. Lett.* **84**, 1398 (2000).
2. R. Alkofer and L. v. Smekal, *Phys. Rep., in press* [hep-ph/0007355].
3. C. D. Roberts and S. M. Schmidt, *Prog. Part. Nucl.Phys.* **45**, S1 (2000).
4. P. Watson and R. Alkofer, *Phys. Rev. Lett.* **86**, 5239 (2001).
5. R. Alkofer, L. v. Smekal and P. Watson in *Dynamical aspects of the QCD phase transition*, eds. D. Blaschke, G. Burau und S. M. Schmidt (University Press Rostock 2001) [hep-ph/0105142].
6. F. Bonnet et al., *Phys. Rev.* D **62**, 051501 (2000).
7. J. I. Skullerud and A. G. Williams, *Phys. Rev.* D **63**, 054508 (2001).
8. S. Ahlig et al., *Phys. Rev.* D **64**, 014004 (2001).
9. A. Bender, C. D. Roberts and L. v. Smekal, *Phys. Lett.* B **380**, 7 (1996).
10. G. Hellstern, R. Alkofer and H. Reinhardt, *Nucl. Phys.* A **625**, 697 (1997).
11. J. C. Bloch, C. D. Roberts and S. M. Schmidt, *Phys. Rev.* C **60**, 065208 (1999).
12. I. Wetzorke and F. Karsch, [hep-lat/0008008].
13. M. Oettel et al., *Phys. Rev.* C **58**, 2459 (1998).
14. M. Oettel, R. Alkofer and L. v. Smekal, *Eur. Phys. J.* A **8**, 553 (2000). *see also:* M. Oettel and R. Alkofer, *Phys. Lett.* B **484**, 243 (2000). G. Hellstern et al., *Nucl. Phys.* A **627**, 679 (1997).
15. M. Oettel, M. Pichowsky and L. v. Smekal, EPJA **8**, 251 (2000).
16. J. C. Bloch, C. D. Roberts and S. M. Schmidt, *Phys. Rev.* C **61**, 065207 (2000).
17. C. D. Roberts, these proceedings; J. C. Bloch, C. D. Roberts and S. M. Schmidt, in preparation.

5. Hadron Excitations, Confinement and Chiral Symmetry Breaking

EXPERIMENTAL STUDIES OF THE HADRON SPECTRUM

JIM NAPOLITANO
Department of Physics
Rensselaer Polytechnic Institute
Troy, NY 12180-3590, USA
E-mail: napolj@rpi.edu

This talk reviews some recent results and future plans for measurements of the hadron spectrum. These measurements emphasize the mass and width of resonant states with particular quantum numbers. We also address the decay dynamics of some of these states. Shortcomings of the presently existing data set are pointed out, and experiments in preparation are discussed.

1 Introduction

This workshop has concentrated on lepton scattering in general, and theory in particular. This talk will focus on neither, and I will bring (I hope) a different perspective on how to study QCD and hadrons. Specifically, I will discuss hadrons and their excitations, and how spectroscopy can be used to learn more about Strong QCD, that is, QCD in the non-perturbative, low energy regime.

Simply put, spectroscopy experiments seek to identify the appropriate degrees of freedom needed to determine hadron structure. The quark model has been extremely successful, but we know that QCD predicts degrees of freedom beyond quarks. Can we observe gluonic degrees of freedom, for example? How would these manifest themselves in terms of the excitation spectrum and also in the strong decays of hadrons?

Consequently, we identify three general goals of hadron spectroscopy.

- Identify the physical states and their quantum numbers, and measure their **mass and width**.
- Determine their **decay modes** and branching ratios.
- Study the underlying **dynamics** of production and decay.

I will try to give specific examples of each of these in this talk.

These goals confront some significant challenges. In very many reactions of interest, there are many broad, overlapping, interfering states, and it can be difficult to unravel the various contributing quantum numbers. (This is particularly true for baryon spectroscopy.) In fact, it is not uncommon to have states with different quantum numbers in exactly the same region of mass. Furthermore, many important states have complicated, many-particle final state decays.

Partial Wave Analysis (PWA) techniques are used to pursue these goals and to attack the challenges. Briefly, the multiparticle phase space is expanded into a (truncated) basis of angular momentum functions which is intended to describe the reaction matrix element and thereby describe the angular distribution. An extended maximum likelihood technique determines the best fit values of the free parameters consistent with the experimental acceptance (simulated in a Monte Carlo program). For the most part, multiparticle final states are described in the "isobar model" in which all decays are quasi-two body, with subsequent decays of the daughter particles.

An example of this procedure applied to meson spectroscopy, from experiment E852 at Brookhaven National Laboratory, concerns the reaction[1]

$$\pi^- p \to \pi^- \pi^+ \pi^- p \tag{1}$$

carried out with an 18 GeV/c pion beam incident on a liquid hydrogen target. The 3π mesonic final state was represented by the quasi-two-body decays $\rho\pi$, $f_2(1270)\pi$, and $(\pi\pi)_S\pi$, where various parameterizations of the $\pi\pi$ S-wave were compared. The PWA of this reaction, to which we will return in this talk, is shown in the upper panel of Figure 1. (This figure also demonstrates the need for data in photoproduction reactions, which will discuss at the end of the talk.) The 3π invariant mass spectrum is clearly dominated by three quantum numbers, namely $J^{PC} = 1^{++}$, $J^{PC} = 2^{++}$, and $J^{PC} = 2^{-+}$, i.e. the $a_1(1260)$, $a_2(1320)$, and $\pi_2(1670)$ meson excitations. A comparison of the (complex) amplitudes for these quantum numbers demonstrates that they are indeed "resonant", consistent with Breit-Wigner functions.

These techniques can be applied to a number of different reactions. This talk will concentrate on diffractive t-channel reactions (such as reaction 1). Other reactions used to study meson excitations include annihilation reactions (e.g. $\bar{p}p \to n\pi$), "central" production, and J/ψ radiative decay[2].

One important area we will *not* cover is the spectroscopy of excited *baryons*. Several mysteries remain since the early days, including the "missing baryons" problem and the nature of the $P11(1440)$ "Roper" resonance. Until very recently, nearly all the high quality data has been from reactions which rely on the isolated $N\pi$ system in the initial or final state (or both). The advent of high quality photon beams at Jefferson Laboratory and at the University of Bonn have given us data on reactions such as $\gamma p \to p\pi^+\pi^-$ and $\gamma p \to p\pi^0\pi^0$. Partial Wave Analyses of these reactions are currently underway, and results should be forthcoming in the next year.

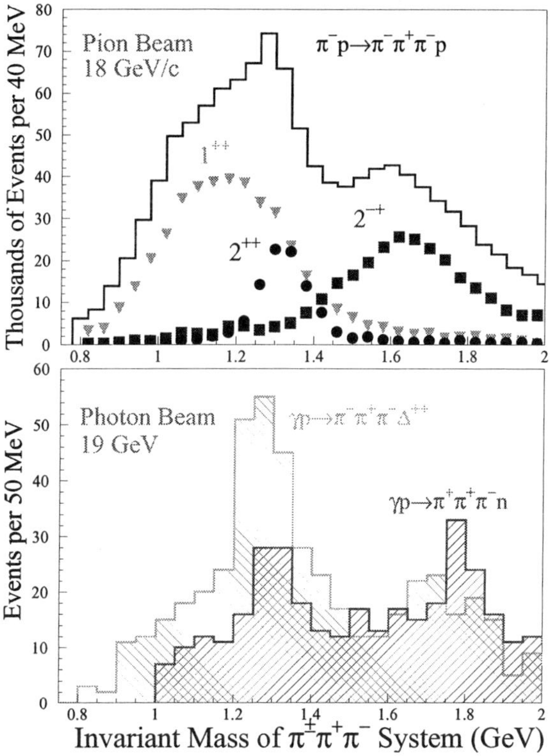

Figure 1. Production of the $\pi^{\pm}\pi^{+}\pi^{-}$ system with pion and photon beams. Pion production, upper panel, is dominated by three meson excitations, namely $a_1(1260)$, $a_2(1320)$, and $\pi_2(1670)$. It also shows a very large number of events, and this allows the quantum numbers of these excitations to be identified using partial wave analysis. Photoproduction, with either positive or negative charge exchange, are seen to be different from pion production, even with severely reduced statistics. Furthermore, they are different from each other. Both reactions seem to show production of $a_2(1320)$ with little indication of the other two states seen in pion production. The reaction $\gamma p \to \pi^{+}\pi^{+}\pi^{-}n$ suggests a new narrow state at ≈ 1775 MeV.

2 New Isoscalar States

Experiment E852 was able to study final states with many photons, and where high statistics were needed to elucidate new states. One example is the spectrum of mesons which decay to $\omega\eta$ as observed in the reaction

$$\pi^{-}p \to \omega\eta n \to \pi^{+}\pi^{-}\pi^{0}\eta n \tag{2}$$

where the π^0 and η are each detected in their decays to two photons, leading to a four photon final state. Such states are necessarily isoscalar, produced by some isovector t-channel exchange process.

E852 has observed[3] the $J^{PC} = 1^{--}$ $\omega(1650)$ in this reaction, as well as a new isoscalar state, the $J^{PC} = 1^{+-}$ $h_1(1595)$, which may be either a radial excitation of the $h_1(1170)$ or perhaps a mixture with states of "hybrid" degrees of freedom. The decay matrix element of the ω leads to a distinctive

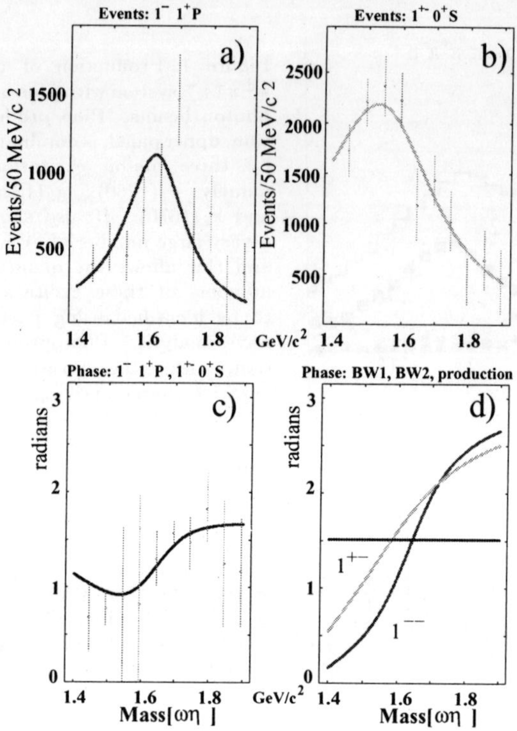

Figure 2. Results of a partial wave analysis (PWA) for the $\omega\eta$ system, produced in the reaction $\pi^- p \to \omega\eta n$ at 18 GeV/c. Panels (a) and (b) show the intensities for two particular partial waves, for $J^{PC} = 1^{--}$ and 1^{+-} respectively. The intensities are fit to Breit-Wigner resonance shapes. Panel (c) shows the phase difference between these two amplitudes, along with the (simultaneous) fit to the same Breit-Wigner parameters as in (a) and (b). Panel (d) shows the phase dependences separately, including a (nearly flat) "production" phase that is included in the fit.

correlation between the three daughter pions, and this was used in the PWA to isolate the omega signal from the underlying combinatoric background.

The results of the PWA are shown in Fig. 2. The maximum likelihood fit is performed independently in each 50 MeV/c^2 bin of $\omega\eta$ invariant mass, and the result is plotted as a point with an error bar. A relatively large number of waves are included in the fit, but Fig. 2 includes only the $J^{PC} = 1^{--}$ wave arising from a P-wave ($l = 1$) $\omega\eta$ combination produced with $m = 1$ and with positive naturality; and the $J^{PC} = 1^{+-}$ wave arising from an S-wave combination, also with positive naturality.

Figure 2 also shows the result of a mass-dependent fit to two Breit-Wigner curves which yield the parameters of the $\omega(1650)$ and $h_1(1595)$ resonances. The fit is carried out simultaneously with the intensities of the two partial waves (upper panels) and the phase difference between the two (complex) amplitudes (lower left). The bottom right shows the two Breit-Wigner phase curves independently, along with the "production" phase which is also fit. For this example, the production phase is a constant, nearly independent of mass.

3 Mesons with Exotic Quantum Numbers

Observation of "exotic" quantum numbers, that is, quantum numbers that cannot arise from a $q\bar{q}$ description, is unequivocal evidence for degrees of freedom beyond the quark model. Consequently, such states have been sought for quite some time. Indeed, one of the early strong successes for the quark model was that these states were *not* observed, and it has in fact proven very difficult to isolate them unambiguously.

E852 has reported evidence for two exotic $J^{PC} = 1^{-+}$ states, namely[a] the $\hat{\rho}(1400)$ and the $\hat{\rho}(1600)$. Figure 3 shows this evidence for[4] $\hat{\rho}(1400) \to \eta\pi$ and for[1] $\hat{\rho}(1600) \to \rho\pi$. *In both cases these are small signals underneath much larger conventional meson excitations, and there has been some controversy.* However, the $\hat{\rho}(1400)$ has been observed by the Crystal Barrel[5] in the reaction $\bar{p}n \to \eta\pi^-\pi^0$, and E852 has confirmed[6] the $\hat{\rho}(1600)$ decaying to $\eta'\pi^-$ where it accounts for roughly half of the observed cross section.

4 Decay of the $b_1(1235)$ Meson

Models of QCD must reflect the underlying field theory, and this leads to some mechanism for quark (and gluon) confinement. Strong decays of mesons come about through the creation of $q\bar{q}$ pairs, and these are manifestations of confinement in that they represent the mechanism by which the original $q\bar{q}$ are separated to large distances. Consequently, studies of strong decays of mesons provide some insight to the underlying dynamics of QCD and confinement.

Experiment E852 has carried out a detailed measurement[7] of one such decay, namely $b_1(1235) \to \omega\pi$ produced in the reaction

$$\pi^- p \to \omega\pi^- p \to \pi^-\pi^+\pi^0\pi^- p \tag{3}$$

with $\pi^0 \to \gamma\gamma$. In particular, the measurement focused on the relative D- and S-wave amplitudes for the final state $\omega\pi$. The experiment determined the relative magnitudes of these two amplitudes, as well as their relative phases. Previous measurements of the magnitudes led to inconsistent results. The relative phase had never been measured before.

[a] Two naming conventions for states with exotic quantum numbers have appeared in the literature. We prefer using the "hat" to signify a state with the "wrong" C-parity. A different choice is to use a symbol identified by I, P, and C, and subscript the spin. Hence an isovector $J^{PC} = 1^{-+}$ state may be called either a $\hat{\rho}$ or a π_1.

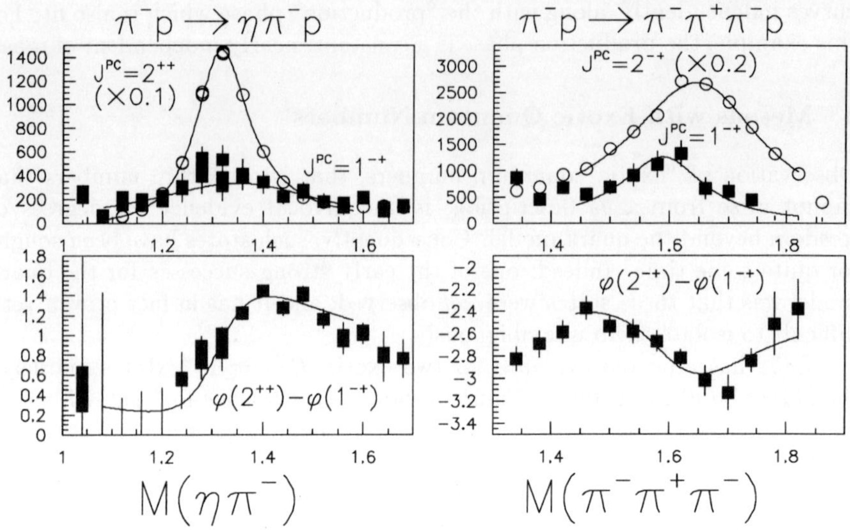

Figure 3. Evidence for mesons with exotic quantum numbers $J^{PC} = 1^{-+}$ from BNL experiment E852. The left side shows evidence for a state at 1400 MeV, decaying to $\eta\pi^-$. The right side, a state at 1600 MeV, decaying to $\pi^-\pi^+\pi^-$. In each case, the top plot compares the exotic wave with a more dominant conventional meson excitation, the $a_2(1320)$ on the left and the $\pi_2(1670)$ on the right. The bottom plot shows the phase motion between the two waves. The solid lines are simultaneous fits to Breit-Wigner resonance shapes for the intensity and the phase motion for the exotic and conventional state in each case.

Calculations of $b_1 \to \omega\pi$ (and other axial vector decays to pseudoscalar+vector systems) have been carried out in the quark model[8] and in the framework of Dyson-Schwinger equations[9]. The results vary depending on a number of adjustable parameters. The relative phase of the D- and S-wave amplitudes is nonzero due to final state interactions, and these have been calculated as well[10].

E852 carried out a partial wave analysis of Reaction 3. They used a technique which fixed the relative magnitude and phase of the D- and S-wave amplitudes in the various relevant partial waves, followed by a grid search to determine the best values of the magnitude ratio and phase and phase difference along with errors. Systematic errors were studied by including different combinations of partial waves, as well as the background beneath the $\omega \to \pi^-\pi^+\pi^0$ signal. They found $|D/S| = 0.275 \pm 0.016 \pm 0.044$ and $\phi(D-S) = 0.118 \pm 0.057 \pm 0.050$ where the first error is statistical and the second is systematic. These values compare well with predictions of the amplitude ratio, and reasonably well with predictions of the phase difference.

5 Meson Photoproduction and Hall D

Although hadron spectroscopy has been an active field for more than three decades, only in the past few years has clear evidence emerged for states beyond the quark model. Technological advances have allowed us to build sophisticated experiments, and to accumulate and analyze large data volumes. These are needed for the sort of detailed partial wave analyses carried out by E852 and the Crystal Barrel, among others.

Most of the relevant spectroscopic data has come from peripheral pion scattering. In these reactions, states are produced via excitation of the (spin-0) pion. It is natural to expect a very different tower of excitations in peripheral reactions of a spin-1 hadronic beam. The photon provides such a beam, since its high energy peripheral hadronic reactions proceed through vector dominance. That is, peripheral photoproduction should provide excitations built on (spin-1) ρ, ω, or ϕ ground states.

Refer again to Fig. 1, where we have already discussed the results from a pion beam shown in the upper plot. The lower is for a photon beam of about the same energy[11]. An immediate conclusion is that the tower of excitations in peripheral photoproduction is indeed very different from that obtained with a pion beam. This is obvious simply from the shape of the invariant mass distributions. What appears to be the $a_2(1320)$ dominates the lower masses in photoproduction but is just one relatively small component in pion production. Furthermore, there is evidence for a relatively narrow high mass structure in photoproduction, but no such structure at all using the pion beam.

Of course, there is another major difference between the top and bottom plots in Fig. 1. There are more than three orders of magnitude more events for the experiment done with the pion beam. This results directly from the much lower intensity photon beams that have been available, as well as the inherently lower cross section for photoproduction. Furthermore, this large number of events is necessary for a thorough partial wave analysis, such as the one which is able to decompose the major contributing excited states to the upper plot in Fig. 1. Indeed, it is this superior statistical precision that allowed E852 to reliably extract the weak exotic signal shown on the right panel of Fig. 3.

A new experiment is being mounted at Jefferson Laboratory to study meson spectroscopy using peripheral photoproduction. This experiment, dubbed the "Hall D Project"[12,13], will use a linearly polarized photon beam with energy in the range 8 to 9 GeV, produced using coherent bremsstrahlung from the 12 GeV electron accelerator beam. A significant aspect of the project is the accelerator upgrade from its current beam energy maximum of 5.5 GeV. The photon beam intensity will exceed those previously available by more than four orders of magnitude, and the experimental facility is especially designed for large and continuous detector acceptance and efficiency, incorporating both charged and neutral particle detection.

Acknowledgments

Many thanks to the organizers for an outstanding conference. I was privileged to have insightful conversations with several people including Tim Londergan, Craig Roberts, Adam Szczepaniak, and Peter Tandy. I would also like to thank my colleagues on E852, in particular Gary Adams, John Cummings, Alex Dzierba, Mina Nozar, and Dennis Weygand. This work was supported by the U.S. National Science Foundation.

References

1. G.S. Adams, *et al*, *Phys. Rev. Lett.* **81**, 5760 (1998); S. Ostrovidov, *et al*, *Phys. Rev.* D, to be published.
2. S. Godfrey and J. Napolitano, *Rev. Mod. Phys.* **71**, 1411 (1999).
3. P. Eugenio, *et al*, *Phys. Lett.* B **497**, 190 (2001).
4. D.R. Thompson, *et al*, *Phys. Rev. Lett.* **79**, 1630 (1997); S.U. Chung, *et al*, *Phys. Rev.* D **60**, 092001 (1999).
5. A. Abele, *et al*, *Phys. Lett.* B **423**, 175 (1998).
6. E.V. Ivanov, *et al*, *Phys. Rev. Lett.* **86**, 3977 (2001).
7. Mina Nozar, *An Analysis of the* $\omega\pi^-$ *System in the Reaction* $\pi^- p \to \pi^-\pi^+\pi^0\pi^- p$ *at 18 GeV/c*, Ph.D. Thesis, Rensselaer Polytechnic Institute (2000).
8. E.S. Ackleh, *et al*, *Phys. Rev.* D **54**, 6811 (1996).
9. J.C.R. Bloch, *et al*, *Phys. Rev.* D **60**, 111502 (1999).
10. T. Barnes, *et al*, *Phys. Rev.* D **63**, 025204 (2001).
11. G. Condo, *et al.*, *Phys. Rev.* D **41**, 3317 (1990); G. Condo, *et al.*, *Phys. Rev.* D **48**, 3045 (1993); G. Condo, *et al.*, *Phys. Rev.* D **43**, 2787 (1991).
12. Dzierba, A., Meyer, C., and Swanson, E., *American Scientist* **88**, 406 (2000).
13. Dzierba, A., and Isgur, N., *CERN Courier* **September** (2000).

THE CHARACTER OF GOLDSTONE BOSONS

M.B. HECHT,*C.D. ROBERTS* AND S.M. SCHMIDT†

Physics Division, Argonne National Laboratory, Argonne IL 60439, USA

†*Institut für Theoretische Physik, Universität Tübingen, D-72076 Tübingen, Germany*

A succinct review of the QCD gap equation and dynamical chiral symmetry breaking; their connection with Bethe-Salpeter equations and resolving the dichotomous nature of the pion; the calculation of the pion's valence-quark distribution; and first results for the π-exchange contribution to the $\gamma N \to \omega N$ cross-section, which is important in the search for missing nucleon resonances.

1 Introduction

In the strong interaction spectrum the pion is identified as both a Goldstone mode, associated with Dynamical Chiral Symmetry Breaking (DCSB), and a bound state composed of u- and d-quarks. This dichotomy is remarkable because, while $m_\rho/2 \sim m_N/3 \sim 350\,\text{MeV} =: M_q$, the constituent quark mass, $m_\pi/2$ is only $\approx 0.2 M_q$; i.e., the pion is much lighter than comparable bound states. M_q is the "effective mass" of quarks in a hadron; and the ratio of this to the renormalisation-group-invariant u- and d-current-quark masses ($\hat{m} \sim 10\,\text{MeV}$) indicates the magnitude of the effect of nonperturbative dressing on light-quark propagation characteristics. The peculiar nature of the pion can be expressed by the question: "How does one form an almost-massless bound state from very massive constituents *without* fine-tuning?" Answering this question; i.e., resolving and understanding the dichotomy of the pion, is a key to unravelling the quark-gluon substructure of QCD's Goldstone mode.

2 QCD's Gap Equation

The insightful study of dynamical symmetry breaking can be pursued using a *gap equation*, which in QCD is the Dyson-Schwinger equation (DSE)[1,2] for the dressed-quark propagator:

$$S(p)^{-1} = Z_2\,(i\gamma\cdot p + m_{\rm bare}) + Z_1 \int_q^\Lambda g^2 D_{\mu\nu}(p-q)\frac{\lambda^a}{2}\gamma_\mu S(q)\Gamma^a_\nu(q,p)\,. \quad (1)$$

In this equation: $D_{\mu\nu}(k)$ is the renormalised dressed-gluon propagator; $\Gamma^a_\nu(q;p)$ is the renormalised dressed-quark-gluon vertex; $m_{\rm bare}$ is the Λ-dependent current-quark bare mass that appears in the Lagrangian; and

$\int_q^\Lambda := \int^\Lambda d^4q/(2\pi)^4$ represents mnemonically a *translationally-invariant* regularisation of the integral, with Λ the regularisation mass-scale. In Eq. (1), $Z_1(\zeta^2,\Lambda^2)$ and $Z_2(\zeta^2,\Lambda^2)$ are the quark-gluon-vertex and quark wave function renormalisation constants, which depend on the renormalisation point, ζ, and the regularisation mass-scale, as does the mass renormalisation constant

$$Z_m(\zeta^2,\Lambda^2) = Z_4(\zeta^2,\Lambda^2)/Z_2(\zeta^2,\Lambda^2), \tag{2}$$

with the renormalised mass given by $m(\zeta) := m_{\rm bare}(\Lambda)/Z_m(\zeta^2,\Lambda^2)$.

The solution of Eq. (1) has the form

$$S(p)^{-1} = i\gamma\cdot p\, A(p^2,\zeta^2) + B(p^2,\zeta^2) = \frac{1}{Z(p^2,\zeta^2)}\left[i\gamma\cdot p + M(p^2,\zeta^2)\right], \tag{3}$$

where the functions $A(p^2,\zeta^2)$, $B(p^2,\zeta^2)$ express the effects of quark-dressing induced by the quark's interaction with its own gluon field. Equation (1) must be solved subject to a renormalisation [boundary] condition, and in QCD it is practical to impose the requirement that at a large spacelike ζ^2

$$\left.S(p)^{-1}\right|_{p^2=\zeta^2} = i\gamma\cdot p + m(\zeta)\,. \tag{4}$$

The dressed-quark mass function: $M(p^2,\zeta^2) = B(p^2,\zeta^2)/A(p^2,\zeta^2)$, is actually independent of the renormalisation point; i.e., it is a function only of $p^2/\Lambda^2_{\rm QCD}$. At one loop order in perturbation theory the running mass

$$m(\zeta) = M(\zeta^2) = \frac{\hat{m}}{\left(\frac{1}{2}\ln\left[\zeta^2/\Lambda^2_{\rm QCD}\right]\right)^{\gamma_m}}\,,\; \gamma_m = 12/(33-2N_f)\,, \tag{5}$$

where $\hat{m}$ is the renormalisation point independent current-quark mass, and the mass renormalisation constant in Eq. (2) is $Z_m(\zeta^2,\Lambda^2) = \left[\alpha(\Lambda^2)/\alpha(\zeta^2)\right]^{\gamma_m}$. (At one-loop in Landau-gauge, $Z_2 \equiv 1$.)

The chiral limit is unambiguously defined by[3] $\hat{m} = 0$. In this case there is no perturbative contribution to the scalar piece of the quark self energy; i.e., $B(p^2,\zeta^2) \equiv 0$ at every order in perturbation theory.

3 Dynamical Chiral Symmetry Breaking

DCSB is the appearance of a $B(p^2,\zeta^2) \not\equiv 0$ solution of Eq. (1) in the chiral limit, which is impossible in perturbation theory. Hence here, as in all studies of dynamical symmetry breaking, a nonperturbative analysis of the gap equation is required. For that one needs a systematic, symmetry-preserving truncation scheme, which goes beyond the weak coupling expansion that yields perturbation theory.

The leading term in one such scheme[4] is the renormalisation-group-improved rainbow approximation to the gap equation:[5]

$$S(p)^{-1} = Z_2\,(i\gamma\cdot p + m_{\rm bare}) + \int_q^{\Lambda} \mathcal{G}((p-q)^2)\, D_{\mu\nu}^{\rm free}(p-q)\frac{\lambda^a}{2}\gamma_\mu S(q)\frac{\lambda^a}{2}\gamma_\nu\,, \tag{6}$$

where $D_{\mu\nu}^{\rm free}(k)$ is the free gauge boson propagator and $\mathcal{G}(k^2)$ is an "effective coupling". In QCD $\mathcal{G}(k^2) = 4\pi\alpha(k^2)$ for $k^2 \gtrsim 1\,{\rm GeV}^2$, where $\alpha(k^2)$ is the strong running coupling constant. However, the behaviour of $\mathcal{G}(k^2)$ for $k^2 < 1\,{\rm GeV}^2$; i.e., at infrared length-scales ($\gtrsim 0.2\,{\rm fm}$), is currently unknown.

The effective coupling is a phenomenological mnemonic for the contracted product of the dressed-gluon propagator and dressed-quark-gluon vertex, which forms the kernel in Eq. (1). The infrared strength of the interaction can then be characterised by the interaction-tension[6]

$$\sigma^\Delta := \frac{1}{4\pi}\int_{\Lambda_{\rm QCD}^2}^{\Lambda_{\rm pQCD}^2} dk^2\, k^2\,\left[\Delta(k^2) - \Delta(\Lambda_{\rm pQCD}^2)\right] \tag{7}$$

with $\Delta(k^2) = \mathcal{G}(k^2)/k^2$, and $\Lambda_{\rm pQCD} = 10\,\Lambda_{\rm QCD}$, the boundary above which perturbation theory is unquestionably valid.

Studies show[7] that DCSB does not occur for $\sigma^\Delta \lesssim 0.5\,{\rm GeV}^2 \sim 9\Lambda_{\rm QCD}^2$, which means that the existence of a dynamically generated, $B \not\equiv 0$ solution of Eq. (1) is impossible without a significant infrared enhancement of the effective interaction. Reproducing observable phenomena requires[6,8] $\sigma^\Delta \gtrsim 4\,{\rm GeV}^2 \sim 70\,\Lambda_{\rm QCD}^2$; i.e., a ten-fold enhancement over the critical value. The origin of this enhancement, whether in the gluon or ghost vacuum vacuum polarisation, or elsewhere, is currently unknown but is actively being sought (see, e.g., Refs. [9,10]). Nevertheless, its characterisation via a one-parameter model yields an efficacious description of light meson observables.[11]

The dynamical generation of a dressed-quark mass in the chiral limit is necessarily accompanied by the formation of a vacuum quark condensate:[3]

$$-\,\langle\bar{q}q\rangle_\zeta^0 = Z_4(\zeta^2,\Lambda^2)\, N_c {\rm tr_D} \int_q^{\Lambda} S^0(q,\zeta)\,, \tag{8}$$

where ${\rm tr}_D$ identifies a trace over Dirac indices only and the superscript "0" indicates the quantity was calculated in the chiral limit. This condensate is gauge-parameter and cutoff independent. In the model summarised in Ref. [11], an explanation of hadron observables *requires* $\langle\bar{q}q\rangle_{\zeta=1\,{\rm GeV}}^0 = -(0.241\,{\rm GeV})^3$. The condensate is analogous to a Cooper pair density and this value corresponds to $1.8\,{\rm fm}^{-3}$. Considering, simply for illustration, the condensate to be a sea of close-packed spheres then this density corresponds to a Cooper pair radius of 0.76 fm, which is just 15% more than the pion's charge radius and 13% less than the nucleon's radius. Clearly the scale set by this condensate is fundamental in QCD.

4 The Goldstone Boson

We began with a question: "How does one form an almost-massless bound state from very massive constituents *without* fine-tuning?" Bound states of a dressed-quark and dressed-antiquark; i.e., mesons, are described by the homogeneous Bethe-Salpeter equation (BSE). (The Bethe-Salpeter amplitude obtained as the solution of this equation is analogous to the wave function in quantum mechanics and provides similar insights into the nature of the bound state.[12]) It was proven in Ref. [3] that the flavour nonsinglet pseudoscalar meson BSE admits a massless solution if, and only if, the QCD gap equation supports DCSB. This is an expression of Goldstone's theorem.

That proof establishes a number of corollaries. One is a collection of quark-level Goldberger-Treiman relations, which establish that the pseudoscalar meson Bethe-Salpeter amplitude necessarily has pseudovector components. It is these components that are responsible for the asymptotic $1/Q^2$ behaviour of the electromagnetic pion form factor.[13] Another is a mass formula, valid for all flavour nonsinglet pseudoscalar mesons, *independent* of the current-quark mass of the constituents.[14] This single mass formula unifies aspects of QCD's light- and heavy-quark sectors, and provides[15] a qualitative understanding of recent results from lattice simulations for the current-quark-mass-dependence of pseudoscalar meson masses.[16] DCSB and the large value of the vacuum quark condensate also explain the large π-ρ mass difference. An heuristic demonstration of this is provided by the bosonisation procedure described in Ref. [17]. In the chiral limit, for the flavour nonsinglet mesons, there is an exact cancellation between the condensate-driven mass-term in the quark-loop piece of the effective action and that in the auxiliary field piece, while for the vector mesons, because $\{\gamma_5, \gamma_\mu\} = 0$, the terms add.

Confinement is the failure to observe coloured excitations in a detector. It is related to the effective interaction and interaction-tension described in Sec. 3. If the interaction-tension is large enough to explain hadron observables then the analytic properties of the dressed-quark propagator are very different from those of a free fermion; e.g., the propagator no longer has a Källén-Lehmann representation. The absence of such a representation is a sufficient condition for confinement.[1,2] From this perspective, the existence of flavour nonsinglet pseudoscalar Goldstone modes is a consequence of confine-

Table 1. Fit parameters for the calculated u valence-quark distribution function, Eq. (9).

scale (GeV)	A_u	η_1	η_2	ϵ_u	γ_u
0.54	11.24	1.43	1.90	2.44	2.54
2.0	4.25	0.97	2.43	1.82	2.46

ment: one can have Goldstone modes without confinement, but one cannot have confinement in the chirally symmetric theory without the appearance of Goldstone modes.

5 Pion's Valence-quark Distribution

The quark-gluon substructure of the pion is also expressed in its parton distribution functions. They can be measured in πN Drell-Yan[18] and deep inelastic scattering[19] (using the Sullivan process) but cannot be calculated in perturbation theory. The valence-quark distribution function has been calculated[20] using the DSE model that unified the small- and large-Q^2 behaviour of the electromagnetic pion form factor.[13] This covariant calculation of the relevant "handbag diagrams" indicates that, at a resolving scale of $q_0 = 0.54\,\mathrm{GeV} = 1/(0.37\,\mathrm{fm})$, valence-quarks with an active mass of 300 MeV carry 71% of the pion's momentum, and yields a numerical result for the distribution function that is pointwise well-described by

$$x\, u_V^\pi(x;q_0) = A_u\, x^{\eta_1}\, (1-x)^{\eta_2}\, (1-\epsilon_u \sqrt{x} + \gamma_u x)\,, \tag{9}$$

with the parameter values listed in Table 1. This parametrisation, with the parameter values listed, provides an equally good fit to $x\, u_V^\pi(x; 2\,\mathrm{GeV})$ obtained by applying the three-flavour, leading-order, nonsinglet renormalisation group evolution equations. The low moments of the distribution calculated at this scale (2 GeV) are [20]

	$\langle x \rangle$	$\langle x^2 \rangle$	$\langle x^3 \rangle$
Calc.	0.24	0.098	0.049
Exp.	0.24 ± 0.01	0.10 ± 0.01	0.058 ± 0.004
Latt.	0.27 ± 0.01	0.11 ± 0.3	0.048 ± 0.020

(10)

with the experimental results from Ref. [18] and the lattice results from Ref. [22].

The value of $\eta_2 \simeq 2$ obtained in the DSE calculation is consistent with a perturbative QCD analysis:[21] $u_V^\pi(x) \sim (1-x)^2$ at $x \simeq 1$. That prediction, however, conflicts with the currently available Drell-Yan data,[18] which

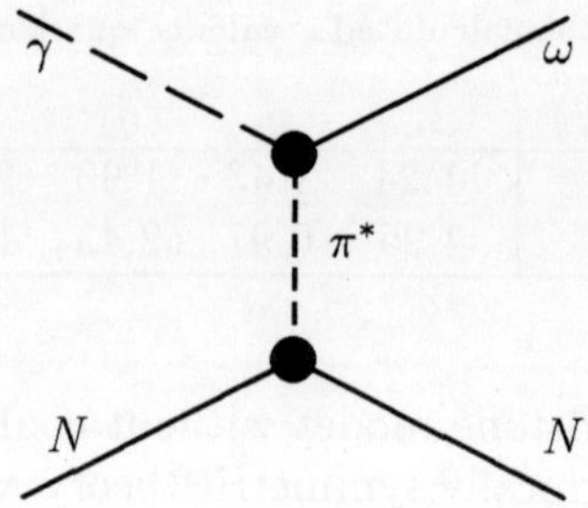

Figure 1. π-exchange contribution to the ω-photoproduction amplitude.

suggests that $u_V^\pi(x) \approx (1-x)$ for $x \simeq 1$. This experimental result, if true, will be a profound challenge to QCD, even questioning whether the strong interaction Lagrangian contains a vector interaction.

We note that the agreement between the moments calculated using the DSE result and those inferred from experiment, see Eq. (10), indicates plainly that the low moments are insensitive to the large-x behaviour of the distribution; i.e., the valence region: even the *sixth* moments of these differently shaped distributions disagree by only 32%. Hence pointwise calculations of the distribution functions are necessary to probe the valence region.

6 ω-meson Photoproduction

Constituent quark models predict many nucleon resonances that are hitherto unobserved. That may be because they only couple weakly to the πN channel, which has been used to search for them. Experiments at facilities such as JLab are therefore seeking the "missing resonances" in electromagnetic reactions. Of these, ω-photoproduction is particularly useful because its non-resonant reaction mechanisms are thought to be well understood.[23] Hence a comparison between experimental data and the cross-section calculated using these mechanisms alone provides a sound base from which to search for nucleon resonance contributions.

Meson exchange models[23] suggest that the leading contribution to the photoproduction cross-section at low energies and forward angles is given by the off-shell-π-exchange diagram depicted in Fig. 1. Here the filled circles represent meson-nucleon and meson-photon form factors, which are customarily parametrised in the calculation of observables.

However, these form factors can be *calculated* when one has a reliable quark-level description of hadrons.[24] We illustrate that using the model of Ref. [24], with the *Ansatz* for the nucleon's Fadde'ev amplitude updated as

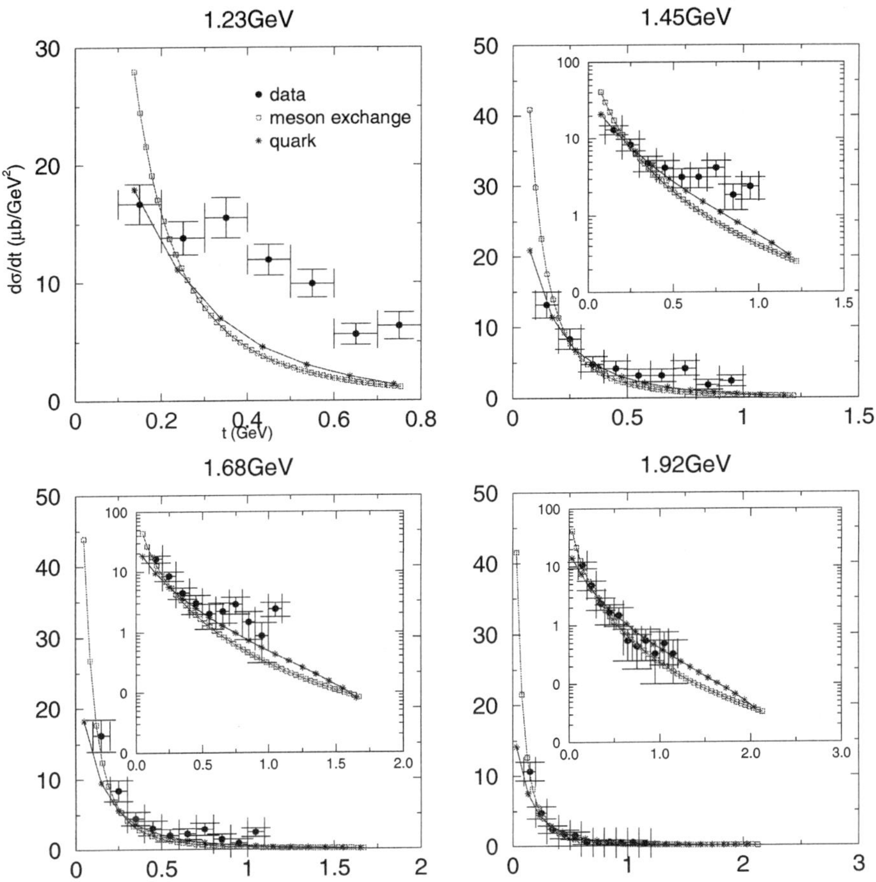

Figure 2. π-exchange contribution to the $\gamma N \to \omega N$ cross-section obtained using the model of Refs. [24,25]. (Panels are labelled by the incident photon energy.) $*$, long-dashed line, our calculation (no parameters were varied to obtain this result); □, short-dashed line, the meson-exchange model calculation of Ref. [23] (only the t-channel π-exchange contribution is shown); data from Ref. [26].

described in Ref. [25], to obtain a *parameter-free* prediction of the contribution from the π-exchange process to the ω-photoproduction cross-section. The results are illustrated in Fig. 2. Much of the discrepancy at large angles owes to our neglect, in this preliminary calculation, of the contribution from s- and u-channel nucleon diagrams. Including them is part of an ongoing programme.

7 Epilogue

The DSEs are maturing as an efficacious nonperturbative tool in hadron physics and in this application they provide crucial information about the long-range part of the QCD interaction.[12] The primary elements in the approach are the Schwinger functions: the dressed-propagators and -vertices. They can also be calculated using other techniques, such as effective field theory and lattice quantisation methods. There is therefore great scope for forging relationships between these approaches, thereby providing a tool whose domain of reliable application is larger than that of its individual parts. One concrete example is the confirmation in recent lattice-QCD simulations[27] of the bulk features of the dressed-quark propagator that have long been suggested by DSE studies.[11,15] Making that agreement quantitative will hone our understanding of the long-range part of the interaction and, e.g., make possible a DSE-built bridge between lattice-QCD and the Q^2-evolution of the electromagnetic pion form factor.[28]

Acknowledgments

CDR gratefully acknowledges the hospitality of the staff at the Special Research Centre for the Subatomic Structure of Matter at the University of Adelaide, and also financial support provided by the Centre. This work was also supported by: the Deutsche Forschungsgemeinschaft under project no. SCHM 1342/3-1; and the US Department of Energy, Nuclear Physics Division, under contract no. W-31-109-ENG-38.

References

1. C.D. Roberts and S.M. Schmidt, *Prog. Part. Nucl. Phys.* **45**, S1 (2000).
2. R. Alkofer and L. von Smekal, "The infrared behavior of QCD Green's functions: Confinement, dynamical symmetry breaking, and hadrons as relativistic bound states," hep-ph/0007355.
3. P. Maris, C.D. Roberts and P.C. Tandy, *Phys. Lett.* B **420**, 267 (1998).
4. A. Bender, C.D. Roberts and L. Von Smekal, *Phys. Lett.* B **380**, 7 (1998).
5. P. Maris and C. D. Roberts, *Phys. Rev.* C **56**, 3369 (1997).
6. C.D. Roberts, "Continuum strong QCD: Confinement and dynamical chiral symmetry breaking," nucl-th/0007054.
7. F.T. Hawes, C.D. Roberts and A.G. Williams, *Phys. Rev.* D **49**, 4683 (1994); F.T. Hawes, P. Maris and C.D. Roberts, *Phys. Lett.* B **440**, 353 (1998).

8. P. Maris and P.C. Tandy, *Phys. Rev.* C **60**, 055214 (1999).
9. A.I. Alekseev and B.A. Arbuzov, *Mod. Phys. Lett.* A **13**, 1747 (1998); L. von Smekal, A. Hauck and R. Alkofer, *Annals Phys.* **267**, 1 (1998) [Erratum-ibid. **269**, 182 (1998); D. Atkinson and J.C.R. Bloch, *Mod. Phys. Lett.* A **13** (1998) 1055.
10. D.B. Leinweber, J.I. Skullerud, A.G. Williams and C. Parrinello [UKQCD Collaboration], *Phys. Rev.* D **60**, 094507 (1999) [Erratum-ibid. D **61**, 079901 (1999)]; A. Cucchieri, *Phys. Rev.* D **60**, 034508 (1999).
11. P. Maris, "Continuum QCD and light mesons," nucl-th/0009064.
12. J.C.R. Bloch, Yu.L. Kalinovsky, C.D. Roberts and S.M. Schmidt, *Phys. Rev.* D **60**, 111502 (1999).
13. P. Maris and C.D. Roberts, *Phys. Rev.* C **58**, 3659 (1998).
14. P. Maris and C.D. Roberts, "QCD bound states and their response to extremes of temperature and density," nucl-th/9806005; M.A. Ivanov, Yu.L. Kalinovsky and C.D. Roberts, *Phys. Rev.* D **60**, 034018 (1999).
15. M.B. Hecht, C.D. Roberts and S.M. Schmidt, "Contemporary applications of Dyson-Schwinger equations," nucl-th/0010024.
16. K. C. Bowler *et al.* [UKQCD Collaboration], *Phys. Rev.* D **62**, 054506 (2000).
17. J. Praschifka, C.D. Roberts and R.T. Cahill, *Phys. Rev.* D **36**, 209 (1987).
18. J. S. Conway *et al.*, *Phys. Rev.* D **39**, 92 (1989).
19. R. J. Holt and P. E. Reimer, "Sturcture of the Goldstone bosons," nucl-ex/0010004; K. Wijesooriya, *et al.*, "The $H(e,e'n)X$ Reaction and the Pion Structure Function," proposal no. JLab PR01-110.
20. M.B. Hecht, C.D. Roberts and S.M. Schmidt, *Phys. Rev.* C **63**, 025213 (2001); M. B. Hecht, C. D. Roberts and S. M. Schmidt, "Dyson-Schwinger equations: Aspects of the pion," nucl-th/0011005.
21. S.J. Brodsky, M. Burkardt and I. Schmidt, *Nucl. Phys.* B **441**, 197 (1995).
22. C. Best *et al.*, *Phys. Rev.* **D 56**, 2743 (1997).
23. Y. Oh, A. I. Titov and T. S. Lee, *Phys. Rev.* C **63**, 025201 (2001).
24. J.C.R. Bloch, C.D. Roberts and S. M. Schmidt, *Phys. Rev.* C **61**, 065207 (2000).
25. M.B. Hecht, C.D. Roberts and S.M. Schmidt, "Neutron electric dipole moment: Constituent-dressing and compositeness," nucl-th/0101058.
26. F.J. Klein, PhD. thesis, Bonn University (1996); F.J. Klein, *et al.*, [SAPHIR Collaboration], πN *Newslett.* **14**, 141 (1998).
27. J. I. Skullerud and A. G. Williams, *Phys. Rev.* D **63**, 054508 (2001).
28. P. Maris and P. C. Tandy, *Phys. Rev.* C **62**, 055204 (2000); J. Volmer *et al.* [The JLab F(pi) Collaboration], *Phys. Rev. Lett.* **86**, 1713 (2001).

CONFINEMENT FROM COULOMB GAUGE QCD AND EXOTIC PHENOMENOLOGY

ADAM P. SZCZEPANIAK

Physics Department and Nuclear Theory Center
Indiana University, Bloomington, IN 47405
E-mail: aszczepa@indiana.edu

I describe how confinement and constituent representations may arise in QCD in the Coulomb gauge. Implications for studies of phenomena related to low energy gluonic excitations, in particular hybrid mesons, are discussed.

1 Motivation

One of the major unsolved mysteries of QCD has to do with establishing the presence of gluonic degrees of freedom at low energies. Gluons play a major role in high energy processes, carrying both momentum and spin of hadrons. At low energies the situation is much more obscure. Even though it is natural to expect that gluons are important contributors to hadronic structure, there is as yet no clear measure of the gluonic content of hadrons. Low energy processes are usually described in terms of valence degrees of freedom. In the constituent quark model or the bag model the gluons appear indirectly, for example as the source of the confining interaction. These valence models have been quite successful when dealing with static properties such as magnetic moments and charge distributions, but in many cases their limitations hint that degrees of freedom may be missing. For example, even some of the simplest reactions, like elastic $\pi - \pi$ scattering reveal structures which are much richer than what is expected from a simple quark model.

An unambiguous identification of nonvalence degrees of freedom may come from spectroscopy of hadrons with exotic quantum numbers. Since only mesons can be exotic one should search for resonances with J^{PC}'s which are outside the $Q\bar{Q}$ classification, normally giving $\mathbf{J} = \mathbf{L}+\mathbf{S}$, $P = (-1)^{L+1}$, $C = (-1)^{L+S}$, *i.e.* search for states with 0^{--}, 0^{+-},1^{-+}, *etc.* It should be noted that this, often referred to as *naive*, classification can be quite rigorous if QCD is quantized in a physical gauge, *e.g.* the Coulomb gauge. As long as L, S, P and C belong to a group of kinematical generators, *i.e.* do not depend on the interactions, they can be used to classify the constituent Fock space basis. In this case the Fock space expansion of an exotic meson will not contain the $Q\bar{Q}$ component.

2 Properties of low energy gluonic excitations

There are many features of the Coulomb gauge which make it particularly suitable for building the constituent, Fock space representation of hadrons. As Gauss's law is part of the Hamiltonian, no additional constraints on the Fock space need to be imposed. The Hamiltonian depends only on physical degrees of freedom, the quark fields, ψ, and the two transverse components of the gauge fields, $\mathbf{A}$, satisfying $\nabla \cdot \mathbf{A} = 0$. This means the hadronic wave functions, $\Psi_h = \Psi_h[\psi, \mathbf{A}]$, are given as solutions of the Schrödinger equation,

$$H_{QCD}^{\nabla \cdot \mathbf{A}=0} \Psi_h = E_h \Psi_h, \tag{1}$$

and have direct probabilistic interpretation. The Coulomb gauge minimizes spurious relativistic effects and thus it is ideal for studies of nonrelativistic systems. Finally, as we will now discuss, confinement emerges already at the mean field, BCS level. Starting from the mean field vacuum one can construct a quasiparticle basis of constituent quark, and constituent gluon excitations and use it to represent a hadronic state. The mean field vacuum is constructed using the variational principle which, as shown by Zwanziger [1], is justified as long as the gauge fields are restricted to the fundamental modular region containing a single Gribov copy. This restriction makes the Faddeev-Popov operator positive, which then constraints the instantaneous Coulomb interaction between color charges.

The mechanisms for confinement in the Coulomb gauge can be illustrated by considering the static $Q\bar{Q}$ system. The relevant parts of the Hamiltonian contain [2],

$$H_{static} = K_g + V_c, \tag{2}$$

where K_g is the gluon kinetic operator and V_c is the Coulomb operator. If expanded in powers of g, the Coulomb operator contains an infinite number of terms to all orders in g. If the Hamiltonian is Weyl ordered there are additional terms usually referred to as the V_1 and V_2 terms by Schwinger, and Christ and Lee [3]. There is also a gluon mass term in K_g which is an artifact of regularization. Renormalization also requires that all operator products are regularized, for example by point-splitting or by smearing the operators over a distance $x \sim 1/\Lambda$. The details of renormalization were discussed in Ref. [4,5].

For static quarks the β function arises from H_{stat} [2]. In particular, to lowest order there are two contributions, one from the $O(g^4)$ part of V_c and the other from iteration of the $O(g^2)$ part of V_c. This is shown in Fig. 1, and the result is

$$\beta(g) = (g^3/(4\pi)^2)\beta_0 = (g^3/(4\pi)^2)[-N_c/3 + 4N_c] \tag{3}$$

Figure 1: $O(g^4)$ contributions to the static $Q\bar{Q}$ potential, a) anti-screening, b) screening.

Iteration of the $O(g^2)$ term (Fig.1b) produces a standard screening correction which is negative since it comes form a second order perturbation on the bare energy. The main positive anti-screening term, $\tilde{\beta}_0 = 4N_c$ comes from the instantaneous, triple gluon vertex, in this case involving one transverse and two Coulomb-gluons (Fig.1a).

Summing up the ladder of self energies shown in Fig.1a results in the running coupling,

$$\alpha(k) = \frac{\alpha(\Lambda)}{1 + \frac{\alpha(\Lambda)\tilde{\beta}_0}{4\pi}\log(k^2/\Lambda^2)}. \tag{4}$$

The running coupling becomes IR singular at a finite k^2, given by $k^2 = \tilde{\Lambda}^2_{QCD} \equiv \Lambda^2 \exp(-4\pi/\alpha(\Lambda)\tilde{\beta}_0)$. The increase of $\alpha(k)$ for small k is the first indication of confinement. In the Coulomb gauge $4\pi\alpha(k)/k^2$ is directly related to the (3-dimensional) Fourier transform of the static $Q\bar{Q}$ potential. Since in the quenched approximation the potential is expected to be well behaved at large distances, the singularity in the running coupling at $k^2 > 0$ indicates that simple ladder summation of the perturbative anti-screening part may be overestimating the coupling. If, however, the dispersion of gluons in the vacuum is not given by the free field relation, $\omega_g(k) = k$ (true only for the perturbative vacuum) then it is possible for the IR singularity to be weaker and appear at $k^2 \leq 0$. The modification of the dispersion relation arises if a variational vacuum with nontrivial gluon distribution is used instead of the perturbative vacuum. For example with a dispersion relation which is finite at small k, $\omega_g(0) \equiv m_g > 0$, the running coupling will approximately be given by

$$\alpha(k) = \frac{\alpha(\Lambda)}{1 + \frac{\alpha(\Lambda)\tilde{\beta}_0}{4\pi}\log((k^2 + m_g^2)/\Lambda^2)}. \tag{5}$$

Then, if m_g is set equal to Λ_{QCD}, the IR singularity will occur at $k^2 = 0$, and furthermore if,

$$8\pi m_g^2 = 3\tilde{\beta}_0 b, \tag{6}$$

with b being the string tension, the low momentum dependence of the running coupling will be consistent with a linearly rising potential for large $Q\bar{Q}$ separation, $V_{Q\bar{Q}}(r) \to br_{Q\bar{Q}}$.

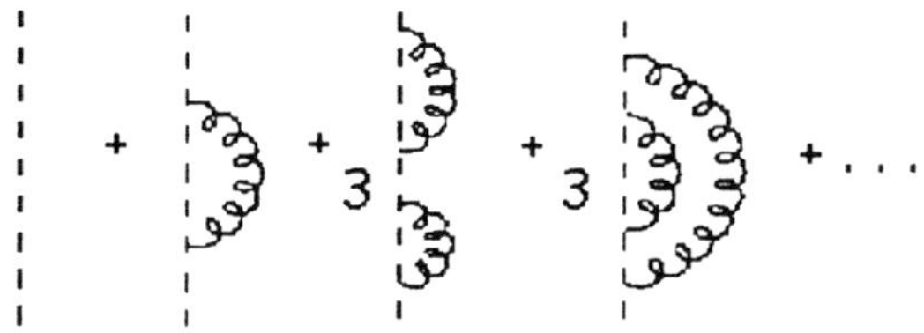

Figure 2: First three terms of a rainbow-ladder expansion of the static $Q\bar{Q}$ ground state potential

It is worth noting that since the ladder summation does not involve iterating the Hamiltonian, the static confining potential emerges solely from the vacuum expectation value of the Hamiltonian in a variational, nonperturbative vacuum. This vacuum ansatz then leads to a non-perturbative dispersion for the gluons which in a rigorous approach should be treated selfconsistently with the Coulomb kernel. [1,4,5,6] However evaluating the exact expectation value of V_c even in a simple, mean-field-BCS-Gaussian ansatz for the vacuum is rather complicated. It can be shown that the dominant IR contribution seems to come from the rainbow-ladder series shown in Fig. 2 [5,6].

The next step in this approach is to take the variational vacuum and build the complete Fock space from quasi-quarks and quasi-gluons. The single-particle energies are modified from the free-field values, $E_g(k) = \omega_g(k) = k$ and $E_Q = \omega_Q(k) = \sqrt{k^2 + m_0^2}$ due to self-energies arising from interactions with the quark and gluon mean fields in the vacuum ansatz. Due to the long-range nature of the interactions these self energies are IR singular, which is another manifestation of confinement – colored states have infinite energy and decouple from the physical spectrum. In color singlet hadrons, however, the residual interaction between quasiparticles enables to cancel these IR divergences from self-energies and produce a finite hadron mass spectrum.

Expanding an eigenstate of the full Hamiltonian, Eq. (1), in the Fock space constructed from the variational ground state has obvious advantages. After cancellation of the IR divergences the average kinetic energy of quasi-particles in a color singlet bound state are of the order of hundreds of MeV which makes truncation to a few (possibly, valence) Fock space components justifiable. Furthermore, selfconsistent treatment of the vacuum structure and the quasiparticle spectrum enables one to maintain chiral symmetry (the vacuum ansatz reflects spontaneous breaking of chiral symmetry) and therefore keep the Goldstone nature of the pion. Recently it has been shown how the constituent quark model description of the vector-pseudoscalar meson mass difference usually attributed to a hyperfine interaction, arises, for low quark masses from the interplay between dynamical chiral symmetry breaking and the hyperfine term and how it matches smoothly onto the $1/m_Q$ behavior for heavy-light mesons [7] .

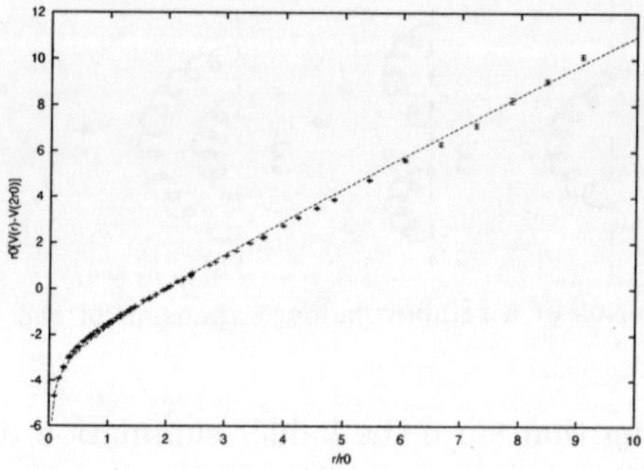

Figure 3: Static $Q\bar{Q}$ ground state potential obtained from Coulomb gauge QCD compared to lattice calculations.

Returning to the gluon sector, the gluon dynamics can be simplified if quarks are static. For a static quark-antiquark pair separated by a distance $r_{Q\bar{Q}}$ the eigenstates of the Hamiltonian can be classified according to: the angular momentum along the $Q\bar{Q}$ axis, $\Lambda = 0, 1, 2, \cdots = \Sigma, \Pi, \Delta, \cdots$, symmetry under reflection in the plane containing the $Q\bar{Q}$ axis, $Y = \pm 1$ and symmetry under combined space reflection and charge conjugation $PC = +1, -1 = g, u$. The eigenvalue problem, Eq. (1) then becomes,

$$H_{QCD}^{\nabla\cdot\mathbf{A}}|\Lambda_{PC}^{Y}, r_{Q\bar{Q}}\rangle = E_{\Lambda_{PC}^{Y}}(r_{Q\bar{Q}})|\Lambda_{PC}^{Y}, r_{Q\bar{Q}}\rangle, \tag{7}$$

and in the quenched approximation the eigenvalues represent the adiabatic $Q\bar{Q}$ potentials in the presence of excited gluonic configurations. For gluons in a ground state the $E_{\Sigma_g^+}(r_{Q\bar{Q}})$ surface is identical to the static $Q\bar{Q}$ potential discussed earlier. The result for $V_{\Sigma_g^+}(r_{Q\bar{Q}})$ obtained from the Coulomb gauge QCD studies, as outlined above, is show in Fig. 3[5] in comparison with the lattice data[8].

The higher adiabatic potentials have also been recently calculated on the lattice[8] and show that the first excited gluon corresponds to $\Lambda = 1$ and $PC = -1$; *i.e* the Π_u configuration and the Π_g ($PC = +1$) configuration is several hundred MeV higher. It is worth noting that this is at odds with the naive expectation. If gluons are considered as $J^{PC} = 1^{--}$ plane waves then one might expect the Π_g, $PC = +1$ to be the ground state configuration.

There are two models which can accommodate this pattern of gluonic levels. In the bag model gluons are treated as noninteracting inside a cavity, and for a given quark-antiquark separation the discrete energy levels come from imposing a boundary condition on the magnetic fields at the bag surface. This has the effect of lowering the $J^{PC} = 1^{+-}$, transverse electric mode with respect to the $J^{PC} = 1^{--}$, transverse magnetic [9] and therefore leads to a correct ordering with $E_{\Pi_u} < E_{\Pi_g}$. Another popular approach, the flux tube model, is based on the idea that the gluon field near the $Q\bar{Q}$ axis can be represented as a collection of beads interacting via simple, nonrelativistic harmonic oscillator forces [10]. This model borrows from the strong coupling limit of QCD – the beads are representing the link fields and the simple interaction represents small vibration of the string of links around the strong coupling equilibrium position of a straight line along the $Q\bar{Q}$ axis. Under parity the displacement of the string in the plane transverse to the $Q\bar{Q}$ changes sign, thus it corresponds to intrinsic negative parity, but under charge conjugation links reverse their orientation but do not change sign and therefore correspond to positive intrinsic charge conjugation. In the flux tube model the lowest gluon mode will then also have $PC = -1$ and the model leads to a correct ordering of the two lowest excited adiabatic surfaces. This is due, however, to a different choice for the degrees of freedom to represent gluons than in the bag model. The flux tube-like form of low energy gluon modes is also consistent with the expected long-range behavior of the spin-dependent $Q\bar{Q}$ potentials. These arise from mixing between the $Q\bar{Q}$ and hybrid, $Q\bar{Q}g$ states and are induced by the electric and/or magnetic field operators [11].

A third popular model is based on the constituent picture, where gluons are described as a localized color charge density [12]. A two body interaction between V_{Q-g} and $V_{\bar{Q}-g}$ can be derived from the QCD Hamiltonian, and produces a series of bound gluon states localized near the quark-antiquark pair [13]. This is because the attraction from V_{Q-g} and $V_{\bar{Q}-g}$ overcomes the repulsion from the octet $V_{Q\bar{Q}}$. However, as one might expect the orbital, s-wave, positive parity gluon configuration has lower energy then the p-wave orbital. Thus, if one takes that the *constituent* gluon has $J^{PC} = 1^{--}$ the model leads to *opposite* ordering of the Π_u and Π_g surfaces in disagreement with the lattice and the other two models. The resolution of this difference may require inclusion of the three-body interaction between the quark-antiquark pair and the transverse constituent gluon arising from the $O(g^4)$ part of V_c.

In the past few years a lot of effort has been put into experimental searches of low energy gluonic excitations. The E852 BNL experiment has reported two exotic candidates, one with mass $M \sim 1.4$ GeV seen in the reaction $\pi^- p \to \eta\pi^- p$ and another at $M \sim 1.6$ GeV in the $\pi^+\pi^-\pi^-$ and $\eta'\pi$ final states [14]. In future the most promising exotic mesons searches may be performed at the Hall D at JLab. In Hall D the exotic mesons will be searched for in the reaction $\vec{\gamma} p \to XN$ using $10^7\gamma/s$ linearly polarized, real photons with $E_{lab} = 8-9$ GeV. There are theoretical arguments for why photoproduction may be a preferred

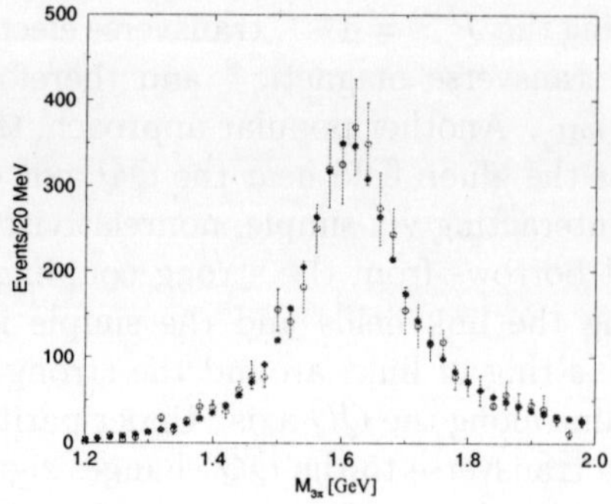

Figure 4: Comparison between MC generated and PWA extracted exotic mesons signals

mechanism to access the exotic mesons as compared to hadronic production with pseudoscalar meson beams. At the quark level, the models discussed above predict that the low lying exotic mesons have the $Q\bar{Q}$ pair in a spin-1 configuration which then directly couples to a real photon as opposed to the π beam. The enhancement of exotic production with photon beams can also be demonstrated using standard Regge phenomenology [15].

One of most promising reactions for the $J^{PC} = 1^{-+}$ exotic meson searches is $\gamma p \to \pi^+\pi^+\pi^- n$. At low-t it is expected to be dominated by the well understood one-pion-exchange production mechanism and since the charged 3π system has $G = -1$ and $I = 1$ the p-wave resonance in $(\pi^+\pi^-)\pi^+ \sim (\rho^0\pi^+)_{l=1}$ channel corresponds to the isovector $J^{PC} = 1^{-+}$ exotic. Furthermore, in the 1 GeV $< M_{3\pi} <$ 2 GeV mass range the 3π mass spectrum has only three, well established, regular $Q\bar{Q}$ resonances, the a_1, a_2 and π_2, which can be used as benchmark signals.

Monte Carlo simulations for this reaction have been performed with physical input parameters for the known resonances and with a small, $O(1\%)$ admixture of the exotic wave. The comparison between generated and extracted, via partial wave analysis, exotic signal is shown in Fig. 4 and corresponds to an event sample equivalent to a few days of Hall D operation.

3 Summary

Mapping the spectrum of low energy gluonic excitations is crucial for understanding of the dynamical mechanisms of strong QCD, *e.g.* confinement, dynamical chiral symmetry breaking and light quark production. The existing phenomenological models begin to converge on the quantum numbers of the low lying exotic mesons, although more input from first principles lattice QCD and/or fixed gauge Hamiltonian methods is needed to establish the nature of the gluonic degrees of freedom. An extensive search for exotic mesons will soon begin with the new experiments in Hall D at the 12 GeV upgraded JLab and will provide the data necessary to address the fundamental issues of QCD.

Acknowledgments

This work was supported by a DOE grant, DE-FG02-87ER40365.

1. D. Zwanziger, *Nucl. Phys.* B **485**, 185 (1997); A. Cucchieri and D. Zwanziger, *Phys. Rev. Lett.* **78**, 3814 (1997).
2. T. D. Lee, *Particle Physics And Introduction To Field Theory*, (New York, Harwood Academic, 1981).
3. N. H. Christ and T. D. Lee, *Phys. Rev.* D **22**, 939 (1980).
4. A. P. Szczepaniak and E. S. Swanson, *Phys. Rev.* D **62**, 094027 (2000); D. G. Robertson, *et al. Phys. Rev.* D **59**, 074019 (1999).
5. A. P. Szczepaniak and E. S. Swanson, *in preparation.*
6. A. R. Swift, *Phys. Rev.* D **38**, 668 (1988).
7. A. P. Szczepaniak and E. S. Swanosn, hep-ph/0006306 *to appear in Phys. Rev. Lett.*
8. K.J. Juge, J. Kuti, C.J. Morningstar, *Nucl. Phys. Proc. Suppl.* **63**, 326 (1998).
9. T. Barnes, F.E. Close, F. de Viron, J. Weyers, *Nucl. Phys.* B**224**, 241 (1983).
10. N. Isgur, J. Paton, *Phys. Rev.* D**31**, 2910 (1985).
11. A. P. Szczepaniak, E. S. Swanson, *Phys. Rev.* D**55**, 3987 (1997).
12. D. Horn , J. Mandula, *Phys. Rev.* D**17**, 898 (1978).
13. E.S. Swanson, A. P. Szczepaniak, *Phys. Rev.* D**59**, 014035 (1999).
14. E852 Collab., *Phys. Rev. Lett.* **79**, 1630 (1997); *Phys. Rev. Lett.* **81**, 5760 (1998); *Phys. Rev. Lett.* **86**, 3977 (2001).
15. A. P. Szczepaniak, M. Swat, hep-ph/0105329.

REGULATOR FREE DYSON-SCHWINGER EQUATION STUDIES OF NON-PERTURBATIVE FIELD THEORY

AYŞE KIZILERSÜ, ANDREAS W. SCHREIBER, TOM SIZER and ANTHONY G. WILLIAMS

Special Research Centre for the Subatomic Structure of Matter and Department of Physics and Mathematical Physics, Adelaide University, 5005, Australia
E-mail: akiziler@physics.adelaide.edu.au

We describe a regularization-independent method for studying a renormalizable field theory nonperturbatively through its Dyson-Schwinger equations. Using QED_4 as an example, we show how the coupled equations determining the non-perturbative fermion and photon propagators can be written entirely in terms of renormalized quantities, which renders the equations manifestly finite in a regularization-independent manner. As an illustration of the technique, we apply it to a study of the fermion propagator in quenched QED_4 with the Curtis-Pennington electron-photon vertex.

1 Introduction

The study of Dyson-Schwinger equations (DSEs) is a well established tool in non-perturbative field theory. In general, these are divergent integral equations which require regularization. The removal of the regulator and the accompanying renormalization of the theory is usually performed numerically. There are, however, some well-known difficulties associated with this. For example, the most commonly used regulator is a hard UV cut-off on momentum integrals. This regulator violates translational invariance in momentum space and, because of this, will in general do violence to any underlying local gauge symmetries of the theory [1]. Although there are recipes available for restoring Ward-Takahashi identities it is fair to say, however, that this is not entirely satisfactory. On the other hand, dimensional regularization, which does not suffer from the above problem, has turned out to be rather tedious to use from a numerical point of view [2]. Moreover, at least some quantities associated with dynamical chiral symmetry breaking have been shown to exhibit discontinuous behaviour as the limit $D \to 4$ is taken. Hence, from a practical point of view, dimensional regularization is unlikely to play a prominent role in future nonperturbative DSE studies.

In this contribution we report on recent work which is designed to circumvent the above problems by removing the regulator *analytically* rather than *numerically*. Details about this approach may be found in Ref. [3] Here, the Dyson-Schwinger equations are formulated in terms of renormalized quantities only. Although the method is expected to be useful within any renormalizable field theory, we have so far concentrated on studies of QED, on which we report here.

Avoiding regulator-induced breaking of gauge-covariance means that common problems in DSE studies, such as the generation of unwanted longitudinal components in gauge-boson propagators and associated quadratic divergences, are avoided. Moreover, the method has an additional benefit: because it is formulated in terms of renormalized quantities , the dependence on the (large) mass scale introduced by the regulator is traded for the momentum scale μ at which the theory is renormalized *before* performing any numerical calculations. In this way dimensional regularization (or any other regularization which doesn't violate gauge covariance) can be used and, more importantly from a numerical point of view, removal of the regulator means that no longer does one have to solve integral equations involving mass scales of vastly different orders of magnitude (and then, in addition, take a limit in which one of the scales goes to infinity). Rather, the important scales in the problem become scales of 'physical' importance, such as the renormalized mass M_μ and the renormalization scale μ at which this mass is defined. It is therefore to be expected that the dominant contributions to any integrands will be from a relatively small region of momenta.

2 Regularization-Independent Formulation

The dimensionally regularized Dyson-Schwinger equations for the renormalized electron and photon propagators (see Fig. 1) are given by

$$\begin{aligned} S^{-1}(\mu;p) &= Z_2(\mu)S^{0^{-1}}(p) - iZ_1(\mu)e_\mu^2 \int \frac{d^Dk}{(2\pi)^D}\,\Gamma^\alpha(\mu;p,k)\,S(\mu;k)\gamma^\beta\,D_{\alpha\beta}(\mu;q) \\ &\equiv Z_2(\mu)S^{0^{-1}}(p) - Z_1(\mu)\,\bar{\Sigma}(p)\,, \qquad (1) \\ D^{-1}_{\alpha\beta}(\mu;q) &= Z_3(\mu)D^{0\;-1}_{\alpha\beta}(q) + iN_F Z_1(\mu)e_\mu^2 \mathrm{tr}\int \frac{d^Dk}{(2\pi)^D}\Gamma_\alpha(\mu;p,k)S(\mu;k)\gamma_\beta S(\mu;p) \\ &\equiv Z_3(\mu){D^0_{\alpha\beta}}^{-1}(q) + Z_1(\mu)\,\bar{\Pi}_{\alpha\beta}(q)\,. \qquad (2) \end{aligned}$$

Here N_F is the number of species of fermions in the theory and the constants $Z_1(\mu)$, $Z_2(\mu)$ and $Z_3(\mu)$ are the vertex, fermion wave function and photon wave function renormalization constants, respectively. We have assumed, for convenience, that the N_F fermions are identical (i.e. that they have equal mass). $S(\mu;p)$ and $D_{\alpha\beta}(\mu;q)$ are the full fermion and photon propagators renormalized at the momentum scale μ ($S^0(p)$ and $D^0_{\alpha\beta}(q)$ are their bare counterparts), while the full (proper) fermion-photon vertex is $\Gamma_\alpha(\mu;k,p)$. These propagators are usually expressed in terms of scalar functions $Z(\mu^2;p^2)$, $M(p^2)$ and

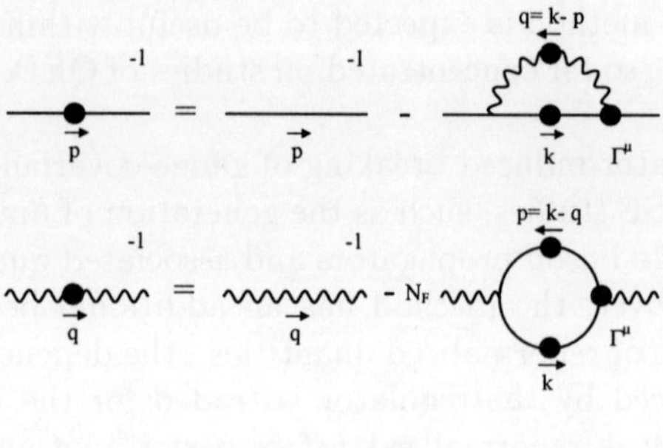

Figure 1: The Dyson-Schwinger equations for the fermion and photon propagators in QED.

$G(\mu^2;q^2)$ through

$$S(\mu;p) = \frac{Z(\mu^2;p^2)}{\not p - M(p^2)},\ -q^2 D_{\alpha\beta}(\mu;q) = G(\mu^2;q^2)\left(g_{\alpha\beta} - \frac{q_\alpha q_\beta}{q^2}\right) + \xi\frac{q_\alpha q_\beta}{q^2} \quad (3)$$

where ξ is the renormalized covariant gauge fixing parameter. The fermion's equation separates into two scalar equations

$$Z^{-1}(\mu^2;p^2) = Z_2(\mu) - Z_1(\mu)\,\bar\Sigma_d(p^2) \quad (4)$$
$$M(p^2)\,Z^{-1}(\mu^2;p^2) = Z_2(\mu)\,m_0 + Z_1(\mu)\,\bar\Sigma_s(p^2) \quad . \quad (5)$$

Here $\bar\Sigma(p) = \not p\ \bar\Sigma_d(p^2) + \bar\Sigma_s(p^2)$, implicit dependence on μ is not shown and the bar over these quantities indicates that we have explicitly separated out the renormalization constants $Z_i(\mu)$. The equivalent equation for the photon wavefunction renormalization $G(\mu^2;q^2)$ is given by

$$G^{-1}(\mu^2;q^2) = Z_3(\mu) + Z_1(\mu)\,\bar\Pi(q^2) \quad , \quad (6)$$

where the definition $\bar\Pi_{\alpha\beta}(q) \equiv -q^2\left(g_{\alpha\beta} - \frac{q_\alpha q_\beta}{q^2}\right)\bar\Pi(q^2)$ is used. Note that, as we are free to use a regulator which respects gauge covariance and will use a vertex which satisfies the Ward-Takahashi identity, $\bar\Pi_{\alpha\beta}(q)$ does not contain spurious quadratic divergences [4] and is transverse as it should be. Hence the term involving the gauge parameter ξ on the left hand side of Eq. (2) precisely cancels the corresponding term involving ξ_0 in $D^0_{\alpha\beta}$ on the right hand side (i.e. $\xi_0 = Z_3\xi$).

The renormalization constants $Z_i(\mu)$ as well as m_0 are fixed by the boundary conditions

$$Z(\mu^2;\mu^2) = 1 \qquad M(\mu^2) \equiv M_\mu \qquad G(\mu^2;\mu^2) = 1 \quad . \quad (7)$$

Because of the Ward-Takahashi identity, which should be satisfied by the vertex Ansatz for Γ_α, one has $Z_1(\mu) = Z_2(\mu)$.

In order to avoid cumbersome notation, we have not explicitly indicated functional dependence on the regulator in Eqs. (1)–(6). Most of the above quantities, i.e. $Z(\mu^2;p^2)$, $M(p^2)$ and $G(\mu^2;q^2)$, as well as $Z_{1,2,3}(\mu)$, are functions of the regulator. Indeed, if one keeps M_μ and e_μ fixed as one removes the regulator, which is the standard textbook prescription, the bare fermion mass m_0 and the bare charge also become regulator dependent.

As one removes the regulator, the integrals on the right hand side of Eqs. (1) and (2), and hence $\bar\Sigma_d(p^2)$, $\bar\Sigma_s(p^2)$ and $\bar\Pi(q^2)$ diverge logarithmically (or, to be more precise, in dimensional regularization they develop singularities at $D = 4$). It is the defining feature of a renormalizable field theory that these divergences may be absorbed into the constants $Z_{1,2,3}(\mu)$ and into the bare mass m_0, rendering finite limits for $Z(\mu^2;p^2)$, $M(p^2)$ and $G(\mu^2;q^2)$. However, we can make use of the p^2 independence of $Z_{1,2,3}(\mu)$ and m_0 in order to eliminate these constants from the above equations. For example, by evaluating Eq. (4) at a second momentum which, for convenience, we take to be $p^2 = \mu^2$, and forming an appropriate difference one obtains

$$Z(\mu^2;p^2) = 1 + Z(\mu^2;p^2)\,\bar\Sigma_d(p^2) - \bar\Sigma_d(\mu^2) \tag{8}$$

where we have made use of the renormalization condition (7). Because the left hand side of this equation is finite as the regulator is removed, the right hand side must also be (even though, of course, the individual terms on the RHS separately diverge). In a similar way one obtains

$$M(p^2) = M_\mu + \left[M(p^2)\bar\Sigma_d(p^2) + \bar\Sigma_s(p^2)\right] - \left[M_\mu\bar\Sigma_d(\mu^2) + \bar\Sigma_s(\mu^2)\right] \tag{9}$$

and

$$G^{-1}(\mu^2;q^2) = 1 + \left[G^{-1}(\mu^2;q^2)\bar\Sigma_d(\mu^2) + \bar\Pi(q^2)\right] - \left[\bar\Sigma_d(\mu^2) + \bar\Pi(\mu^2)\right] \tag{10}$$

We stress that the above manipulations should be possible within any renormalizable field theory, for example in QCD. Of course, in that theory the relevant DSE for the ghosts must also be taken into account.

3 Application to Quenched QED$_4$

For quenched QED Eq. (10) becomes irrelevant and Eqs. (8) and (9) reduce to one-dimensional integral equations. Employing the CP vertex these renor-

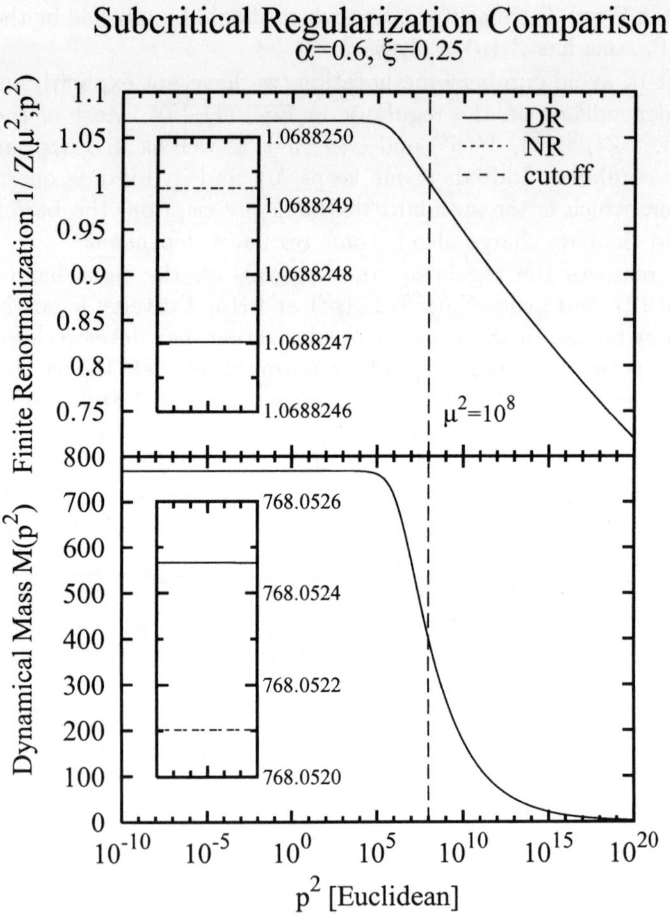

Figure 2: The finite renormalization $1/Z(\mu^2;p^2)$ and dynamical mass $M(p^2)$ for the solution of the fermion DSE for subcritical coupling $\alpha = 0.6$ and gauge parameter $\xi = 0.25$ found from the regularization-independent (NR) method compared with solutions using the modified UV cut-off regulator and dimensional regularization (DR). The DR solution is extrapolated from $D \neq 4$ to $D = 4$ using a cubic fit in $(D-4)$ at each momentum point. All solutions have renormalized mass $m_\mu = 400$ (in arbitrary units) at the renormalization point $\mu^2 = 10^8$.

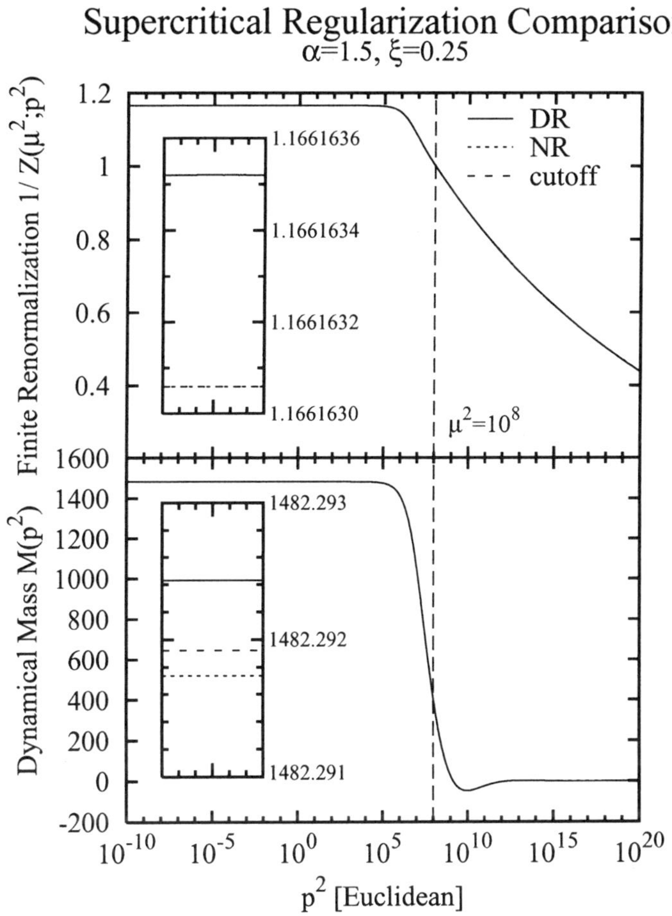

Figure 3: Same as Fig.2, but at supercritical coupling

malized equations become, in Euclidean space,

$$Z(\mu^2;p^2) = 1 - \frac{\alpha\xi}{4\pi}\int_{p^2}^{\mu^2} dk^2 \frac{1}{[k^2+M^2(k^2)]} Z(\mu^2;k^2)$$
$$+\frac{\alpha}{4\pi}\int_0^\infty \frac{dk^2}{[k^2+M^2(k^2)]}\left[Z(\mu^2;p^2)\, I(k^2,p^2) - I(k^2,\mu^2)\right] \quad (11)$$

$$M(p^2) = M_\mu + \frac{\alpha}{4\pi}\int_0^\infty \frac{dk^2}{[k^2+M^2(k^2)]}\left[J(k^2,p^2) - J(k^2,\mu^2)\right.$$
$$\left. + M(p^2)\, I(k^2,p^2) - M_\mu\, I(k^2,\mu^2)\right] \quad (12)$$

where $I(k^2,p^2)$ and $J(k^2,p^2)$ are those of Table 1 of Atkinson et al.'s paper [5], apart from the term proportional to $\xi\left(1+\frac{M^2(k^2)}{k^2}\right)$ in $I(k^2,p^2)$ in their paper which originated from the violation of gauge covariance by their cutoff. The equation for the wavefunction renormalization function (Eq. 11) is the same as Eq. (10) in Ref. [6], however our treatment of the mass function differs from theirs. These equations were solved numerically [3] using $\xi = 0.25$, $\mu^2 = 10^8$ and $m_\mu = 400$ as input. Solutions for $\alpha = 0.6$ (subcritical) and $\alpha = 1.5$ (supercritical) are shown in Figs. 2 and 3 where they are compared to those from previously published studies employing cut-off regularization [7] (with the gauge-covariance modification) or dimensional regularization [2].

One can see the agreement between the regulator-independent and the modified cut-off and dimensionally regularized solutions is excellent. They are indistinguishable on the main figures: the inserts in Figs. 2 and 3 have the same p^2 scale and reveal the remarkable agreement between them in the infrared region. We conclude with some confidence that the three regularization schemes give identical results for the renormalized solutions.

In short, it appears that the regulator-free method outlined here reproduces well established results in quenched QED to high precision and with significantly reduced numerical effort. Having established this, we are now proceeding to the study of unquenched QED_4.

Acknowledgments

This work was supported by the Australian Research Council.

References

1. Z. Dong, H. Munczek, and C. D. Roberts, *Phys. Lett.* B**333**, 536 (1994).
2. A.W. Schreiber, T. Sizer and A.G. Williams, *Phys. Rev.* D**58**, 125014 (1998); V. P. Gusynin, A. W. Schreiber, T. Sizer and A.G. Williams, *Phys. Rev.* D**60**, 065007 (1999).
3. A.Kızılersü, A.W. Schreiber and A.G. Williams, *Phys. Lett.* B**499**, 261 (2001); A.Kızılersü, T. Sizer and A.G. Williams, hep-ph/0101188.
4. K.-I. Kondo, H. Mino and H. Nakatani, *Mod. Phys. Lett. A***17**, 1509 (1992); J. C. R. Bloch and M.R. Pennington, *Mod. Phys. Lett. A***10**, 1225 (1995).
5. D. Atkinson et al., *Phys. Lett.* B**329**, 117 (1994).
6. D.C. Curtis and M.R. Pennington, *Phys. Rev.* D**46**, 2663 (1992).
7. F.T. Hawes and A.G. Williams, *Phys. Rev.* D**51**, 3081 (1995); F.T. Hawes, A.G. Williams and C.D. Roberts, *Phys. Rev.* D**54**, 5361 (1996); F.T. Hawes, T. Sizer and A.G. Williams, *Phys. Rev.* D**55**, 3866 (1997).

DOMAIN-LIKE STRUCTURES IN THE QCD VACUUM, CONFINEMENT AND CHIRAL SYMMETRY BREAKING

A. C. KALLONIATIS

Centre for the Subatomic Structure of Matter, University of Adelaide
South Australia 5005, Australia
E-mail: akalloni@physics.adelaide.edu.au

S. N. NEDELKO [a]

Institüt für Theoretische Physik III, Universität Erlangen-Nürnberg,
Staudtstrasse 7, Erlangen 91058, Germany
E-mail: nedelko@theorie3.physik.uni-erlangen.de

We discuss how the inclusion of singular gauge fields in the partition function for QCD can lead to a domain-like picture for the QCD vacuum by virtue of specific conditions on quantum fluctuations at the singularities. With a simplified model of hyperspherical domain regions with interiors of constant field strength we calculate the basic parameters of the QCD vacuum, the gluon condensate, topological susceptibility, string constant and quark condensate, and briefly discuss confinement of dynamical quarks and gluons.

1 Introduction

Singular gauge fields are most probably unavoidable in nonabelian gauge theories [1]. In the last two decades great effort has gone into clarifying whether or not (and, if yes, then how) such fields provide for confinement and chiral symmetry breaking.

We address this question in a model based on an assumption that singularities in vector potentials are concentrated on hypersurfaces ∂V_j in Euclidean space, in the vicinity of which gauge fields can be divided into a sum of a singular pure gauge S_μ and a regular fluctuation part Q_μ, and where a certain colour vector n_j^a can be associated with ∂V_j. For such configurations not to have infinite action[2] the fluctuation fields charged with respect to n_j must obey specific boundary conditions on ∂V_j. The interiors of these regions thus constitute "domains" V_j. Gauge field modes neutral with respect to n_j^a are not restricted and provide for interactions between domains. In a given domain V_j the effect of fluctuations in neighbouring regions can be seen as an external gauge field $B^a_{j\mu}$ neutral with respect to n_j^a. This enables an approximation in which different domains are assumed to be decoupled from each other but, to compensate, a certain mean field is introduced in domain interiors.

[a]On leave of absence from Bogoliubov LTPh, JINR, Dubna, Russia

Decomposing a general gauge field $A^j_\mu = S^j_\mu + Q^j_\mu$ and demanding finiteness of the classical action we come to the conditions

$$\breve{n}_j Q^{(j)}_\mu = 0,\ \psi = -i\,\not\eta^j e^{i\alpha_j\gamma_5}\psi,\ \bar\psi = \bar\psi i\,\not\eta^j e^{-i\alpha_j\gamma_5},\ x \in \partial V_j, \tag{1}$$

with the adjoint matrix $\breve{n}_j = T^a n^a_j$ appearing in the condition for gluons, and a bag-like boundary condition arising for quarks, with a unit vector $\eta^j_\mu(x) = x_\mu/|x|$ normal to ∂V_j.

To make the model analytically tractable we consider spherical domains with fixed radius R and approximate the mean field in V_j by a covariantly constant (anti-)self-dual configuration with the field strength

$$\hat{\mathcal{B}}^{(j)}_{\mu\nu} = \hat n^{(j)} B^{(j)}_{\mu\nu},\ \tilde B^{(j)}_{\mu\nu} = \pm B^{(j)}_{\mu\nu},\ B^{(j)}_{\mu\nu} B^{(j)}_{\rho\nu} = B^2\delta_{\mu\rho},\ B = \text{const},$$
$$\hat n^{(j)} = t^3\cos\xi_j + t^8\sin\xi_j\ (\xi_j \in \{\frac{\pi}{6}(2k+1),\ k = 0,\ldots,5\}),$$

where the parameter B is the same for all domains and the constant matrix $n^a_j t^a$ belongs to the Cartan subalgebra. It should be stressed that there is no source for this field on the boundary and therefore it should be treated as strictly homogeneous in all further calculations. The homogeneity itself appears here just as an approximation.

Thus the partition function we will deal with can be written

$$\mathcal{Z} = \mathcal{N} \lim_{V,N\to\infty} \prod_{i=1}^{N} \int_V \frac{d^4 z_i}{V} \int_\Sigma d\sigma_i \int_{\mathcal{F}^i_Q} \mathcal{D}Q^i \int_{\mathcal{F}^i_\psi} \mathcal{D}\psi_i \mathcal{D}\bar\psi_i$$
$$\delta[D(\breve{\mathcal{B}}^{(i)})Q^{(i)}]\Delta_{\rm FP}[\breve{\mathcal{B}}^{(i)}, Q^{(i)}] \exp\left\{-S^{\rm QCD}_{V_i}\left[Q^{(i)} + \mathcal{B}^{(i)}, \psi^{(i)}, \bar\psi^{(i)}\right]\right\},$$

where the thermodynamic limit assumes $V, N \to \infty$ but with the density $v^{-1} = N/V$ taken fixed and finite. The measure of integration over parameters characterising domains is defined as

$$\int_\Sigma d\sigma_i \ldots = \frac{1}{48\pi^2} \int_0^{2\pi} d\alpha_i \int_0^{2\pi} d\varphi_i \int_0^\pi d\theta_i \sin\theta_i$$
$$\times \int_0^\pi d\omega_i \sum_{k=0,1} \delta(\omega_i - \pi k) \int_0^{2\pi} d\xi_i \sum_{l=0,1,2}^{3,4,5} \delta(\xi_i - (2l+1)\pi/6)\ldots \tag{2}$$

Here φ_i and θ_i are spherical angles of the chromomagnetic field, ω_i is the angle between the chromomagnetic and chromoelectric fields, ξ_i is the angle in the colour matrix $\hat n_i$, α_i is the chiral angle and z_i is the centre of the domain V_i.

This partition function describes a statistical system of density N/V composed of noninteracting extended domain-like structures, each of which is characterised by a set of internal parameters and whose internal dynamics are represented by the fluctuation fields.

2 Mean Field Correlators

In this model the connected $n-$point correlator

$$\langle B^{a_1}_{\mu_1\nu_1}(x_1)\ldots B^{a_n}_{\mu_n\nu_n}(x_n)\rangle = B^n t^{a_1\ldots a_n}_{\mu_1\nu_1,\ldots,\mu_n\nu_n}\Xi_n(x_1,\ldots,x_n),$$
$$t^{a_1\ldots a_n}_{\mu_1\nu_1,\ldots,\mu_n\nu_n} = \int d\sigma_j\, n^{(j)a_1}\ldots n^{(j)a_n} B^{(j)}_{\mu_1\nu_1}\ldots B^{(j)}_{\mu_n\nu_n},$$

of field strength tensors,

$$B^a_{\mu\nu}(x) = \sum_j^N n^{(j)a} B^{(j)}_{\mu\nu}\theta(1-(x-z_j)^2/R^2),$$

can be calculated explicitly using the measure, Eq. (2). Translation-invariant functions

$$\Xi_n(x_1,\ldots,x_n) = \frac{1}{v}\int d^4z\,\theta(1-(x_1-z)^2/R^2)\ldots\theta(1-(x_n-z)^2/R^2)$$

emerge and can be seen as the volume of the region of overlap of n hyperspheres of radius R and centres $(x_1,\ldots,x_n)$, normalised to the volume of a single hypersphere $v=\pi^2R^4/2$. The functions Ξ_n are continuous and vanish if $|x_i-x_j|\geq 2R$. Correlations in the background field have finite range $2R$. The Fourier transform of Ξ_n is then an entire analytical function and thus correlations do not have a particle interpretation. The statistical ensemble of background fields is not Gaussian since all connected correlators are independent of each other and cannot be reduced to the two-point correlations.

The simplest application of the above correlators gives a gluon condensate density which to this approximation is

$$g^2\langle F^a_{\mu\nu}(x)F^a_{\mu\nu}(x)\rangle = 4B^2.$$

Another vacuum parameter which plays a significant role in the resolution of the $U_A(1)$ problem is the topological susceptibility [3]. To define this we consider first the topological charge density for the colour group $SU(3)$ in the lowest approximation

$$Q(x) = \frac{g^2}{32\pi^2}\tilde{F}^a_{\mu\nu}(x)F^a_{\mu\nu}(x) = \frac{B^2}{8\pi^2}\sum_{j=1}^N\theta[1-(x-z_j)^2/R^2]\cos\omega_j,$$

where $\omega_j \in \{0, \pi\}$ depends on the duality of the j-th domain. For a given field configuration the topological charge is additive

$$Q = \int_V d^4x \; Q(x) = q(N_+ - N_-), \; q = B^2 R^4/16, \; -Nq \leq Q \leq Nq$$

where q is a 'unit' topological charge, namely the absolute value of the topological charge of a single domain, and N_+ (N_-) is the number of domains with (anti-)self-dual field, $N = N_+ + N_-$. The probability of finding the topological charge Q in a given configuration is defined by the distribution

$$\mathcal{P}_N(Q) = \frac{\mathcal{N}_N(Q)}{\mathcal{N}_N} = \frac{N!}{2^N \, (N/2 - Q/2q)! \, (N/2 + Q/2q)!},$$

where $\mathcal{N}_N(Q)$ is the number of configurations with a given charge and $\mathcal{N}_N$ is the total number of configurations. The distribution is symmetric about $Q = 0$, where it has a maximum for N even. For N odd the maximum is at $Q = \pm q$. Averaged topological charge is zero.

The topological susceptibility χ is determined by the two-point correlator of topological charge density, which in the lowest approximation gives

$$\chi = \int d^4x \; \langle Q(x) Q(0) \rangle = \frac{B^4}{64\pi^4} \int d^4x \; \Xi_2(x) = \frac{B^4 R^4}{128\pi^2}. \tag{3}$$

3 Area Law for the Wilson Loop

To zeroth order in fluctuations the Wilson loop is given by the integral

$$W(L) = \lim_{V,N \to \infty} \prod_{j=1}^{N} \int_V \frac{d^4 z_j}{V} \int d\sigma_j \frac{1}{N_c} \mathrm{Tr} \exp \left\{ i \int_{S_L} d\sigma_{\mu\nu}(x) \; \hat{B}_{\mu\nu}(x) \right\}.$$

Note that path ordering in our case is not necessary since the matrices $\hat{n}^k$ are assumed to be in the Cartan subalgebra. Computationally it is convenient to consider a circular contour in the (x_3, x_4) plane of radius L with centre at the origin. To illustrate the steps in the calculation we consider here the case of colour group $SU(2)$, though the final result for $SU(3)$ will be quoted below. For colour $SU(2)$ we have $\hat{n}^k = \epsilon^k \tau_3$, $\epsilon^k = \pm 1$. The thermodynamic limit assumes that the density N/V is fixed. Calculation of the colour trace gives

$$W(L) = \lim_{V,N \to \infty} \left[\int_V \frac{d^4 z_j}{V} \int d\sigma_j \frac{1}{2} \left(e^{i B^j_{\mu\nu} J_{\mu\nu}(z_j)} + e^{-i B^j_{\mu\nu} J_{\mu\nu}(z_j)} \right) \right]^N,$$

$$J_{\mu\nu}(z_k) = \int_{S_L} d\sigma_{\mu\nu}(x) \; \theta(1 - (x - z_k)^2/R^2).$$

We have exploited the property that the integral above does not depend on the index j. As the contour of the Wilson loop is in the (x_3, x_4)-plane, the only nonzero components of $J_{\mu\nu}$ are

$$J_{34} = -J_{43} = \int_{S_L} dx_3\, dx_4\, \theta(1-(x-z)^2/R^2),\ B_{\mu\nu}J_{\mu\nu} = 2BJ_{43}\cos\theta,$$

where θ is the angle between chromoelectric field and the third coordinate axis. After integrating over the spatial orientations of the vacuum field the Wilson loop takes the form

$$W(L) = \lim_{N,V\to\infty}\left[\frac{1}{V}\int_V dz\,\frac{\sin 2BJ_{43}(z)}{2BJ_{43}(z)}\right]^N.$$

Integrating over z and then taking the thermodynamic limit ($N\to\infty$, $v = V/N = \pi^2R^4/2$), gives finally for a large Wilson loop $L \gg R$

$$W(L) = \lim_{N\to\infty}\left[1-\frac{1}{N}U(L)\right]^N = e^{-U(L)},\ U(L) = \sigma\pi L^2 + O(L),$$

$$\sigma = Bf(BR^2),\ f(z) = \frac{2}{\pi z}\left(1-\frac{1}{2\pi z}\int_0^{2\pi z}\frac{dx}{x}\sin x\right),$$

which displays an area law. For $SU(3)$ the only difference is in the function $f(z)$ which turns out to be:

$$f(z) = \frac{2}{3\pi z}\left(3-\frac{\sqrt{3}}{2\pi z}\int_0^{2\pi z/\sqrt{3}}\frac{dx}{x}\sin x - \frac{2\sqrt{3}}{\pi z}\int_0^{\pi z/\sqrt{3}}\frac{dx}{x}\sin x\right).$$

The function f is positive for $z > 0$ and has a maximum for $z = 1.55$. We choose this maximum to estimate the model parameters by fitting the string constant to the lattice result,

$$\sqrt{B} = 947\text{MeV},\ R^{-1} = 760\text{MeV}, \tag{4}$$

and get for the gluonic parameters of the vacuum

$$\sqrt{\sigma} = 420\text{MeV},\ \chi = (197\text{MeV})^4,\ \frac{\alpha_s}{\pi}\langle F^2\rangle = 0.081(\text{GeV})^4,\ q = 0.15. \tag{5}$$

If there were clear separation of the two scales characterising the system (namely if either $\sqrt{B}R \gg 1$ or $\sqrt{B}R \ll 1$) then an approximate treatment of Green's functions for the fluctuation fields would be possible. We observe that there is no separation: $\sqrt{B}R \approx 1$. Thus an approximation based on large or small domains is not justifiable.

Observe also that if B goes to zero then the string constant vanishes. This underscores the role of the gluon condensate in the confinement of static charges. On the other hand, if the number of domains is fixed and the thermodynamic limit is defined as $V, R \to \infty, N = \text{const.} < \infty$, namely if the domains are macroscopically large, then $W(L) = 1$, which indicates the absence of a linear potential between infinitely heavy charges in a purely homogeneous field.

Since we have exactly integrated over background fields the role of finite range of correlation functions is hidden in the above calculation. In order to see this role explicitly one would need to decompose the integrand into an infinite series and integrate term by term. At this step all correlation functions of the background field up to infinite order would be manifest. The arguments about the importance of a fast decay of correlators for confinement of static charges [4] would be seen to apply here via this representation.

4 Confinement of Fluctuation Fields

Confinement of the fluctuational fields can be studied via the analytical properties of their propagators. For the types of fields and boundary conditions we have described here, the propagators of fluctuations can be analytically calculated by reduction of the problems for both quarks and gluons to that of a charged scalar field. The scalar case is essentially equivalent to the problem of a four-dimensional harmonic oscillator with the orbital momentum coupled to the external field, and the general solution for these Green's functions can be found exactly by standard decomposition over hyperspherical harmonics.

Qualitatively it is clear that due to the boundary conditions we are dealing with, the $x-$space propagators of charged fields are defined in regions of finite support where they have integrable singularities so that their Fourier transforms are entire functions in the complex momentum plane. This can be treated as confinement of dynamical fields[5]. A known consequence of entire propagators is a Regge spectrum of relativistic bound states[6], which is physically appealing. A more expansive analysis of confinement in this model will be given elsewhere[7].

5 Quark Condensate

Due to averaging over self- and anti-self-dual configurations and all possible values of the angle α in the partition function chiral symmetry is not broken explicitly. However, as we show below, a nonzero quark condensate arises in the massless limit due to an interplay between the random distribution of domains with self- and anti-self-dual field and the boundary conditions with random value of the chirality violating angle α.

The quark propagator is defined by the equations

$$(i\partial\!\!\!/ - \frac{1}{2}\hat{n}\gamma_\mu B_{\mu\nu}x_\nu - m)S(x,y) = -\delta(x,y), \tag{6}$$
$$i\,\eta\!\!\!/(x)e^{i\alpha\gamma_5}S(x,y) = -S(x,y),\ (x-z)^2 = R^2,$$
$$S(x,y)i\,\eta\!\!\!/(y)e^{-i\alpha\gamma_5} = S(x,y),\ (y-z)^2 = R^2,$$

where $\eta_\mu(x) = (x-z)_\mu/|x-z|$. Substitution

$$S(x,y) = (i\,D\!\!\!\!/ + m)[P_\pm\mathcal{H}_0 + P_\mp O_+\mathcal{H}_1 + P_\mp O_-\mathcal{H}_{-1}],$$
$$O_+ = N_+\Sigma_+ + N_-\Sigma_-,\ O_- = N_+\Sigma_- + N_-\Sigma_+,$$
$$N_\pm = \frac{1}{2}(1 \pm \hat{n}/|\hat{n}|),\ \Sigma_\pm = \frac{1}{2}(1 \pm \vec{\Sigma}\vec{B}/B),\ \hat{B} = |\hat{n}|B,$$

shows that scalar functions $\mathcal{H}_\zeta(x,y)$ should solve the equations:

$$(-D^2 + m^2 + 2\zeta\hat{B})\mathcal{H}_\zeta = \delta(x,y),$$

If we were to look for solutions vanishing at infinity then the Green's function $\mathcal{H}_{-1}$ $(\zeta = -1)$ would be divergent in the massless limit due to zero modes of the Dirac operator. The bag-like boundary conditions remove zero eigenvalues from the spectrum, and the massless limit is regular.

In order to avoid cumbersome calculations and render the role of the former zero modes transparent, we turn to the particular choice $y = z = 0$ and calculate the value of the quark condensate at the centre of the domain. In this case $\mathcal{H}_\zeta$ are functions of x^2 only, and the general solutions for scalar Green's functions take the form $(\mu_\zeta = m^2/2B + \zeta)$

$$\mathcal{H}_\zeta = \Delta(x^2|\mu_\zeta) + C_\zeta\Phi(x^2|\mu_\zeta),\ \Phi(x^2|\mu) = e^{-Bx^2/4}M(1+\mu, 2, Bx^2/2).$$

Here $\Delta(x^2|\mu)$ is scalar propagator which vanishes at infinity with mass $2B\mu$, and Φ_ζ is a solution to the homogeneous equation regular at $x^2 = 0$, expressed in terms of the confluent hypergeometric function. The constants C_ζ can be fit to implement the boundary condition. The terms $m\mathcal{H}_0$ and $m\mathcal{H}_1$ vanish in the massless limit and do not contribute to the condensate. The nontrivial contribution comes from $m\mathcal{H}_{-1}$. The bag-like conditions imply that on the boundary $\mathcal{H}_{-1}$ satisfies a mixed condition, where $f' = df/d|x|$ and the sign $(-)+$ corresponds to an (anti-)self-dual domain,

$$2e^{\mp i\alpha}m\mathcal{H}_{-1} = -2\mathcal{H}'_{-1} - \hat{B}R^2\mathcal{H}_{-1}$$

which leads to the relations

$$\lim_{m\to 0} m\mathcal{H}_{-1} = \frac{e^{\pm i\alpha}}{2\pi^2 R^3} F(\hat{B}R^2/2) e^{-\hat{B}x^2/4}, \ \mathrm{Tr}S(0,0) = \frac{e^{\pm i\alpha}}{2\pi^2 R^3} \sum_{|\hat{n}|} F(\hat{B}R^2/2),$$

$$F(z) = e^z - z - 1 + \frac{z^2}{4} \int_0^\infty dt \ \frac{e^{2t - z(\coth t - 1)/2}}{\sinh^2 t} (\coth t - 1).$$

Note that the term in the propagator with nonzero trace is proportional to the zero mode of the Dirac operator: $\not{D} P_{\mp} O_- \exp(-\hat{B}x^2/4) = 0$.

Averaging this result over the angle α and (anti-)self-dual configurations and taking into account the α-dependence of quark determinant [8] $\det S^{-1} \propto \exp\{\pm iq\alpha\}$, with q being unit topological charge, we get a finite result

$$\langle\bar{\psi}\psi\rangle = -\frac{q}{2\pi^2 R^3 (1+q)} \sum_{|\hat{n}|} F(\hat{B}R^2/2).$$

For B and R fixed by the string constant as in Eq. (4) this is equal to

$$\langle\bar{\psi}\psi\rangle = -(228\mathrm{MeV})^3,$$

which indicates spontaneous breaking of chiral symmetry. Final conclusions require complete calculation including averaging over positions of domains.

Acknowledgments

ACK is supported by the Australian Research Council. We are grateful to Frieder Lenz for fruitful discussions and critical comments, as well as to Garii Efimov, Jan Pawlowski and Lorenz von Smekal for valuable communication.

References

1. V. N. Gribov, *Nucl. Phys.* B **139**, 1 (1978); I. M. Singer, *Comm. Math. Phys.* **60**, 7 (1978).
2. F. Lenz, O. Jahn, *Phys. Rev.* D **58**, 085006 (1998); F. Lenz, S. Worlen, hep-ph/0010099.
3. R.J. Crewther, *Phys.Lett.* B **70**, 349 (1977); G. Veneziano, *Nucl. Phys.* B **159**, 213 (1979); E. Witten, *Nucl. Phys.* B **156**, 269 (1979).
4. G.Dosch, Y.A. Simonov, *Phys. Lett.* B **205**, 339 (1988).
5. H. Leutwyler, *Phys. Lett.* B **96**, 154 (1980).
6. G.V. Efimov, S.N. Nedelko, *Phys. Rev.* D **51**, 174 (1995); G.V. Efimov, G. Ganbold, hep-ph/0103101.
7. S. Nedelko, A. Kalloniatis, *in preparation.*
8. S. Duerr, A. Wipf, *Nucl. Phys.* B **443**, 201 (1995).

GLUEBALL PROPERTIES IN ANISOTROPIC SU(3) LATTICE QCD WITH AN IMPROVED ACTION

NORIYOSHI ISHII
The Institute of Physical and Chemical Research(RIKEN),
2-1 Hirosawa, Wako, Saitama 351-0198, JAPAN

HIDEO SUGANUMA
Tokyo Institute of Technology,
2-12-1 Ohkayama, Meguro, Tokyo 152-8552, JAPAN

HIDEO MATSUFURU
Yukawa Institute for Theoretical Physics, Kyoto University,
Kitashirakawa-Oiwake, Sakyo, Kyoto 606-8502, JAPAN

We study glueball properties at finite temperature using $SU(3)$ lattice QCD at the quenched level with an anisotropic lattice. We use a tree-level Symanzik $O(a^2)$ improved action. We present our preliminary results which shows the slight reduction of the scalar glueball mass near T_c.

1 Introduction

Quenched $SU(3)$ lattice QCD Monte Carlo simulations suggest the existence of a finite temperature deconfinement transition at the critical temperature $T_c \simeq 260$ MeV [1]. At the quenched level, without involving dynamical quark excitations, the lightest physical excitation is the scalar glueball with a mass of 1.5 GeV $\lesssim m_G \lesssim$ 1.7 GeV [2,3], which can be expected to dominate the thermodynamic properties below T_c. The closed packing model, a phenomenological model for the QCD phase transition, provides us with an intuitive picture that the deconfinement transition takes place when "hadrons" excited from the thermal system begin to overlap significantly with one another. However, since the contributions from glueballs are suppressed by a rather strong statistical factor $e^{-m_G/T}$, we may wonder if the total number of glueballs to be excited may be insufficient in quenched QCD. To keep the consistency with the closed packing model, the resolution may be found in either a tremendous expansion of the glueball size or a drastic reduction of the mass [a] near the critical temperature (See Fig.1). Indeed, the lattice result for the correlator of Polyakov loops supports a change in the shape of the $q\bar{q}$ potential at finite temperatures [5,6]. As a consequence, the phenomenological potential model also

[a] The reduction of the glueball mass is also anticipated in the monopole condensation picture of confinement [4].

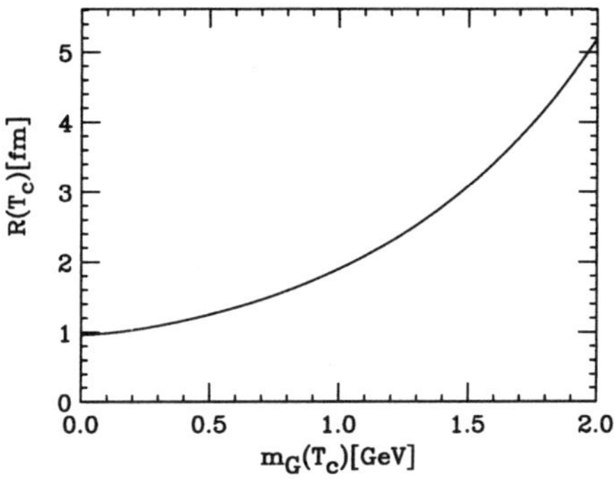

Figure 1: The glueball radius $R(T_c)$, adjusted so as to reproduce $T_c \simeq 260$ MeV in the closed packing model, is plotted against the effective thermal glueball mass $m_G(T_c)$.

suggests a change in the wave function of the glueball, particularly its mass and size. It is therefore important to study the properties of glueballs at finite temperature, although no such studies have been done so far on the lattice. Even for typical hadrons, most studies of mass shifts on the lattice at finite temperature are unfortunately limited to the screening mass , i.e., the correlation along the spatial direction [7], whereas what is really interesting is the direct information on the mass from correlations along the temporal direction. This is because the lattice points available in the temporal direction decrease as the temperature becomes higher, and the extraction of the mass spectrum becomes more difficult. Recently, the use of an anisotropic lattice where the temporal spacing is finer than the spatial one has been introduced [5,10], which enables us to overcome the difficulties mentioned above. In this paper, after an analytical consideration of the closed packing model, we will outline how to study glueball properties at finite temperature by means of the anisotropic lattice and the smearing method. We will present preliminary results in $SU(3)$ lattice QCD with a tree-level Symanzik $O(a^2)$ improved action.

2 The closed packing model for glueballs in quenched QCD

In this section, we analytically investigate the deconfinement phase transition at the quenched level in terms of low-lying scalar glueball properties. Consider the bag model picture of the scalar glueball. In this picture, a sphere of radius R distinguishes the outside from the inside, where the gluonic degrees of freedom appear. As long as the temperature T is low, the number $N(T)$ of thermally excited bags is small. However, as the temperature increases, the number

$N(T)$ of bags increases. Finally, these bags begin to overlap with each other and cover the whole space, and thus the deconfinement phase transition takes place in the closed packing model. The ratio of the volume inside the gluonic bags to the total volume V is given as

$$r_V(T) = \frac{\frac{4\pi}{3}R(T)^3 \times N(T)}{V} = \frac{4\pi}{3}R(T)^3 \int \frac{d^3k}{(2\pi)^3}\frac{1}{e^{E(\vec{k},T)/T}-1}. \quad (1)$$

Now we simplify the dispersion relation to the quasi-free relation $E(\vec{k},T) = \sqrt{\vec{k}^2 + m_{\rm G}^2(T)}$ with $m_{\rm G}(T)$ being the effective glueball mass at temperature T. Since the deconfinement phase transition is expected to take place when $r_V(T) \simeq 1$, the glueball radius $R(T_c)$ near the deconfinement phase transition is estimated as

$$R(T_c) \simeq \left(\frac{4\pi}{3}\int \frac{d^3k}{(2\pi)^3}\frac{1}{e^{\sqrt{\vec{k}^2+m_{\rm G}^2(T_c)}/T_c}-1}\right)^{-1/3}. \quad (2)$$

In Fig.1, the glueball radius $R(T_c)$, adjusted so as to reproduce the deconfinement phase transition at $T_c = 260$ MeV, is plotted as a function of the glueball mass $m_{\rm G}(T_c)$ in the closed packing model. Thus, if $m_{\rm G}$ persists to be 1.7 GeV near T_c, the glueball should have the abnormally huge radius 3.8 fm. Instead, if R remains at a typical hadron size of about 1 fm, the glueball mass should reduce drastically to $m_{\rm G}(T_c) \lesssim 0.5$ GeV.

3 The smearing method in $SU(3)$ lattice QCD

Using $SU(3)$ lattice QCD, we study the glueball correlator

$$G(t) \equiv \langle \tilde{O}(t)\tilde{O}(0)\rangle, \quad \tilde{O}(t) \equiv O(t) - \langle O \rangle, \quad O(t) \equiv \sum_{\vec{x}} O(t,\vec{x}). \quad (3)$$

The subtraction of the vacuum expectation value is necessary for the scalar glueball with $J^{\rm PC} = 0^{++}$. The summation over $\vec{x}$ physically means the momentum projection as $\vec{p} = \vec{0}$. In the case of the scalar glueball, the summand is [2,3,11,12]

$$O(t,\vec{x}) \equiv O_{12}(t,\vec{x}) + O_{23}(t,\vec{x}) + O_{31}(t,\vec{x}), \quad (4)$$

where $O_{ij}(t,\vec{x})$ is the real part of the trace of the plaquette operator which lies within the i-j plain. In the spectral representation, $G(t)$ is expressed as

$$G(t)/G(0) = \sum_{n=1}^{\infty} c_n e^{-E_n t}, \quad c_n \equiv \left|\langle n|\tilde{O}|0\rangle\right|^2 / G(0), \quad G(0) = \sum_{n=1}^{\infty}\left|\langle n|\tilde{O}|0\rangle\right|^2, \quad (5)$$

where E_n denotes the energy of the n-th excited state $|n\rangle$. (In this paper, $|0\rangle$ denotes the vacuum, and $|1\rangle$ denotes the ground-state 0^{++} glueball.) Note that c_n is a positive number with $\sum c_n = 1$, and thus $G(t)/G(0)$ can be expressed as a weighted average of the exponentials $e^{-E_n t}$ with weight c_n. On a finite lattice with spacing a, the simple plaquette operator $O_{ij}(t, \vec{x})$ has a small overlap with the glueball ground state $|G\rangle \equiv |1\rangle$, and the extracted mass from $G(t)$ looks heavier owing to excited-state contamination. This problem of small overlap originates from the fact that the plaquette operator has a smaller "size" of $O(a)$ than the physical size of the glueball. This problem becomes more severe near the continuum limit, i.e., $a \simeq 0$. We thus have to enhance the overlap by improving the glueball operator so as to have approximately the same size as the physical size of the glueball. One of the systematic ways to achieve this is known to be the smearing method[2,11,13,14]. The smearing method is expressed as the iterative replacement of the original spatial link variables $U_i(s)$ by the associated fat link variables, $\bar{U}_i(s) \in SU(3)_c$. The fat link variable $\bar{U}_i(s)$ is defined to be the $SU(3)_c$ element which maximizes

$$\mathrm{ReTr}\left\{ \bar{U}_i^\dagger(s) \left[\alpha U_i(s) + \sum_{j\neq i} \sum_{\pm} U_j(s) U_i(s \pm \hat{j}) U_j^\dagger(s \pm \hat{i}) \right] \right\}, \tag{6}$$

where $U_{-\mu}(s) \equiv U_\mu^\dagger(s - \hat{\mu})$, and α is a real parameter. A schematic illustration of the fat link is given by

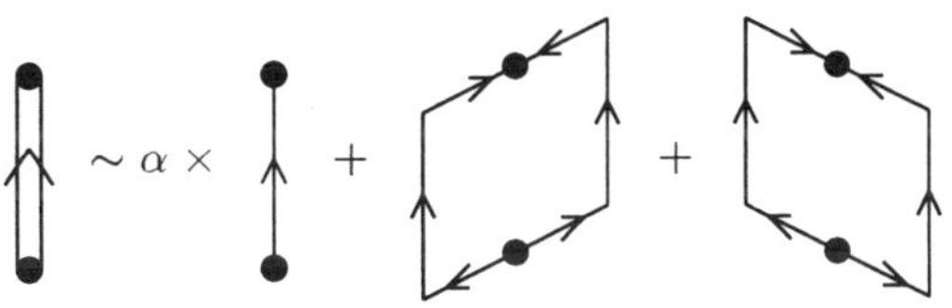

(7)

Note that the summation is only over the spatial direction so as to avoid the nonlocality in the temporal direction. Note also that $\bar{U}_i(s)$ has the same gauge transformation properties as $U_i(s)$. We will refer to the fat link defined in Eq.(6) as the first fat link. The n-th fat link is defined iteratively, i.e., it is obtained by repeating the procedure in Eq.(6) with $(n-1)$ fat links. For the physically extended glueball operator, we use the n-th smeared operator, the plaquette operator which is constructed with the n-th fat link variables. To attribute a size to the n-th smeared operator, we use the linearization and the continuum approximation, and obtain the diffusion equation[8] as

$$\frac{\partial}{\partial n}\phi_i(n;\vec{x}) = D\Delta\phi_i(n;\vec{x}), \quad D \equiv \frac{a_s^2}{\alpha+4}, \quad a_s : \text{spatial lattice spacing.} \tag{8}$$

Here $\phi_i(n;\vec{x})$ denotes the distribution function of the gluon field $A_i(\vec{x},t)$ after n smearing operations. The n-th smeared plaquette located at the origin $\vec{x}=\vec{0}$ physically means the Gaussian extended glueball operator with the distribution

$$\phi_i(n;\vec{x}) = \frac{1}{(4\pi Dn)^{3/2}} \exp\left[-\frac{\vec{x}^2}{4Dn}\right]. \tag{9}$$

Hence the size of the operator is estimated as

$$\sqrt{\langle \vec{x}^2 \rangle} = \sqrt{6Dn} = a_s \sqrt{\frac{6n}{\alpha+4}}. \tag{10}$$

So far, the smearing method was introduced to carry out an accurate mass measurement by maximizing the overlap. However, it can be also used to give a rough estimate of the physical glueball size. In fact, once we obtain the maximum overlap with some n and α, the physical size of the glueball is estimated with Eq.(10).

4 The $SU(3)$ lattice QCD result for glueball mass and size at $T \neq 0$

In this section, we present our preliminary results. To generate the gauge field configurations, we use the tree-level Symanzik $O(a^2)$ improved action [9]

$$\begin{aligned} S_{\rm G} =& \beta \frac{1}{\gamma_{\rm G}} \sum_{s,i<j\leq 3} \Big[c_{11}(1-P_{ij}(s)) + c_{12}(2-R_{ij}(s)-R_{ji}(s)) \Big] \\ &+ \beta\gamma_{\rm G} \sum_{s,i\leq 3} \Big[c_{11}(1-P_{i4}(s)) + c_{12}(2-R_{i4}(s)-R_{4i}(s)) \Big], \end{aligned} \tag{11}$$

where $c_{11} = 5/3$, $c_{12} = -1/12$, $\beta = 4.56$ and $\gamma_{\rm G} = 3.45$. $P_{\mu\nu}(s)$ is the plaquette and the rectangular loop $R_{\mu\nu}(s)$ is defined as $R_{\mu\nu}(s) = \frac{1}{3}{\rm ReTr}\Big(U_\mu(s)U_\mu(s+\hat{\mu})U_\nu(s+2\hat{\mu})U_\mu^\dagger(s+\hat{\nu}+\hat{\mu})U_\mu^\dagger(s+\hat{\nu})U_\nu^\dagger(s) \Big)$. The numerical calculations are performed on lattices of size $16^2 \times 24 \times N_t$, $N_t = 28$ ($T = 0.87T_c$) and $N_t = 96$ ($T \simeq 0$) [5,10]. These parameters generate a spatial lattice spacing of $a_s = 0.12$ fm and a temporal lattice spacing of $a_t = 0.03$ fm. We have used 500 gauge field configurations for $N_t = 28$ and 150 for $N_t = 96$. Throughout this section, we take the smearing parameter to be $\alpha = 2.3$.

In Fig.2, the overlap $C \equiv \left|\langle G|\tilde{O}|0\rangle\right|^2 / G(0)$ and $\chi^2/N_{\rm DF}$ are plotted against the iteration number of the smearing $N_{\rm smear}$ $(= n)$. The overlap becomes maximum with reasonable χ^2 as $\chi^2/N_{\rm DF} \lesssim 1$ in the region $10 \leq N_{\rm smear} \leq 24$. In this region, as is shown in Fig.3, the glueball mass, which is extracted with

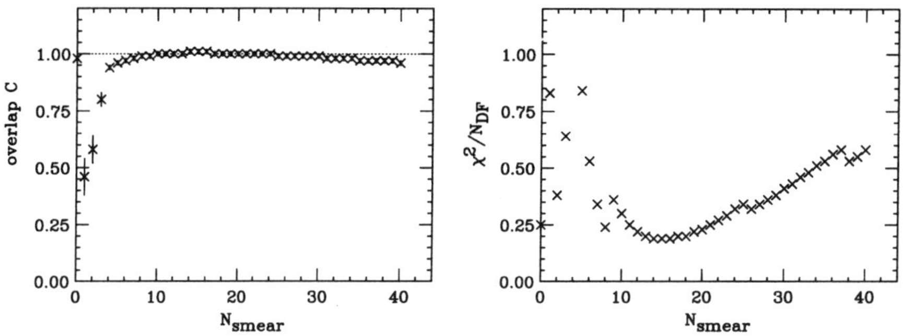

Figure 2: The overlap C is plotted against $N_{\rm smear}$ at $T = 0$ (left figure). $\chi^2/N_{\rm DF}$ is plotted against $N_{\rm smear}$ at $T = 0$ (right figure).

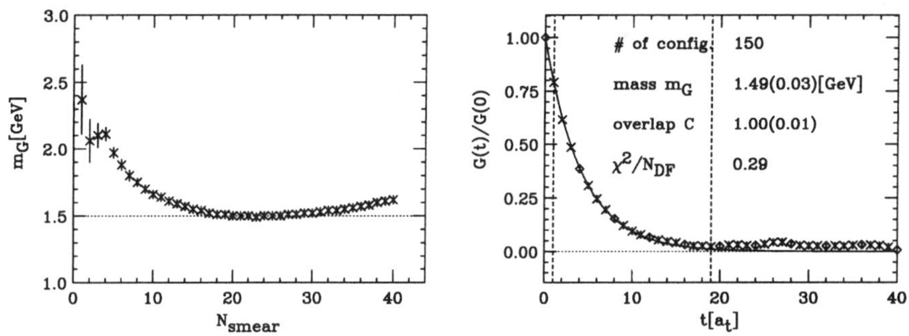

Figure 3: The glueball mass $m_{\rm G}$ is plotted against $N_{\rm smear}$ (left figure). The normalized glueball correlator $G(t)/G(0)$ at $T = 0$ is plotted at $N_{\rm smear} = 23$ (right figure). The curve denotes the best single-exponential fit from the region between the vertical dotted lines.

a single-exponential fit, takes its minimum as $m_{\rm G} = 1.49$ GeV at $N_{\rm smear} = 23$. This indicates the achievement of the ground-state glueball. We thus obtain $m_{\rm G} = 1.49$ GeV at $T = 0$. The glueball size R is estimated from Eq.(10) to be 0.37 fm $\lesssim R \lesssim 0.57$ fm from the maximal overlap region with $10 \le N_{\rm smear} \le 24$. We show also the glueball correlator $G(t)$ at $T = 0$ for a smearing number $N_{\rm smear} = 23$ in Fig.3. Note that, if we used the isotropic lattice, we could only use the points indicated by the diamonds, which would be insufficient to get the mass.

At finite temperature, we use the hyperbolic cosine fitting as $G(t)/G(0) = C(e^{-m_{\rm G} t a_t} + e^{-m_{\rm G}(N_t - t)a_t})$, where N_t denotes the number of the temporal lattice size, and a_t the temporal lattice spacing. In Fig.4, the overlap C and $\chi^2/N_{\rm DF}$ for $T = 0.87T_c$ are plotted against $N_{\rm smear}$. In the whole region, we

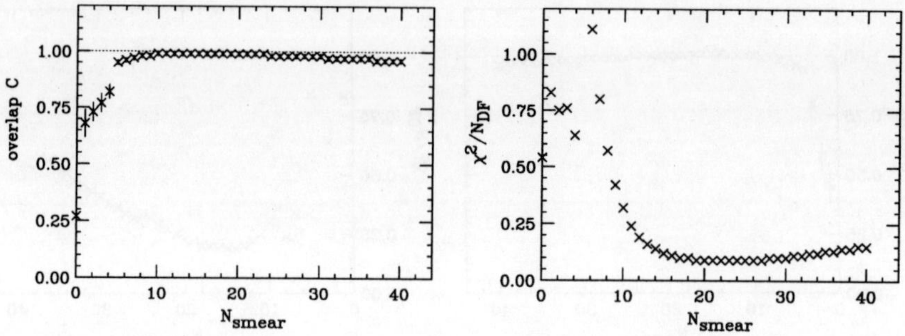

Figure 4: The overlap C is plotted against $N_{\rm smear}$ at $T = 0.87T_c$ (left figure). $\chi^2/N_{\rm DF}$ is plotted against $N_{\rm smear}$ at $T = 0.87T_c$ (right figure).

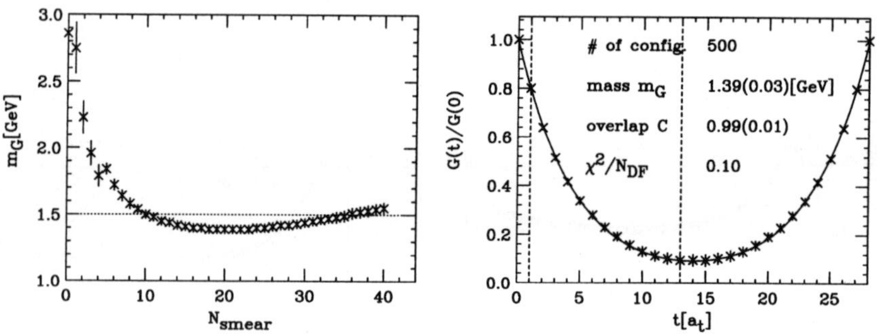

Figure 5: The fitted glueball mass at $T = 0.87T_c$ is plotted against $N_{\rm smear}$ (left figure). The normalized glueball correlator $G(t)/G(0)$ at $T = 0.87T_c$ is plotted for $N_{\rm smear} = 23$ (right figure). The curve denotes the best hyperbolic cosine fit in the interval indicated by the vertical dotted lines.

have reasonable χ^2 as $\chi^2/N_{\rm DF}\lesssim 1$. The overlap becomes maximal in the region $10 \le N_{\rm smear} \le 23$. In this region, as shown in Fig.5, the glueball mass becomes a minimum as $m_{\rm G} = 1.39$ GeV in the interval $18 \le N_{\rm smear} \le 23$. We thus obtain $m_{\rm G} = 1.39$ GeV at $T = 0.87T_c$. The glueball size R is also estimated from Eq.(10) as $0.37 \text{ fm} \lesssim R \lesssim 0.56$ fm at $T = 0.87T_c$, from the maximal overlap region with $10 \le N_{\rm smear} \le 23$. We also show the glueball correlator $G(t)$ at $T = 0.87T_c$ for $N_{\rm smear} = 23$ in Fig.5.

5 Summary and discussions

We have applied the closed packing model to quenched QCD to derive the relation between the mass and the size of the glueball so as to reproduce the deconfinement phase transition at $T_c = 260$ MeV. We have then studied the mass and the size of the scalar glueball at finite temperature by using $SU(3)$ lattice QCD at the quenched level with an anisotropic lattice and the smearing method. To generate gauge field configurations, we have used a tree-level Symanzik $O(a^2)$ improved action with a lattice size of $16^2 \times 24 \times N_t$, $N_t = 96 (T = 0)$ and $N_t = 28 (T = 0.87T_c)$. As for the glueball size $R(T)$, no drastic change was observed between 0.37 fm $\lesssim R(T = 0) \lesssim 0.57$ fm and 0.37 fm $\lesssim R(T = 0.87T_c) \lesssim 0.56$ fm. We obtained the glueball mass to be $m_G = 1.49$ GeV at $T = 0$ and $m_G = 1.39$ GeV at $T = 0.87T_c$. This result may suggest a slight reduction of the glueball mass at finite temperature, which would provide support for our conjecture discussed in Sect. 1 and Sect. 2. However, this reduction does not seem as drastic as is suggested by the closed packing model for quenched QCD. At any rate, to draw a definite conclusion, more statistics are necessary.

Acknowledgements

The calculations have been performed on the NEC SX-5 in Osaka University. The authors thank Dr. T. Umeda for useful comments and discussions.

References

1. G. Boyd, J. Engels, F. Karsch, E. Laermann, C. Legeland, M. Lutgemeier and B. Petersson, *Nucl. Phys.* B **469**, 419 (1996).
2. C. Moningstar and M. J. Peardon, *Phys. Rev.* D **60**, 034509 (1999).
3. J. Sexton, A.Vaccarino, D. Weingarten, *Phys. Rev. Lett.* **75**, 4563 (1995).
4. H. Ichie, H. Suganuma and H. Toki, *Phys. Rev.* D **52**, 2944 (1995).
5. H. Matsufuru, Y. Nemoto, H. Suganuma, T. T. Takahashi and T. Umeda, Proceedings of *Quantum Chromodynamics and Color Confinement*, edited by H. Suganuma et al. (World Scientific, 2001) 246.
6. O. Kaczmarek, F. Karsch, E.Laermann and M. Lutgemeier, *Phys. Rev.* D **62**, 034021 (2000).
7. C. DeTar and J. B. Kogut, *Phys. Rev. Lett.* **59**, 399 (1987).
8. N. Ishii, H. Suganuma and H. Matsufuru, in preparation.
9. K. Symanzik, *Nucl. Phys.* B **226**, 187 (1983).
10. T. R. Klassen, *Nucl. Phys.* B **533**, 557 (1998).
11. H. J. Rothe, *Lattice Gauge Theories*, (World Scientific, 1992) 1.
12. I. Montvay and G. Münster, *Quantum Fields on a Lattice*, (Cambridge Monographs on Mathematical Physics, 1994) 1.
13. APE Collaboration, M. Albanese et al., *Phys. Lett.* B **192**, 163 (1987).
14. T. T. Takahashi, H. Matsufuru, Y. Nemoto and H. Suganuma, *Phys. Rev. Lett.* **86**, 18 (2001).

DECONFINING BY WINDING

RALF HOFMANN
Max-Planck-Institut für Physik,
Werner Heisenberg-Institut,
Föhringer Ring 6, 80805 München
E-mail: ralfh@mppmu.mpg.de

A model for the quantum effective description of the vacuum structure of thermalized SU(3) Yang-Mills theory is proposed. The model is based on Abelian projection leading to a Ginzburg-Landau theory for the magnetic sector. The possibility of topologically non-trivial, effective monopole fields in the deconfining phase is explored. These fields are assumed to be Bogomol'nyi-Prasad-Sommerfield saturated solutions along the compact, Euclidean time dimension. Accordingly, a gauge invariant interaction for the monopole fields is constructed. Motivated by the corresponding lattice results the vacuum dynamics is assumed to be dominated by the monopole fields. A reasonable value for the critical temperature is obtained, and the partial persistence of non-perturbative features in the deconfining phase of SU(3) Yang-Mills theory, as it is measured on the lattice, follows naturally.

1 Abelian projection, monopole trajectories, and field theoretic description

Quantum Chromodynamics (QCD) and SU(N) Yang-Mills (YM) theory are asymptotically free (regime of large momenta)[a], and they exhibit a growth of the coupling toward the infrared, perturbatively. The latter is a *necessary* condition for a successful explanation of the fact that the non-collective propagation of color charges over large distances has not been observed experimentally. Due to lattice results this is commonly believed to be caused by the peculiarities of the gauge boson interactions. Therefore, we restrict ourselves to the discussion of pure gluodynamics, that is, SU(N) YM theory. A promising attempt was made by Mandelstam and 't Hooft in the late seventies and early eighties [1]. Roughly, they proposed to view the QCD and YM vacuum as a dual superconductor which forces the chromo-electric flux between a pair of largely separated test charges into a tube characterized by a constant tension σ. The corresponding potential grows linearly with separation, and hence color charges are confined. How can one obtain a condensate of magnetic charges responsible for this? The idea is that due to a presumably educated gauge fixing the low-energy degrees of freedom of SU(N) Yang-Mills

[a]For QCD this statement is only true if the number of active quark flavors is does not exceed a critical value set by the number of colors.

become transparent. Imposing a gauge condition invariant under the maximal Abelian subgroup $U(1)^{N-1}$, the emergence of chromo-magnetic monopoles can be observed. For example, demanding that the homogeneously transforming, hermitian field strength component $F_{23}(x)$ be diagonal after gauge fixing is no constraint for a gauge transformation$\in U(1)^{N-1}$. Moreover, for a given configuration A_μ there may be points in space where the eigenvalues of F_{23} are degenerate. At such a location the entire gauge group is unconstrained by the gauge condition. With respect to *Abelian* components of the gauge field A_μ one encounters radially directed magnetic flux with the magnetic field becoming singular at the singularity of the gauge fixing procedure and along a half-line starting there. A magnetic monopole together with its Dirac string is recovered. However, the statement that at the instant x^0 there sits a magnetic monopole at point $\vec{x}$ is a *highly gauge variant* and arbitrary one. Instead of diagonalizing $F_{23}(x)$ under gauge transformations we could have chosen to diagonalize $F_{12}(x)$. Generically, this would have caused the gauge singularities to be distributed in a different way. This is a serious problem, and one only can hope that the dynamics itself proliferates the gauge *invariant* existence of monopoles at low resolution. Another unanswered question arises in view of the monopole interpretation using Abelian components of the gauge fields A_μ: What justifies the projection onto Abelian components? The only hint that this procedure describes the low-resolution physics properly comes from the lattice where the string tension indeed was found to be saturated by the projected dynamics and, even better, by the monopole dynamics alone [5,6,7]. Analytical insight into this miracle may be obtained if the vacuum at low resolution can be shown to be governed by the dynamics of an adjoint Higgs model with the gauge field being essentially pure gauge [8].

For now let us be pragmatic. Point-like monopoles in space correspond to line-like trajectories in spacetime. It can be shown that a summation over these trajectories for monopole charge $|Q| = 1$ in the partition function leads to a scalar field theory for the monopole sector [2]. However, working with the projected gauge field alone the theory necessarily contains non-local terms which one would like to avoid [2]. The introduction of a twin set of gauge fields can cure this [2,3]. Projecting QCD, one obtains a theory of electrically and magnetically charged matter fields and sets of (N-1) electric and magnetic abelian gauge fields $\vec{A}_\mu$ and $\vec{B}_\mu$, respectively. A *phenomenological* self-interaction of the magnetic monopole fields is assumed to yield a spontaneously generated condensate breaking the $U(1)^{N-1}$ symmetry [4]. As a result, the set of gauge fields $\vec{B}_\mu$, interacting with the magnetic charges [b], becomes

[b] This sector is a dual Ginzburg-Landau theory.

massive, and a constant string-tension σ emerges.

2 The case of finite temperature

It has been shown on the lattice (for example [12]) that a thermalized SU(3) Yang-Mills theory undergoes a deconfinement phase transition at critical temperatures $T_c^{lat} \sim 200 - 300$ MeV. As explained above the starting point for an analytical investigation of this phenomenon is the dual Ginzburg-Landau theory given by the following Lagrangian [4]

$$\mathcal{L}_{DGL} = -\frac{1}{4}(\partial_\mu \vec{B}_\nu - \partial_\nu \vec{B}_\mu)^2 + \sum_{k=1}^{3} \left\{ \left| (i\partial_\mu - g\vec{\varepsilon}_k \cdot \vec{B}_\mu)\phi_k \right|^2 - V_k(\phi_k, \bar{\phi}_k) \right\} . \tag{1}$$

In contrast to the electric, Abelian gauge group the magnetic $U(1)^2$ (with gauge field $\vec{B}_\mu = (B_\mu^3, B_\mu^8))$ is believed to be spontaneously broken by nonvanishing VEVs of the monopole fields ϕ_k. In Eq. (1) $g\vec{\varepsilon}_i$ denote the effective magnetic charges with $\vec{\varepsilon}_1 = (1, 0)$, $\vec{\varepsilon}_2 = (-1/2, -\sqrt{3}/2)$, $\vec{\varepsilon}_3 = (-1/2, \sqrt{3}/2)$, and the fields ϕ_k satisfy the constraint $\sum_{k=1}^{3} \arg \phi_k = 0$ [4].

The potential $V \equiv \sum_{k=1}^{3} V_k(\phi_k, \bar{\phi}_k)$ is introduced *phenomenologically* to account for the self-interaction of the monopole fields ϕ_k. In the framework of the dual Ginzburg-Landau theory the deconfinement phase transition at finite temperature has been first discussed in Ref. [10]. In Ref. [11] the critical temperature T_c of the deconfinement phase transition was determined as the point where a thermal one-loop effective potential, calculated in Euclidean spacetime with compactified time dimension of size $\beta \equiv 1/T$, starts possessing an absolute minimum with vanishing monopole VEV's. For the sake of renormalizability and the desired feature of spontaneous breaking of the magnetic gauge symmetry the potential V for the monopole fields was chosen to be Higgs-like: $\lambda \sum_{k=1}^{3} \left(\bar{\phi}_k \phi_k - v^2\right)^2$.

However, since the Abelian description of gluodynamics can only be valid up to a certain resolution Λ_b[c] one may wonder whether a renormalizable monopole interaction is imperative. Introducing a pair of external, oppositely charged, static, electric color-charges and integrating out the electric and (massive) magnetic gauge bosons and the monopole fluctuations about their VEV's in quadratic approximation a critical temperature $T_c \sim 500$ MeV was obtained from the corresponding effective potential. This is too high. The result is not surprising since the effective potential was calculated by abusing perturbation theory ($\lambda \sim 25$ from a fit to the string tension σ!).

[c] After all perturbative QCD works well for large momenta.

Our approach is therefore different. We view the monopole potential to be a quantum effective potential already. With the lattice-motivated assumption that the vacuum dynamics is *dominated by the monopole fields* it is sufficient to look for solutions to the classical equations of motion of the monopole sector. The potential is then constructed such that
1) it is gauge invariant under the magnetic $U(1)^2$,
2) it admits topologically non-trivial, BPS saturated solutions along the compact, Euclidean time dimension [d], and
3) it allows for *one* topologically trivial, non-vanishing VEV for the monopole fields at zero temperature.
Thereby, the requirement of BPS saturation derives from the fact that we are interested in a vacuum description. The corresponding fields must then saturate the lowest bound for the Euclidean action. From **1)** and **2)** it follows that the "square root" $V_k^{1/2}$, defined as $V_k(\bar{\phi}_k\phi_k) \equiv V_k^{1/2}(\bar{\phi}_k)V_k^{1/2}(\phi_k)$, must have a single pole at $\phi_k = 0$ [13,14,16]. **3)** enforces an analytical part in order to obtain finite VEV's at zero temperature (stabilization). Furthermore, this analytical part can only be a power $\sim \phi^N$ since a genuine polynomial would introduce non-degenerate vacua. From **1)** it finally follows that

$$V_k(\phi_k, \bar{\phi}_k) == \lim_{N\to\infty} \left\{ \frac{\Lambda^6}{\bar{\phi}_k\phi_k} + \kappa^2\Lambda^{-2(N-2)}(\bar{\phi}_k\phi_k)^N - 2\,\kappa\Lambda^{5-N}\frac{1}{\bar{\phi}_k\phi_k}\mathrm{Re}\,\phi_k^{N+1} \right\} \tag{2}$$

Thereby, Λ is a mass-parameter, and κ is some dimensionless coupling constant. Considering the rhs of Eq. (2) at finite N, the potential explicitly breaks the magnetic $U(1)^2$ gauge symmetry, $\vec{B}_\mu \to \vec{B}_\mu + 1/g\,\partial_\mu\vec{\theta}(x)$, $\phi_k \to e^{-i\vec{\epsilon}_k\cdot\vec{\theta}(x)}\phi_k$, down to Z_{N+1}^2 due to the term $\mathrm{Re}\,\phi_k^{N+1}$ [16]. Only in the limit of large N is the gauge symmetry restored. For the purpose of qualitative illustration Fig. 1 shows the potential V_k. Note that at $T = 0$ this effective potential V indeed does not allow for fluctuations of the fields ϕ_k due to its infinite curvature at $\phi_k = \Lambda$. The BPS equations, corresponding to the potential of Eq. (2), read

$$\partial_\tau\phi_k = \bar{V}_{1/2}^k(\bar{\phi}_k)\ , \quad \partial_\tau\bar{\phi}_k = V_{1/2}^k(\phi_k)\ . \tag{3}$$

The rhs are only fixed up to phase factors $e^{i\delta}$, $e^{-i\delta}$, respectively. From Eq. (2)

---

[d] Recall, that in the Euclidean description finite temperature is implemented by compactifying the time dimension.

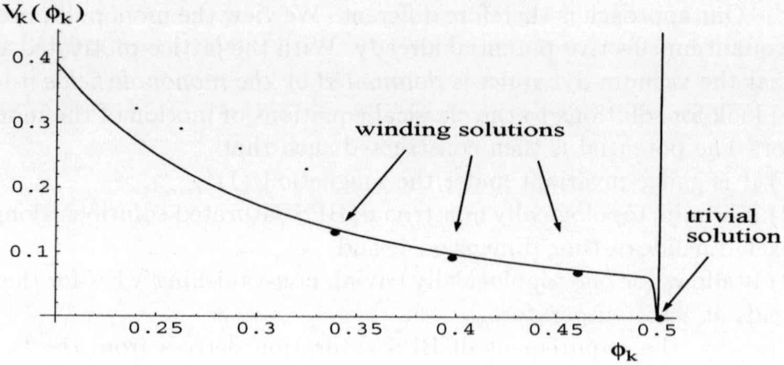

Figure 1. The monopole potential V_k as a function of real ϕ_k for $\kappa = 1$, $\Lambda = 0.5$, and $N = 500$. Indicated are the potential energy densities of topologically trivial and nontrivial solutions to the BPS equations.

we have for *periodic* solutions

$$V^k_{1/2}(\phi_k) = \begin{cases} \pm i\frac{\Lambda^3}{\phi_k}\,, & (|\phi_k| < \Lambda) \\ 0\,, & (|\phi_k| = \Lambda) \\ \infty\,, & (|\phi_k| > \Lambda) \end{cases} \,. \tag{4}$$

3 BPS saturated winding solutions

For the case $|\phi_k| < \Lambda$ periodic solutions ($\phi_k(0) = \phi_k(\beta)$) to Eqs. (3) subject to Eq. (4) have been discussed in Refs. [13,14] within the framework of supersymmetric theories. For a compact, Euclidean time dimension of length $\beta \equiv 1/T$ the set of topologically distinct solutions is

$$\phi_k^{n_k}(\tau) = \sqrt{\frac{\Lambda^3\beta}{2|n_k|\pi}}\,\mathrm{e}^{2n_k\pi i\frac{\tau}{\beta}}\,, \quad (n_k \in \mathbf{Z})\,. \tag{5}$$

Thereby, the sign of n_k corresponds to the choice of phase in the BPS equations. Due to the phase constraint the sets of solutions $(\phi_1^{n_1}, \phi_2^{n_2}, \phi_3^{n_3})$ can be labelled by the integers n_1 and m, which are both odd *or* even. The winding numbers n_2, n_3 are then given as $-n_1/2 \mp m/2$, respectively. The gauge function $\vec{\theta}_{n_1,m}(\tau)$, transforming to the unitary gauge $\mathrm{Re}\,\phi_k(\tau) > 0$, $\mathrm{Im}\,\phi_k(\tau) = 0$, reads

$$\vec{\theta}_{n_1,m}(\tau) = \frac{2\pi}{\beta}\,\tau \begin{pmatrix} n_1 \\ \frac{m}{\sqrt{3}} \end{pmatrix} \,. \tag{6}$$

Note that this non-periodic function leaves the periodicity of the gauge field $\vec{B}_\mu$ intact.

4 Deconfinement phase transition

Eqs. (3) admit the solution $\phi_k^0 \equiv \Lambda$ with winding number and energy density zero, and the winding solutions of the previous section. We identify the set $(\phi_1^0, \phi_2^0, \phi_3^0)$ with the confining vacuum at low temperatures. In this regime the spectrum consists of glueballs with masses that are much larger than the prevailing temperatures [9]. The thermal equilibrium between the vacuum "medium" and its excitations is essentially realized at pressure zero, which indeed is observed on the lattice [18]. Above the deconfinement phase transition the ground state "medium" readily emits and absorbs (almost) free gluons under the influence of the heat bath like a black body emits and absorbs photons. We identify the genuine winding set of lowest potential energy density[e], represented by $(n_1, m) = (2, 0)$, with the ground state just above the deconfinement transition. In this picture the confining vacuum is a perfect thermal insulator up to the transition, where its structure drastically changes by the absorption of an amount of latent heat per volume [18] equal to the gap $\Delta\varepsilon^{(2,0)}$ between the potential V of the zero winding and the lowest genuine winding set. We estimate the critical temperature T_c of this transition by assuming the deconfining vacuum to behave like an incompressible, static fluid with traceless energy-momentum-tensor [17] (no scale anomaly) which is in thermal equilibrium with an ideal gas of gluons. In this case we have the equation of state $\varepsilon = 3p$ for pressure p and energy density ε for both the vacuum (vac) and the gluon gas (gg). The equilibrium condition of equal pressures, $p_{vac} = p_{gg}$, then takes the following form [17]

$$\Delta\varepsilon^{(2,0)} = \frac{8}{15}\pi^2 T_c^4 \,, \quad \text{where} \quad \Delta\varepsilon^{(2,0)} = 8\pi\Lambda^3 T_c \,. \tag{7}$$

Using Eqs. (7) and the condition $p_{vac} = p_{gg}$, we obtain $T_c = \left(\frac{15}{\pi}\right)^{\frac{1}{3}} \Lambda \sim 1.68\,\Lambda$. Adopting the value $\Lambda = 0.126$ GeV for the monopole condensate at $T = 0$ from Refs. [9,11], this yields $T_c = 0.212$ GeV which is compatible with the lattice results of $T_c = 0.2 - 0.3$ GeV.

---

[e] By "genuine" we mean that each of the solutions ϕ_k is winding. Note, however, that there are also semi-winding sets, for example $(n_1, m) = (1, 1)$, which contain two winding fields and one field of winding number zero. Since none of the monopole fields ϕ_k should be singled out it is natural to assume that either all ϕ_k are winding or none at all.

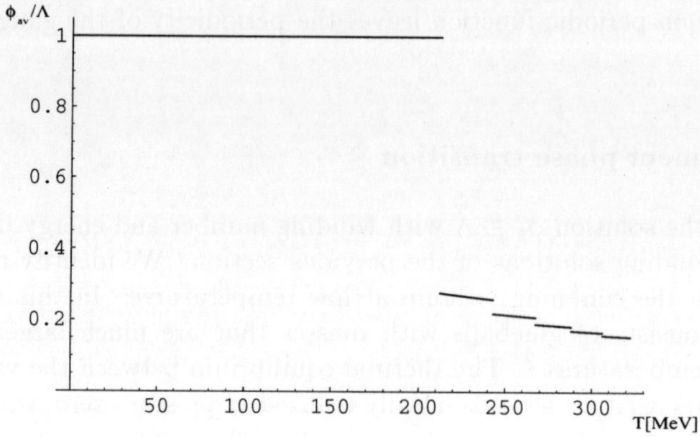

Figure 2. The average monopole field $\phi_{av}(T)$ in units of $\phi_{av}(T=0)$. For further explanation see text.

5 The deconfining phase

Transforming the winding solution of Eq. (5) to the unitary gauge by means of Eq. (6), we obtain the following expression for the individual condensate, corresponding to winding n_k: $\phi_k^{n_k} = \frac{\Lambda^{3/2}}{\sqrt{2\pi|n_k|T_l}} = \frac{\Lambda}{\sqrt{(30\pi^2 l)^{1/3}|n_k|}}$ [f]. For the configuration $(n_1, m) = (2, 0)$ we obtain $\phi_{av}^{(2,0)} \sim 0.28\,\Lambda$. In Fig. 2 the evolution of the average monopole condensate is depicted up to $T \sim 0.34$ GeV. In unitary gauge the monopole fields are real, and the spontaneous breakdown of the U(1)^2 magnetic gauge symmetry then transparently generates a mass m_B for the field $\vec{B}_\mu$ which scales linearly with the average monopole condensate, $m_B^{(n_1,m)} = \sqrt{3}\, g\, \phi_{av}^{(n_1,m)}$. In Refs. [9,11] the value $m_B \sim 0.5$ GeV was determined at $T = 0$, yielding a magnetic gauge coupling of $g = 2.3$.

How does the string tension σ evolve? An analytical expression for σ in terms of g, m_B, and m_ϕ was derived in Ref. [9] by calculating the potential between two heavy and static quarks of opposite color charge, separated by a distance R. Thereby, an effective Lagrangian for the dynamics of the electric $U(1)^2$ gauge field $\vec{A}_\mu$ is obtained by integrating out the massive, magnetic vector field $\vec{B}_\mu$ and by assuming that the monopole fields do not fluctuate about their VEVs. Applying convenient gauge fixing terms and introducing external, static color currents, which couple to $\vec{A}_0$, one obtains an effective action. From this action one extracts the potential energy $V_{\bar{q}q}(R)$ and reads

[f] $l \equiv \sum_k |n_k|$.

off as $\sigma = 2\pi m_B^2/(3g^2)\ln\left[1+m_\phi^2/m_B^2\right]$. Since *static* test charges were assumed finite temperature does not necessitate any new considerations. The appearance of the monopole mass $m_\phi > m_B$ under the logarithm in the expression for σ is due to its role as an ultraviolet cutoff for the integration over transverse momenta in the expression for the linear part of the potential $V_{\bar{q}q}(R)$. This is justified by the observation that $m_B \sim 0$ inside the flux tube of radius $\rho \sim m_\phi^{-1}$ in a type II dual superconductor (see Ref. [11] and Refs. therein). Since we work with effective fields ϕ_k we cannot asign a mass to the fluctuations of the monopole fields about the background of a classical ground state. We may, however, assume that ρ^{-1} scales with temperature in the same way as m_B does. Since the dependence of σ on the cutoff is logarithmic the result is not dramatically sensitive to the exact T dependence of ρ^{-1}, which we will make explicit by distinguishing the following cases:
(a) $\rho^{-1}(0) = \rho^{-1}(T) = 1.26$ GeV [9] $\Rightarrow$

$$\frac{\sigma(T_c)}{\sigma(0)} \sim (0.28)^2\frac{\ln(1+(1.26)^2/(0.5\times 0.28)^2)}{\ln(1+(1.26)^2/(0.5)^2)} \sim 0.19\ . \tag{8}$$

(b) $\rho^{-1}(T) \propto m_B(T) \Rightarrow$

$$\frac{\sigma(T_c)}{\sigma(0)} \sim (0.28)^2 \sim 0.08\ . \tag{9}$$

Hence, the discrepency amounts to a factor 2 for these two extreme cases.

6 Summary and discussion

We considered a thermalized dual Ginzburg-Landau theory modelling hot SU(3) Yang-Mills theory. From the postulate that in contrast to the confining phase the ground state of the deconfining phase is characterized by topologically nontrivial, BPS saturated solutions to the classical equations of motion of the monopole sector and the uniqueness of the vacuum at $T=0$ we deduced the corresponding, gauge invariant interaction. As a consequence, the average monopole field undergoes a drastic decrease to about 1/4 of its zero temperature value across the phase boundary. Scaling linearly with the monopole condensate, the same applies to the mass of the magnetic vector fields generated by the spontaneous breaking of the magnetic $U(1)^2$ gauge symmetry. Working with the zero temperature value for the monopole condensate of Refs. [9,11], we obtain a reasonable critical temperature $T_c \sim 212$ MeV in contrast to the effective potential calculation of Ref. [11], where $T_c \sim 500$ MeV, and the more realistic value of $T_c \sim 200$ MeV was obtained by introducing an ad hoc T-dependence of the dimensionless coupling constant λ.

Depending on the assumption about the T-dependence of the flux tube radius ρ the string tension σ decreases to about 1/12..1/5 of its value at $T = 0$ across the phase boundary which is compatible with lattice measurements [18]. To determine T_c we assumed a *free* gluon gas although close to the critical T_c the lattice indicates a pressure in the deconfining phase which is sizably smaller than the Stefan-Boltzmann asymptotics. However, since the leading term in an expansion of p in (mass scale)/T is quartic in T we expect T_c to be robust against changes in the subleading terms[g]. The presence of non-perturbative effects is described by non-vanishing values of the monopole condensate ϕ_{av} and the string tension σ, and the model predicts a slow decrease of these quantities. On the lattice and in effective potential calculations the deconfinement phase transition has been determined to be of first order [11,18,19]. In contrast, SU(2) Yang-Mills theory exhibits a second order phase transition [19]. The model of the present work can be easily adapted to the dual Ginzburg-Landau theory describing maximal abelian gauge fixed SU(2) Yang-Mills theory. The qualitative results would then be the same. It is clear that we cannot resolve this difference since by definition the order parameter $|\phi_{av}|$ is always discontinuous. However, the model gives a reason for the residual, non-trivial structure of monopole condensation in SU(2) and SU(3) Yang-Mills theories at high temperatures, it predicts a reasonable value for the critical temperature T_c, and it incorporates the low-temperature limit in the form of a constant solution to the BPS equations. The existence of a non-trivial structure of monopole condensation above the phase transition has been pointed out in Ref. [20] a long time ago based on lattice investigations.

Acknowledgments

The author would like to thank the organizers of "Lepton Scattering, Hadrons and QCD" for a very interesting workshop and a stimulating atmosphere.

References

1. S. Mandelstam, *Phys. Rep.* C**23**, 245 (1976). G. 't Hooft, *Nucl. Phys.* B**190**, 455 (1981).
2. K. Bardakci and S. Samuel, *Phys. Rev.* D**18**, 2849 (1978).
3. D. Zwanziger, *Phys. Rev.* D**3**, 880 (1970).
4. T. Suzuki, *Prog. Theor. Phys.* **80**, 929 (1988); **81**, 752 (1989). S. Maedan and T. Suzuki, *Prog. Theor. Phys.* **81**, 229 (1989).

[g]If there were no subleading terms T_c would be given by a cubic root.

5. H. Shiba and T. Suzuki, *Phys. Lett.* **B351**, 519 (1995).
6. K. Yamagishi, S. Kitahara and T. Suzuki, *JHEP* **0002**, 012 (2000).
7. M. N. Chernodub, S. Fujimoto, S. Kato, M. Murata, M. I. Polikarpov and T. Suzuki, hep-lat/0006025, to appear in *Phys. Rev.* D.
8. P. Suranyi, hep-lat/0102009.
   R. Hofmann, hep-ph/0103279.
9. H. Suganuma, S. Sasaki and H. Toki, *Nucl. Phys.* **B435**, 207 (1995).
10. H. Monden, T. Suzuki and Y. Matsubara, *Phys. Lett.* **B294**, 100 (1992)
11. H. Ichie, H. Suganuma and H. Toki, *Phys. Rev.* **D52**, 2944 (1995).
12. T. Hashimoto *et al.*, *Phys. Rev.* **D42**, 620 (1990).
13. G. Dvali and M. Shifman, *Phys. Lett.* **B454**, 277 (1999).
14. X. Hou, A. Losev and M. Shifman, *Phys. Rev.* **D61**, 085005 (2000).
15. T. Schafer and E. V. Shuryak, *Rev. Mod. Phys.* **70**, 323 (1998).
16. R. Hofmann, *Phys. Rev.* **D62**, 105021 (2000), Erratum-ibid. **D63**, 049901 (2001).
17. J. Cleymans, E. Nykänen and E. Suhonen, *Phys. Rev.* **D33**, 2585 (1986).
18. B. Beinlich *et al.*, *Eur. Phys. J.* **C6**, 133 (1999).
19. F. Karsch, "THE DECONFINEMENT TRANSITION IN FINITE TEMPERATURE LATTICE GAUGE THEORY", Invited talk given at the 'Enrico Fermi' Int. School of Physics, Varenna, Italy, Jun 26-Jul 6, 1984.
20. M. L. Laursen and G. Schierholz, *Z. Phys.* **C38**, 501 (1988).

CHARMONIUM, GLUEBALLS AND EXOTIC HYBRIDS IN A RELATIVISTIC MANY-BODY APPROACH

STEPHEN R. COTANCH

Department of Physics, North Carolina State University, Raleigh, NC 27695-8202, USA

E-mail: cotanch@ncsu.edu

Utilizing conventional many-body techniques, an effective QCD Hamiltonian formulated in the Coulomb gauge is approximately diagonalized. This approach preserves chiral symmetry yet generates constituents with dynamical mass confined through a linear potential with slope specified by lattice gauge theory. The experimental meson and lattice glueball spectra are reasonably described and exhibit Regge trajectories consistent with observation including the Pomeron. A fully relativistic, three quasiparticle calculation for hybrid mesons reinforces alternative theoretical models and predicts the lightest hybrid states are near but above 2 GeV. This strongly suggests the recently observed $J^{PC} = 1^{-+}$ exotics at 1.4 and 1.6 GeV are of a different, perhaps four quark, structure. Charmonium predictions now resolve the overpopulation of J/Ψ states relative to observation. Extensions and preliminary results for low energy $\pi\pi$ scattering are also discussed.

1 Introduction

In many areas of physics the quintessential treatment of the quantum mechanical bound state problem is the use of many-body techniques to diagonalize an effective Hamiltonian. In the last few years this approach has been successfully applied[1,2,3] to hadron structure implementing the established BCS, TDA and RPA methods. A fundamental feature of this formulation is that it unifies the quark and gluon sectors and is amenable to systematic improvement through controllable approximations. The purpose of this paper is to report further extensions and applications of this comprehensive approach.

This paper is organized into five sections. In the next section the model Hamiltonian is specified along with the effective interaction and model parameters. Section 3 reports approximate diagonalizations of the Hamiltonian for 2 and 3 constituent hadrons– heavy mesons, higher spin glueballs and charmed and exotic hybrids. This section also details a glueball Regge trajectory consistent with the Pomeron as well as a Pomeron exchange model analysis of recent $p(\gamma, \phi)p$ data. New, preliminary applications to 4 quark systems are presented in section 4 for low energy $\pi - \pi$ scattering. Here contact is made with the Beth-Salpeter/Dyson-Schwinger approach to calculate the scattering amplitude at zero momentum yielding results first derived by Weinberg[4]. Finally, conclusions are summarized in Section 5.

2 Effective Hamiltonian

The essence of this model is an effective Coulomb gauge Hamiltonian for both the quark and gluon sectors

$$H = \int d\mathbf{x}\, \psi^\dagger(\mathbf{x}) \left[-i\, \boldsymbol{\alpha} \cdot \boldsymbol{\nabla} + \beta\, m_q\right] \psi(\mathbf{x}) + \frac{1}{2} \int d\mathbf{x} \left[\mathbf{E}^a(\mathbf{x}) \cdot \mathbf{E}^a(\mathbf{x}) \right. \\ \left. + \mathbf{B}^a(\mathbf{x}) \cdot \mathbf{B}^a(\mathbf{x})\right] + \frac{1}{2} \int d\mathbf{x} d\mathbf{y}\, \rho^a(\mathbf{x}) V(\mathbf{x}, \mathbf{y}) \rho^a(\mathbf{y}) , \qquad (1)$$

entailing current quark masses, m_q, fields, ψ, color charge density $\rho^a(\mathbf{x}) = \psi^\dagger(\mathbf{x}) T^a \psi(\mathbf{x}) + f^{abc} \mathbf{A}^b(\mathbf{x}) \cdot \mathbf{E}^c(\mathbf{x})$, and gluon fields $\mathbf{A}^a$ (vector), $\mathbf{E}^a$ (electric), and $\mathbf{B}^a$ (magnetic) with color index a. The leading QCD canonical interactions and confinement are represented by the potential $V = -\frac{\alpha_s}{r} + \sigma r$. The parameters are the string tension, $\sigma = .135\, GeV^2$, from lattice gauge calculations, $\alpha_s = .2$ and standard quark masses ($m_u = m_d = 5\, MeV, m_s = 150\, MeV, m_c = 1200\, MeV$).

To find the approximate ground state of H a BCS rotation is performed to a new, but unknown quasiparticle basis. Then by minimizing the vacuum expectation of H with respect to this rotation, non-linear gap equations are generated for both quark and gluon quasiparticles energies. Using these dynamically "dressed" constituent degrees of freedom, hadron wavefunctions and masses are obtained by solving the TDA and RPA equations. Consult refs.[1,2,3] for further details.

Results from approximately diagonalizing this Hamiltonian for mesons, hybrids and glueballs are now presented. It is significant to note that this approach contains only a minimal number of pre-determined parameters.

3 Charmed Mesons, Glueballs and Exotic Hybrids

As detailed in an earlier publication[2], the RPA operator, but not the TDA, commutes with the chiral charge and therefore preserves chiral symmetry and a Goldstone boson pion. Consequently for light quark mesons, such as the π and η, significant RPA-TDA differences were numerically calculated. However, for hadrons, especially glueballs, with mass above 1 GeV these two approaches are essentially identical. Accordingly, all results presented in this section were obtained using the computationally simpler TDA.

Figure 1 summarizes a recent TDA calculation for the J/Ψ charmonium system, $c\bar{c}$. This system is believed to be gluon rich and since previous quark model treatments underpredicted the number of states relative to observation it has been argued that this anomalous overpopulation was evidence for hy-

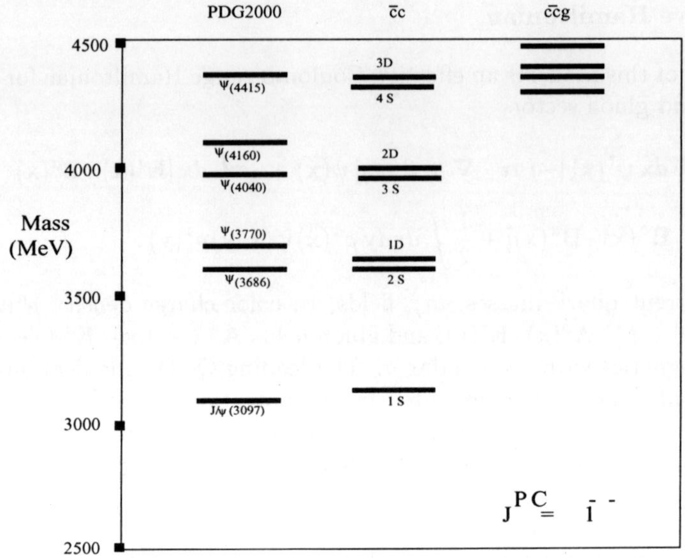

Figure 1. Charmonium spectrum.

brid states, $c\bar{c}g$. However, as shown in Fig. 1, by simply including D wave $c\bar{c}$ excitations the number of theoretical and experimental states is now comparable. Ironically, by also including gluonic excitations (see discussion below), $c\bar{c}g$, the theoretical "underpopulation" problem is now an "overpopulation" as there are now 11 predicted states but only 6 observed. Clearly, density of states arguments for exotic hybrids must be carefully made.

In Fig. 2 results from a glueball calculation for higher spin are compared to lattice measurements[5,6,7,8]. Because these TDA results do not yet include the quark sector, comparison is only made to quenched lattice calculations.

An extensive TDA analysis of the meson spectrum has revealed Regge trajectories with good agreement with the Chew-Frautschi trajectory, $\alpha(t) = .55 + 0.9\, GeV^{-2} t$, using the same model parameters given above. Interestingly, the predicted glueball spectra also reveals a Regge trajectory that is similar to the Pomeron, defined as the J^{++} trajectory with the largest intercept. From high energy cross section fits the Pomeron is given by $\alpha_P(t) = 1.08 + 0.25\, GeV^{-2} t$. Figure 3 compares the Pomeron to the TDA and lattice results.

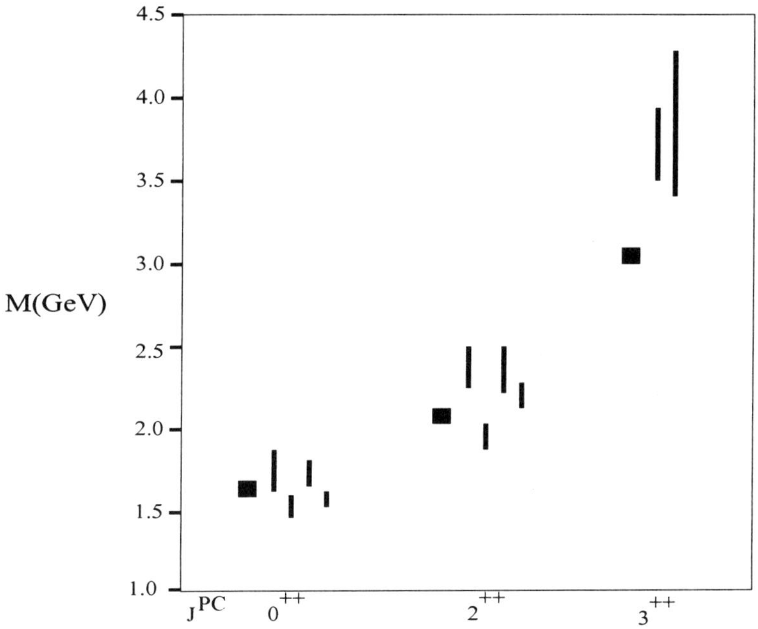

Figure 2. Hamiltonian TDA (boxes) and lattice (bars) glueball spectra.

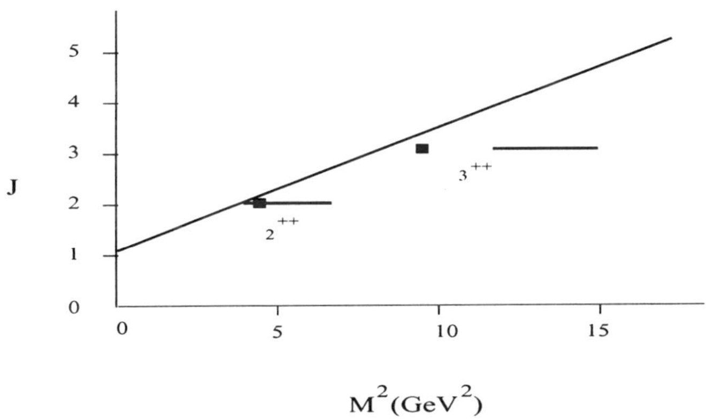

Figure 3. TDA (boxes) and lattice (bars) spectra compared to the Pomeron (solid line).

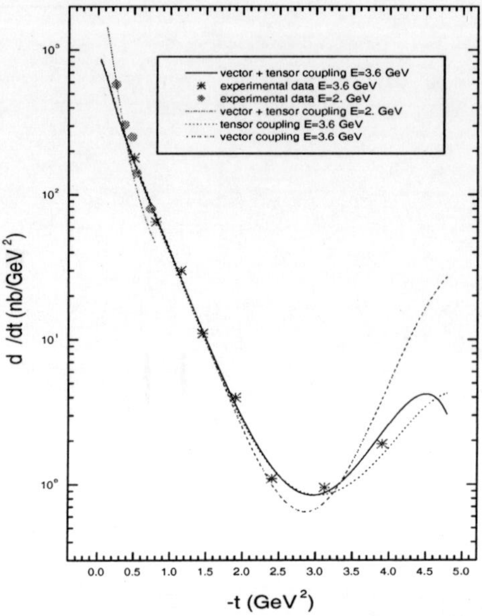

Figure 4. Theory and data for p(γ,ϕ)p.

Finally, an analysis of recent ϕ photoproduction data[9] using a Pomeron exchange model[10] provides further support for a $J^{PC} = 0^{++}$ glueball mass near 1.7 GeV. This is shown in Fig. 4 where calculations are compared to data for a variety of photon energies. The low t predictions are very sensitive (not shown) to the t channel exchange Pomeron mass and a clear preference for mass around 1.7 GeV, in contrast to the previous 1.3 GeV parameterization, emerges.

Our last key hadron structure result addresses exotic hybrid mesons having quantum numbers not possible for $q\overline{q}$ states. Using the quark and gluon quasiparticle gap solutions, a three-body hybrid state has been formulated, $q\overline{q}g$, which now entails color octet quark states. Because the TDA hybrid meson equation now involves solving a formidable nonlocal, 12 dimensional integral equation, the hybrid masses where determined variationally using the Monte Carlo code VEGAS. The results, shown in Fig. 5, are compared to alternative models and recent $J^{PC} = 1^{-+}$ exotic states observed at BNL

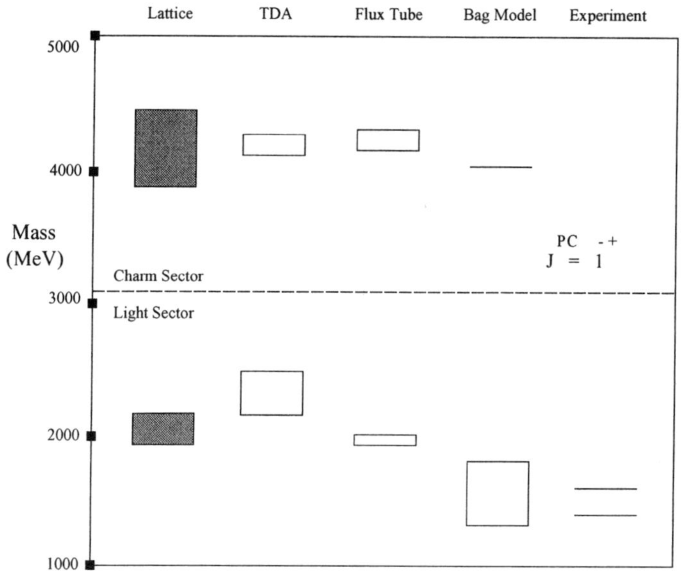

Figure 5. Hybrid meson calculations.

by the E852 collaboration[11]. With the exception of the vintage bag model predictions, the different theoretical approaches (lattice, flux tube and TDA) are in general agreement and predict the lightest exotic hybrid to have mass near, but above, 2 GeV. This is significantly higher than the observed states, indicating these states are more likely to have a four quark, $q\bar{q}q\bar{q}$, structure. Because octet quarks experience a repulsive force, color singlet states will be lighter and meson-meson molecules, such as a πa_1 resonance, are likely candidates.

4 Low energy π π Scattering

The final topic concerns the ambitious task of calculating hadron scattering amplitudes within a relativistic field theoretical quark formulation. Here a brief synopsis is given of a recent analysis[12] of $\pi\pi$ scattering at low energies near threshold. Using the axial Ward identity and a Schwinger-Dyson (SD), Bethe-Salpeter (BS) formulation in the rainbow-ladder approximation[13,14], ref. 12 has formally derived the low energy $\pi\pi$ S- matrix and rigorously reproduced Weinberg's result[4] for the isospin 0 and 2 amplitudes

$$A_0 = i\,\frac{2\,s - M_\pi^2}{2f_\pi^2},\quad A_2 = i\,\frac{-s + 2M_\pi^2}{2f_\pi^2}\ . \tag{2}$$

At threshold, $s = 4M_\pi^2$, $t = u = 0$, and the well known isospin 0 and 1 scattering lengths are immediately reproduced, $a_0 = +7L/2$ and $a_2 = -L/2$ ($L = m_\pi^2/8\pi f_\pi^2$), respectively, as well as the Adler zero in the chiral limit. Direct numerical evaluation of the BS equations, again in the rainbow-ladder approximation, yields $A_0 = 17.1$ and $A_2 = -4.54$ in good agreement with Weinberg's values of 15.7 and -4.49, respectively.

Lastly, ref. 12 also details a formal connection between the Hamiltonian RPA treatment and the BS formulation in the instantaneous limit. For point pions, Weinberg's results is again directly derived in the RPA. Utilizing the chiral symmetry preservation of the RPA, the same result can be obtained for finite size pions but a numerical calculation is required.

5 Conclusion

Summarizing, the approximate diagonalization of an effective QCD Hamiltonian using the BCS, TDA, and RPA many-body techniques is an attractive and effective approach to hadron structure. The RPA preserves chiral symmetry and enables the chiral governance of the pion to naturally emerge in both bound state and scattering applications. The method is extremely powerful for few constituent systems and provides dynamical insight regarding gluonic degrees of freedom, especially for glueballs and hybrid mesons. Since this formalism entails controllable approximations, the method is amenable to systematic improvement by including higher Fock state components. This will permit study of molecular ($q\bar{q}q\bar{q}$) mesons, baryons, including the nucleon strangeness content (pentaquark systems), and even dibaryons (six quark systems). Finally, the approach also appears promising for scattering analysis and further applications are in progress.

Acknowledgments

The author is especially grateful to his many collaborators who spearheaded the topics in this talk: P. Bicudo (Pomeron), F. J. Llanes-Estrada (mesons and hybrids), P. Maris (BS/SD $\pi\pi$ calculation), E. Ribeiro ($\pi\pi$ RPA-BS/SD formulation), A. Szczepaniak (glueballs) and A. Talkachou (ϕ photoproduction). This work is supported in part by grants DOE DE-FG02-97ER41048 and NSF INT-9807009.

References

1. A. Szczepaniak, E. S. Swanson, C.-R. Ji, and S. R. Cotanch, *Phys. Rev. Lett.* **76**, 2011 (1996).
2. F. J. Llanes-Estrada and S. R. Cotanch, *Phys. Rev. Lett.* **84**, 1102 (2000).
3. F. J. Llanes-Estrada and S. R. Cotanch, *Phys. Lett.* B **504**, 15 (2001).
4. S. Weinberg, *Phys. Rev. Lett.* **17**, 616 (1966).
5. C. J. Morningstar and M. Peardon, *Phys. Rev.* D **60**, 034509 (1999).
6. UKQCD Collaboration, G. S. Bali *et al.*, *Phys. Lett.* B **309**, 378 (1993).
7. H. Chen, J. Sexton, A. Vaccarino, and D. Weingarten, *Nucl. Phys.* B **34**, 357 (1994).
8. M. Teper, hep-th/9812187 (1998).
9. E. Anciant *et al.*, *Phys. Rev. Lett.* **85**, 4682 (2000).
10. R. A. Williams, *Phys. Rev.* C **57**, 223 (1998).
11. G. S. Adams et al. (E852 Collaboration), *Phys. Rev. Lett.* **81**, 5760 (1998); D. R. Thompson et al. (E852 Collaboration), *Phys. Rev. Lett.* **79**, 1630 (1997).
12. P. Bicudo, S. R. Cotanch, F. J. Llanes-Estrada, P. Maris, E. Ribeiro, and A. Szczepaniak, in preparation.
13. P. Maris and C. D. Roberts, *Phys. Rev.* C **56**, 3369 (1997).
14. P. Maris and P. C. Tandy, *Phys. Rev.* C **60**, 055214 (1999).

6. Small-x Physics and Nuclear Medium Effects

SMALL X PHYSICS AND THE INITIAL CONDITIONS IN HEAVY ION COLLISIONS

ALEX KRASNITZ
UCEH, Universidade do Algarve,
Campus de Gambelas, P-8000 Faro, Portugal.

RAJU VENUGOPALAN
Physics Department and RIKEN-BNL Research Center,
Brookhaven National Laboratory, Upton, NY 11973, USA

At very high energies, the high parton densities (characterized by a semi-hard saturation scale Λ_s) ensure that parton distributions can be described by a classical effective field theory with remarkable properties analogous to those of spin glass systems. This color glass condensate (CGC) of gluons also provides the initial conditions for multi-particle production in high energy nuclear collisions. In this talk, we briefly summarize recent theoretical and phenomenological progress in the CGC approach to small x physics. In particular, we discuss recent numerical work on the real time gluodynamics of partons after a nuclear collision. The implications of this work for the theoretical study of thermalization in nuclear collisions and on the phenomenological interpretation of results of the recent RHIC experiments are also discussed.

1 Introduction

Why is small x physics relevant to heavy ion collisions? The canonical picture of how matter is produced and evolves in a high energy nuclear collision was articulated by Bjorken [1]. The "valence" components of the nuclear wavefunction (partons carrying a large fraction, per nucleon, of the nuclear momentum) interact weakly and populate primarily the fragmentation regions at large projectile and target rapidities. It is the "wee" virtual excitations of the nuclear ground state, gluons (mostly) and sea quarks, carrying a small fraction of the nuclear longitudinal momentum, that populate the central rapidity region, eventually producing the $\sim$ 1000 hadrons that were measured in one unit of pseudorapidity at RHIC. If the nucleus is viewed as a superposition of Fock states, these wee partons belong to a state with a large number of gluons and sea quarks–each of these can therefore have only a small fraction of the nuclear momentum [2]. The higher the energy of the nucleus, the larger the number of wee partons. In a nuclear collision, this Fock state is "destroyed" and the wee partons go on–shell creating the large energy densities in the relatively large space–time volumes associated with quark gluon plasma formation. Understanding the possible formation and dynamical evolution of this partonic state therefore requires understanding small x physics at a deep level.

In the perturbative [3] QCD approach to small x physics, the "cloud" of gluons is generated by Bremsstrahlung of a ladder of soft gluons off harder "valence" partons. The parton density grows rapidly with decreasing x at a rate which, if extrapolated to $x \to 0$, would violate the Froissart (unitarity) bound. This seems unlikely because at sufficiently small x the Bremsstrahlung approach must break down. When the density of partons becomes large, parton recombination and screening effects become important, and may perturbatively "saturate" the rapid growth of parton distributions at small x [4].

Since the density of partons is large, one can formulate the small x problem as a classical effective field theory (EFT) [5] [a]. The small x partons described by the EFT constitute a *color glass condensate* (CGC) [7,8]. They are a glass because their properties are analogous to a spin glass. They form a condensate because they belong to a state with high occupation number of order $1/\alpha_S$.

The CGC contains a dimensionful scale, the saturation scale $\Lambda_s(x, A)$. It is proportional to the gluon density per unit area and grows as $\Lambda_s^2 \sim A^{1/3}/x^\delta$, where $\delta \approx 0.2 - 0.3$ is the power that controls the rise in parton distributions in a nucleon. If $\Lambda_s \gg \Lambda_{QCD}$, weak coupling methods are applicable and one can compute the parton distributions in the classical EFT. At RHIC, one can estimate $\Lambda_s = 2$ GeV, while at LHC, $\Lambda_s = 4$ GeV –weak coupling methods are therefore only marginally applicable at the former [9].

This talk is organized as follows. In section 2, we discuss the color field of a single nucleus. We will advertise the impressive recent theoretical developments in this approach. We will also discuss the striking phenomenological success of saturation models in describing the HERA data. In the following section, we will discuss the application of these ideas to nuclear scattering. Results from non–perturbative numerical simulations of the initial stages of nuclear collisions will be presented. In section 4, we will briefly confront saturation models against the recent RHIC data. We end in section 5 with an outlook describing relevant projects in progress.

2 The color field of a nucleus: QCD is a color glass condensate at high energies

The classical EFT approach to small x physics has been discussed extensively in the literature [5,10,8]. It contains a dimensionful scale Λ_s equal to the color charge squared per unit area of the valence sources. The EFT, formulated in the infinite momentum frame, in light cone gauge, is a (nearly) two-dimensional theory with a structure mathematically analogous to that of a disordered sys-

[a] The approach described here is by no means the only approach. For a review of the recent status of different approaches to small x physics, see Ref. [6].

tem of Ising spins in a random magnetic field [11,7]. That's where the color glass comes from [b]. Further, since the occupation number of the gluonic fields is large–the field strength squared $F_{\mu\nu}^2 \sim 1/\alpha_S$–what we have is a color glass condensate (CGC) [8]. A virtual photon (from an energetic leptonic probe) would therefore indeed "see" a colored glass on the time scales of its interaction.

The classical theory is solvable and classical correlation functions have been computed [12]. The gluon number has the Weizsäcker–Williams distribution $\Lambda_s^2/k_t^2/\alpha_S$ for $k_t >> \Lambda_s$ and (due to an infinite resummation in Λ_s/k_t) saturates as the distribution $\ln(\Lambda_s/k_t)/\alpha_S$. This suggests that the typical momentum of the gluons is peaked around Λ_s. Recently, Lam and Mahlon have done a lot of work elucidating the nature of the classical distributions when strict color neutrality is imposed [13].

When quantum corrections are included, a Wilsonian renormalization group (RG) picture emerges. The structure of the classical field remains intact while the scale Λ_s grows. This is because including hard gluons (harder than the x scale of interest) in the source increases the typical size of color fluctuations. The scale $\Lambda_s(x)$ is then given by the gluon density per unit area,

$$\Lambda_s^2(x) = \frac{A^{1/3}}{x^\delta}. \tag{1}$$

Here $\delta(\Lambda_s)$ denotes the power of the rise in the *nucleon* gluon distribution. This equation must be solved self-consistently for Λ_s at each x.

In particular, the weight function that gluonic correlation functions have to be averaged over satisfies a non-linear RG equation [10]. There has been very important work recently in understanding and *solving* this RG equation [8]. (The correlation functions obtained from this functional equation are identical to those obtained independently in different approaches by Balitsky and Kovchegov [14]. For numerical solutions of these, see Ref. [15].) The equations have been solved in a mean field approximation–for large transverse momenta ($k_t >> \Lambda_s$) the structure of the source is the simple Gaussian one of the McLerran–Venugopalan model, while at small transverse momenta, ($k_t << \Lambda_s$) it is entirely different. Much work remains to be done to uncover the complex structure of these equations–their phenomenological implications for heavy ion collisions and for deeply inelastic scattering [16] are tremendous.

Simple saturation models which incorporate the scale Λ_s have been applied to study the HERA data. With only three parameters, they fit the entire range of the HERA inclusive data (for $x < 0.01$ and for all Q^2 up to 450 GeV^2). The

[b]Unlike the original model of Parisi and Sourlas, which fails in describing real spin glasses, this theory, on account of its peculiar quantum structure, may indeed be a glass. This remains to be proven though.

model, with the same parameters, reproduces the HERA diffractive and (with some caveats) vector meson production data [17].

3 Classical approach to nuclear collisions: shattering the color glass condensate

In a nuclear collision the color glass shatters, producing a large number of gluons. The full quantum description of how this happens is a very difficult problem which has no solution yet. The standard pQCD factorization ("mini-jets") approach to particle production is not of much help here because it is most unreliable at small x and small Q^2. Inevitably, the better pQCD based models of nuclear scattering include saturation ideas in some form [18].

However, one can make quite a bit of headway in the classical EFT approach. Since the classical fields of the two nuclei are known, they provide the initial conditions for the evolution of the gluon fields produced in the nuclear collision. The small x fields are described by the classical Yang-Mills equations

$$D_\mu F_{\mu\nu} = J_\nu \tag{2}$$

with the random sources on the two light cones: $J_\nu = \sum_{1,2} \delta_{\nu,\pm}\delta(x_\mp)\rho_{1,2}(r_t)$. The two signs correspond to two possible directions of motion along the beam axis z. As shown by Kovner, McLerran and Weigert (KMW) [19], low-x fields in the central region of the collision obey sourceless Yang-Mills equations (this region is in the forward light cone of both nuclei) with the initial conditions in the $A_\tau = 0$ gauge given by

$$A^i = A^i_1 + A^i_2; \qquad A^\pm = \pm\frac{ig}{2}x^\pm[A^i_1, A^i_2]. \tag{3}$$

Here the pure gauge fields $A^i_{1,2}$ are solutions of (2) for each of the two nuclei in the absence of the other nucleus.

In order to obtain the resulting gluon field configuration at late proper times, one needs to solve (2) with the initial condition (3). Since the latter depends on the random color source, averages over realizations of the source must be performed. KMW showed that in perturbation theory the gluon number distribution by transverse momentum (per unit rapidity) suffers from an infrared divergence. A reliable way to go beyond perturbation theory is to re-formulate the EFT on a lattice by discretizing the transverse plane. (Boost invariance in pseudorapidity is assumed, which reduces the theory to a $2+1$–dimensional theory.) The resulting lattice theory can then be solved numerically [c]. We shall

[c] For alternative analytical approaches, see the work of Balitsky [20] and of Kovchegov [21].

not dwell here on the details of the lattice formulation, which is described in detail in Ref. [22].

We have simplified the problem considerably since Λ_s and the linear size L of the nucleus are the only parameters in the problem. We can write any dimensionful physical quantity q as $\Lambda_s^d f_q(\Lambda_s L)$, where d is the dimension of q. All the non-trivial physical information is contained in the dimensionless function $f_q(\Lambda_s L)$ [7]. We can estimate the values of the product $\Lambda_s L$ which correspond to key collider experiments. Assuming Au-Au collisions, we take $L = 11.6$ fm (for a square nucleus!) and estimate the standard deviation Λ_s to be 2 GeV for RHIC and 4 GeV for LHC [9]. Also, we have approximately $g = 2$ for energies of interest. The rough estimate is then $\Lambda_s L \approx 120$ for RHIC and $\Lambda_s L \approx 240$ for LHC.

Our simulations were performed for the SU(2) gauge group–all results for physical quantities are based on extrapolations to SU(3). A straightforward computation is of the initial energy per unit transverse area per unit rapidity, deposited in the central region by the colliding nuclei [23]. At late times, this is given by the non–perturbative formula

$$\frac{1}{\pi R^2}\,\frac{dE_t^i}{d\eta}|_{\eta=0} = \frac{1}{g^2}\, f_E\, \Lambda_s^3\,, \tag{4}$$

where $f_E \equiv f_E(\Lambda_s R) = 0.21$–$0.26$. Using this formula, and assuming, in accordance with Ref. [25], the $(N_c^2-1)/N_c$ dependence of the energy on the number of colors N_c, we arrive at the values of 2700 GeV and of 25000 GeV for the *initial* transverse energy per unit rapidity at RHIC and at LHC, respectively [23]. We can also estimate a formation time τ_f in our picture. For RHIC (LHC), we have $\tau_f = 0.3$ fm (0.13 fm). This then enables one to estimate the initial energy density to be 66.5 GeV/fm^3 (1.3 TeV/fm^3) at RHIC (LHC).

The number and distribution of produced gluons are of considerable interest as initial data for possible evolution of the gluon gas towards thermal equilibrium [25,26,27]. Strictly speaking, a particle number is only well-defined in a free-field theory, and there is no unique extension of this notion to a general interacting case. For this reason, we use two different generalizations of the particle number to an interacting theory, each having the correct free-field limit. We verify that the two definitions agree in the weak-coupling regime corresponding to late proper times in the central region [24]. Our first definition is straightforward. We impose the Coulomb gauge condition in the transverse plane: $\vec{\nabla}_\perp \cdot \vec{A}_\perp = 0$ and and determine the momentum components of the resulting field configuration. Our second definition is based on the behavior of a free-field theory under relaxation. This "cooling" technique allows a gauge invariant determination of the total particle number. Unfortunately, it cannot be used to determine the momentum distributions leaving the Coulomb gauge determination as the only one available.

We obtain

$$\frac{1}{\pi R^2}\frac{dN}{d\eta}|_{\eta=0} = \frac{1}{g^2} f_N \Lambda_s^2 \,, \tag{5}$$

where $f_N \equiv f_N(\Lambda_s L) = 0.13$ (0.15) for RHIC (LHC) energies. The agreement between the cooling and Coulomb techniques at the larger values of $\Lambda_s L$ relevant for RHIC and LHC is excellent. It is not as good at the smaller values: in general, the cooling number is more reliable. Naively extrapolating our results to SU(3), we find for *Au-Au* central collisions at RHIC energies, $dN/d\eta \sim 950$ for RHIC and $dN/d\eta \sim 4300$ for LHC energies.

Our results for the number distribution, computed in Coulomb gauge, are as follows. We have verified that for large $k_\perp$ our numerically obtained multiplicity agrees with the lattice analogue of the perturbative expression. At smaller $k_\perp$, the distribution softens and converges to a constant value, unlike its perturbative counterpart. Notably, this qualitative change of the distribution occurs at $k_\perp \sim \Lambda_s$. We tried to quantify this non-perturbative behavior by fitting the distribution to a variety of physically motivated functional forms. Surprisingly, we find that the shape of the distribution is closely reproduced by the two dimensional Bose-Einstein form $n(k) = A/(\exp(\beta\omega_k) - 1)$, with the inverse temperature β of the order of 1 in units of Λ_s, and with ω_k corresponding to a free *massive* dispersion relation, with the mass of the order of $0.1\Lambda_s$. This is a remarkable result for a purely classical theory, whose meaning, beyond providing us with a convenient parametrization, is not yet clear.

We have studied the number distribution at different proper time slices. Our numerical results show clearly that the size of the mass gap decreases as the square root of the proper time–this is what one would expect for instance for a screening mass. Further, though the peak of the distribution softens, its slope remains unchanged–the inverse two dimensional "temperature" β is a constant as a function of proper time. Thus the particle number is essentially unchanged–it increases slightly due to the "decay" of the mass gap.

Interestingly, all of these results (as well as the Λ_s/k_t form of the distribution for $k_t < \Lambda_s$) are consistent with the theoretical analysis of Baier et al. [26] as communicated to us by Son [28]. Baier et al. have studied the importance of $2 \rightarrow 3$ processes in driving the system to equilibrium. The particular feature of their work, distinguishing it from previous ones, is the inclusion of the Landau–Pomeranchuk–Migdal (LPM) effect that results in the suppression of hard gluon radiation in QCD [29]. In their kinetic theory analysis, for proper times $\tau < 1/\alpha_S^{3/2}$, the occupation number f is greater than unity. This is the overlap region between our classical simulations and kinetic theory and it is gratifying to see that they agree. This agreement further suggests that LPM interference effects are automatically included in the classical simulation.

What we have discussed above are the *initial* energy and number distributions of gluons produced in nuclear collisions. The subsequent evolution of the system (when the occupation number f falls below unity) is beyond the scope of the classical approach–it does however provide the initial conditions for the evolution. The approach to equilibrium in the McLerran–Venugopalan model was first investigated by Mueller [25]. He initially considered only elastic scattering and derived a Landau–type equation for the single particle distributions. This equation has since been solved numerically and the approach to equilibrium studied quantitatively [27]. One finds that elastic processes are very inefficient in driving the system to equilibrium (see also Ref. [30]). Baier et al. have argued that $2 \to 3$ processes are not suppressed relative to $2 \to 2$, and that they may actually be more efficient. For a more detailed discussion of thermalization, we refer the reader to their paper and to Son's talk at Quark Matter 2001 [28].

4 The CGC confronts RHIC data

Data from RHIC were presented at Quark Matter 2001 and some of it has already been published. How does the CGC picture fare when confronted with the data? The predictions of various models for the RHIC multiplicity have been summarized nicely by Eskola in his QM2001 talk [31] (see also Ref. [32]). Our prediction in Ref. [24] does quite well. This however was fortuitous since from Eq. 5 it is clear that the multiplicity depends very sensitively on the saturation scale–the latter clearly has at least a 10% uncertainty.

A more sensitive test is the centrality dependence of the multiplicity. Kharzeev and Nardi plotted the charged particle multiplicity per participant pair at pseudorapidity $\eta = 0$ versus the number of participants. They find that both the CGC and Glauber models give nearly identical predictions [33]. These in turn agree well with the data [34].

A consequence of the CGC picture is that $< p_t^2 > \propto dN/dy/\pi R^2$, as is suggested by Eq. 5. This dependence has been verified recently for the E735 $p\bar{p}$ data at the Tevatron as well as the UA(1) data [35]. It also works for the NA49 Heavy Ion CERN SPS data. First comparisons with the RHIC data are also promising [36].

In Ref. [24], we found that the momentum distributions of the produced gluons were nearly universal functions that scaled as $f \equiv f(p_t/\Lambda_s)$. Whether the RHIC data reflects this can be easily checked because Λ_s should be different for different centralities. Juergen Schaffner–Bielich has checked that the RHIC data for charged particles satisfies this property by rescaling the p_t scale by $p_t \to \lambda p_t$, where λ was an apparently arbitrary scaling constant. What is truly remarkable however is that the scaling constant is not arbitrary–it corresponds closely to the ratio of the saturation scales at the two centralities. If this scaling constant is used instead to determine Λ_s at the different centralities and if then the number of charged particles per participant pairs at $\eta = 0$ is plotted versus the number of participants (a la Kharzeev–Nardi), it agrees well with the data [36].

So far so good. However, the ratio of $R_i = E_t/N_{\rm gluon} \sim 1.5\Lambda_s$ computed by us [23,24] is too large relative to the data [37]. This is not necessarily a problem because R_i is not a conserved quantity and depends on the complex dynamics of the space–time evolution. Our simulation only computes the *initial* value of this ratio. Both number changing processes a la Baier et al. and the expansion of the system should decrease this ratio. If the former is the dominant process, one might be throwing the baby out with the bath water because the α_S dependence of the multiplicity might change. A more mundane possibility is that the function f_E which was computed in SU(2) might be significantly lower for SU(3). That's one reason to press ahead with SU(3) simulations of the problem. As pointed out by Dumitru in his QM2001 talk, R_i is a sensitive discriminant of the space–time dynamics in various models [30]. Detailed space–time simulations are in progress [38].

Tests of the CGC scenario in the coming years include the energy dependence of Λ_s, event by event fluctuations, and heavy quark production among others. We have recently computed the likelihood of the formation of CP–odd domains in the classical approach [39].

5 Outlook

The theoretical foundations of the classical approach to nuclear scattering clearly need to be on a firmer foundation. This is especially pressing in light of recent theoretical advances in small x physics [8]. There are several practical improvements that need to be done independently of the above. The extension to SU(3) and relaxing the strict boost invariance assumption are on the top of the agenda. A related interesting problem is: does the hot gluonic matter thermalize? At least asymptotically, this problem should have a solution. Recent attempts to address this problem are encouraging but much work (hopefully aided by new ideas) remains. Finally, as more and more RHIC data pour in over the years, we expect that they will provide a sensitive test of the ideas discussed here. We live in interesting times!

Acknowledgements

R. V.'s research was supported by DOE Contract No. DE-AC02-98CH10886. The authors acknowledge support from the Portuguese FCT, under grants CERN/P/FIS/1203/98 and CERN/P/FIS/15196/1999.

References

1. J. D. Bjorken, *Phys. Rev.* D **27**, 140 (1983).
2. J. D. Bjorken, J. B. Kogut and D. E. Soper, *Phys. Rev.* D **3**, 1382 (1971).
3. E. A. Kuraev, L. N. Lipatov and V. S. Fadin, *Sov. Phys. JETP* **45**, 199 (1977); I. I. Balitsky and L. N. Lipatov, *Sov. J. Nucl. Phys.* **28**, 822 (1978).
4. L. V. Gribov, E. M. Levin and M. G. Ryskin, *Phys. Rept.* **100**, 1 (1983); A. H. Mueller and J. Qiu, *Nucl. Phys.* B **268**, 427 (1986).
5. L. McLerran and R. Venugopalan, *Phys. Rev.* D **49**, 2233 (1994); D **49**, 3352 (1994); D **50**, 2225 (1994); D **59**, 094002 (1999).
6. R. Venugopalan, *Pramana* **55**, 73 (2000).
7. R. V. Gavai and R. Venugopalan, *Phys. Rev.* D **54**, 5795 (1996).
8. E. Iancu, A. Leonidov and L. McLerran, hep-ph/0011241; hep-ph/0102009; hep-ph/0103032.
9. M. Gyulassy and L. McLerran, *Phys. Rev.* C **56**, 2219 (1997).
10. J. Jalilian-Marian, A. Kovner, A. Leonidov and H. Weigert, *Nucl. Phys.* B **504**, 415 (1997); J. Jalilian-Marian, A. Kovner and H. Weigert, *Phys. Rev.* D **59**, 014015 (1999).
11. G. Parisi and N. Sourlas, *Phys. Rev. Lett.* **43**, 744 (1979).
12. J. Jalilian-Marian, A. Kovner, L. McLerran and H. Weigert, *Phys. Rev.* D **55**, 5414 (1997); Y. V. Kovchegov, *Phys. Rev.* D **54**, 5463 (1996).
13. C. S. Lam and G. Mahlon, hep-ph/0102337; *Phys. Rev.* D **62**, 114023 (2000); D **61**, 014005 (2000).
14. Y. V. Kovchegov, *Phys. Rev.* D **61**, 074018 (2000); I. Balitsky, *Nucl. Phys.* B **463**, 99 (1996).
15. N. Armesto and M. Braun, hep-ph/0104038; E. Levin and M. Lublinsky, hep-ph/0104108.
16. R. Venugopalan, hep-ph/0102087.
17. K. Golec-Biernat and M. Wusthoff, *Phys. Rev.* D **59**, 014017 (1999); D **60**, 114023 (1999); A. M. Stasto, K. Golec-Biernat and J. Kwiecinski, *Phys. Rev. Lett.* **86**, 596 (2001); A. C. Caldwell and M. S. Soares, hep-ph/0101085.
18. K. J. Eskola, K. Kajantie, P. V. Ruuskanen and K. Tuominen, *Nucl. Phys.* B **570**, 379 (2000).

19. A. Kovner, L. McLerran and H. Weigert, *Phys. Rev.* D **52**, 3809 (1995); D **52**, 6231 (1995).
20. I. Balitsky, hep-ph/0101042.
21. Y. V. Kovchegov, hep-ph/0011252.
22. A. Krasnitz and R. Venugopalan, *Nucl. Phys.* B **557**, 237 (1999).
23. A. Krasnitz and R. Venugopalan, *Phys. Rev. Lett.* **84**, 4309 (2000).
24. A. Krasnitz and R. Venugopalan, *Phys. Rev. Lett.* **86**, 1717 (2001).
25. A. H. Mueller, *Nucl. Phys.* B **572**, 227 (2000); *Phys. Lett.* B **475**, 220 (2000).
26. R. Baier, A. H. Mueller, D. Schiff and D. T. Son, *Phys. Lett.* B **502**, 51 (2001).
27. J. Bjoraker and R. Venugopalan, *Phys. Rev.* C **63**, 024609 (2001).
28. D. T. Son, talk at Quark Matter 2001, Stony Brook, New York, January 2001; private communication.
29. R. Baier, Y. L. Dokshitzer, A. H. Mueller, S. Peigne and D. Schiff, *Nucl. Phys.* B **483**, 291 (1997).
30. A. Dumitru and M. Gyulassy, *Phys. Lett.* B **494**, 215 (2000); hep-ph/0103291.
31. K. J. Eskola, hep-ph/0104058.
32. N. Armesto and C. Pajares, *Int. J. Mod. Phys.* A **15**, 2019 (2000).
33. D. Kharzeev and M. Nardi, nucl-th/0012025.
34. K. Adcox *et al.* [PHENIX Collaboration], nucl-ex/0012008.
35. L. McLerran and J. Schaffner-Bielich, hep-ph/0101133.
36. from discussions with J. Schaffner-Bielich, unpublished.
37. P. Steinberg, Rapporteur talk, Quark Matter 2001, Stony Brook, New York, January 2001.
38. Y. Nara and R. Venugopalan, in progress.
39. D. Kharzeev, A. Krasnitz and R. Venugopalan, in preparation.

LEADING NUCLEON PRODUCTION AT HERA

GARRY LEVMAN

University of Toronto, Department of Physics, 60 St George St, Toronto ON M5S 1A7, Canada

E-mail: levman@physics.utoronto.ca

Experimental results from HERA on the production of leading nucleons in e^+p collisions are reviewed. The data cover a large kinematic range from photoproduction to deep inelastic scattering. The protons and neutrons are produced with low transverse momentum ($p_T^2 \lesssim 0.5$ GeV^2) and carry a large fraction of the incoming proton's energy ($x_L > 0.6$). The data are compared to hadroproduction experiments and to the predictions of particle-exchange models. Some implications of the data for the structure of the pion are also discussed.

1 Introduction

The confinement of partons, the nature of the vacuum, the long range structure of the nucleon, remain puzzles of quantum chromodynamics closely related to the problem of the transition from perturbative to non-perturbative physics.

Leading nucleons have been observed in positron-proton collisions at HERA[1,2,3]. These are protons and neutrons which carry a large fraction of the energy ($x_L \gtrsim 0.2$) of the target proton, but which are produced with small transverse momentum ($p_T \lesssim 1$ GeV). The presence of a leading nucleon in the final state flags a peripheral process associated with long distances, even when short distances are probed by high photon virtuality (Q^2) or high transverse energies. The study of leading nucleon production provides useful information on the relationship between the soft and hard aspects of the strong interaction.

Data from HERA cover a large kinematic regime in Q^2: from photoproduction $Q^2 < 0.02$ GeV^2, through the transition region $0.1 < Q^2 < 0.74$ GeV^2 to deep inelastic scattering (DIS) $Q^2 > 4$ GeV^2. The studies include both semi-inclusive reactions[1,2], $ep \to eNX$ ($N = p$ or n), as well as the semi-exclusive production of jets with high transverse energy[3,4,5], $ep \to eN$ jet jet X.

In analogy with the inclusive structure function $F_2(x, Q^2)$, which is defined through the inclusive ep cross section,

$$\frac{d^2\sigma(ep \to eX)}{dxdQ^2} = \frac{4\pi\alpha^2}{dxQ^4}\left(1 - y + \frac{y^2}{2}\right) F_2(x, Q^2)$$

the semi-inclusive structure function $F_2^{LB(4)}(x, Q^2, x_L, p_T^2)$ is defined through the four-fold differential cross section for leading nucleon production,

$$\frac{d^4\sigma(ep \to eNX)}{dxdQ^2dx_Ldp_T^2} = \frac{4\pi\alpha^2}{xQ^4}\left(1 - y + \frac{y^2}{2}\right) F_2^{LB(4)}(x, Q^2, x_L, p_T^2)$$

where $B = P$ or N for leading protons or neutrons, respectively; x is Bjorken's variable, and y is the inelasticity of the electron in the rest frame of the target proton.

2 Leading proton production

Figure 1 shows the H1 measurement[2] of F_2^{LP} for protons in the kinematical range $2 < Q^2 < 50$ GeV^2, $0.7 < x_L < 0.9$ and $p_T < 0.2$ GeV. In each bin of x and Q^2 the structure function is approximately flat in x_L; however, the level increases with increasing Q^2. The flatness of the x_L distribution for protons is also apparent in the ZEUS data[6] which shows that there is no difference in shape or rate between the DIS and the transition region ($0.1 < Q^2 < 0.8$ GeV^2) data. ZEUS has also studied the angular distribution of leading protons as a function of x_L in the transition region and in DIS. In bins of x_L the data are well described by the form $dN/dp_T^2 \propto \exp(-bp_T^2)$. The values found for the slope parameters b are largely independent of x_L, and are the same in both the DIS and the transition samples. The data indicate that

$$F_2^{LP}(x, Q^2, x_L, p_T^2) \approx 3\exp(-6p_T^2)F_2(x, Q^2)$$

(with p_T in GeV) since approximately 15% of the DIS events have a leading proton with $0.6 < x_L < 0.9$ and $p_T^2 < 0.5$ GeV^2. The positron variables x and Q^2 factorize from the proton variables x_L and p_T, a form of the hypothesis of limiting fragmentation[7].

3 Leading neutron production

Figure 1 shows the H1 measurement of F_2^{LN} at high x_L. In contrast to the leading proton case, the structure function for neutrons decreases with increasing x_L approaching zero at the kinematic limit.

ZEUS has measured the x_L spectrum of neutrons in photoproduction, the transition region and DIS. The data are shown in Fig. 2 as a fraction of

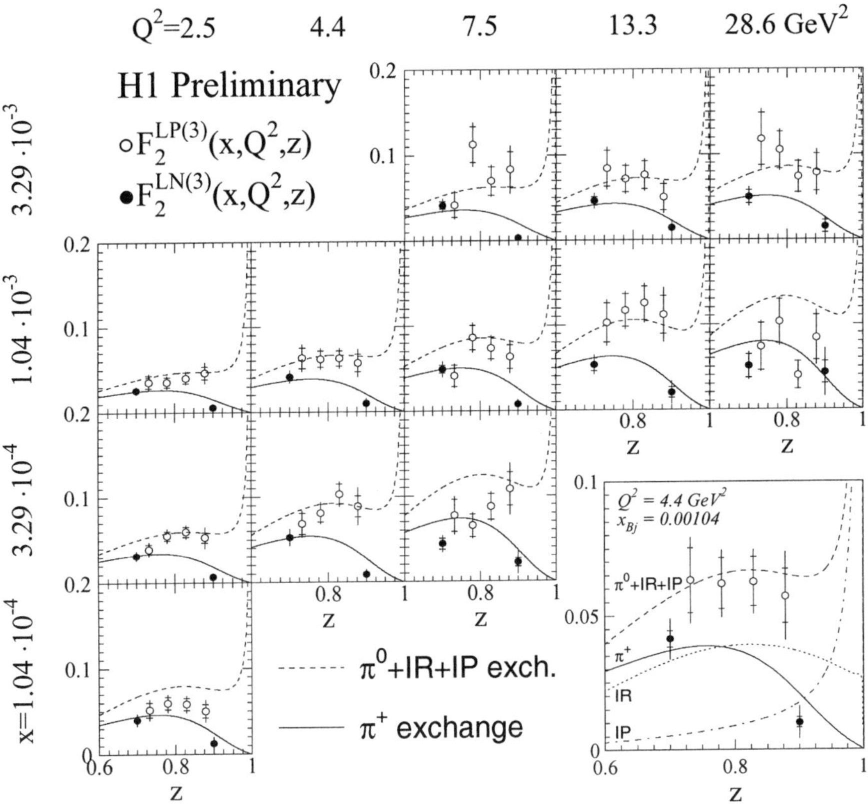

Figure 1. F_2^{LP} and F_2^{LN} as a function of z in bins of x and Q^2. $z \equiv x_L$ is the fraction of the target proton's energy carried by the produced nucleon. The structure functions are integrated over $0 < p_T < 0.2$ GeV. The curves show the result of the H1 particle-exchange model for leading nucleon production. The amount of pion exchange in nucleon production is fixed by neutron production. $I\!R$ represents the remaining meson exchanges contributing to proton production. $I\!P$ is pomeron exchange.

the corresponding total inclusive cross section. The neutron scattering angle satisfies $\theta < 0.8$ mrad ($p_T < 0.66x_L$ GeV). The increase in the p_T range with x_L may account for the rising neutron yield with x_L.

The neutron rate increases by 20-30% between photoproduction and DIS. For DIS the spectra are roughly independent of Q^2, although at low x_L the rate continues to increase slowly while at high x_L the rate decreases. The strong rise in rate between photoproduction and DIS can be attributed to the absorptive rescattering of the produced neutron at low Q^2. For $Q^2 \gtrsim 4$ GeV^2 the photon is point-like and absorptive effects are minimal.

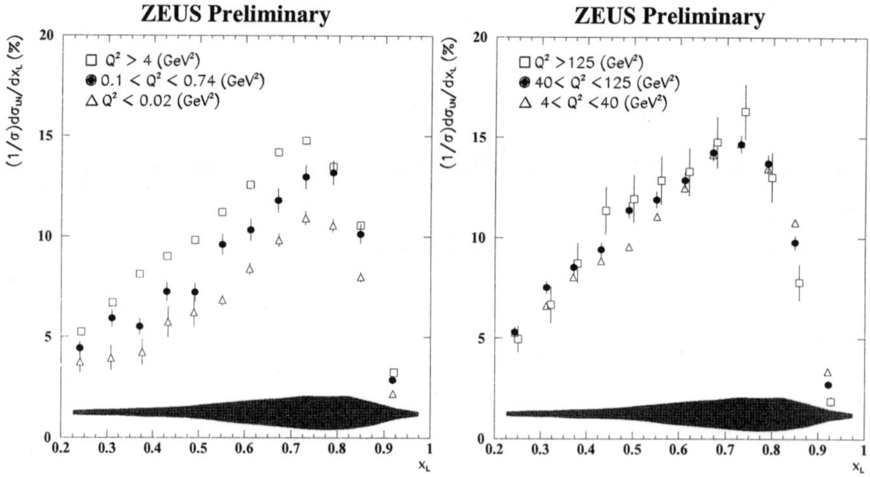

(a) (b)

Figure 2. (a) Energy spectra of neutrons in photoproduction $Q^2 < 0.02$ GeV^2, in the intermediate region $0.1 < Q^2 < 0.74$ GeV^2, and in deep inelastic scattering $Q^2 > 4$ GeV^2. (b) Energy spectrum of neutrons in DIS for three Q^2 selections: $4 < Q^2 < 40$ GeV^2, $40 < Q^2 < 125$ GeV^2, and $Q^2 > 125$ GeV^2. The points in (b) are offset slightly for visibility.

In the peak region, $0.64 < x_L < 0.82$, the neutron yield in DIS is approximately 2.5% independent of Q^2 and x.

4 Comparison to hadroproduction

The hadroproduction of leading nucleons has been studied in experiments[a] at Fermilab and CERN which show that the characteristics of the produced nucleon broadly conform to those of electroproduction at HERA.

Bubble chamber data from Fermilab[9,10] show that protons produced at low p_T in hadron-hadron collisions have flat x_L spectra for $0.6 < x_L < 0.95$, exponentially falling p_T^2 spectra, and production rates comparable to protons in DIS at HERA, where the center of momentum energy is much larger ($W \approx$ 200 GeV compared to 20 GeV). Moreover, the slopes of the p_T^2 distributions are approximately independent of x_L and have values of about 6 GeV^{-2} as in electroproduction.

Both the energy and angular distributions of leading neutrons at HERA are similar to hadroproduced neutrons[16,17,18,19] for $x_L \gtrsim 0.4$; however, the rate of neutrons produced in photoproduction at HERA is roughly half that expected of neutrons produced at Fermilab.

5 Comparison to particle exchange models

Both H1 and ZEUS have compared the predictions of particle-exchange[2,22,23,24] models to their data. In these models the leading nucleon is produced as a result of a virtual exchange at the proton vertex. For protons, the exchange can be either a meson $(f_2, \omega, \pi, \rho, a_2)$ or the pomeron ($I\!P$), the exchange which contributes to diffractive scattering and carries the quantum numbers of the vacuum. Neutron production is simpler in that only isovector mesons (π, ρ, a_2) contribute. In these models

$$F_2^{LB(3)}(x, Q^2, x_L) = \sum_i f_{i/p}(x_L) F_2^i(x/(1-x_L), Q^2) \tag{1}$$

where $i = I\!P, \pi, \rho, \omega, a_2, f_2$, $f_{i/p}$ is the splitting function for $p \to N$ via i exchange, and F_2^i is the structure function of i. As seen in Fig. 1, the H1 x_L distributions are in good agreement with one such model. The ZEUS collaboration finds that such models can describe both the x_L and the p_T^2 distributions of leading protons.

6 The structure of the pion

Pion exchange is expected to dominate in the production of leading neutrons[22,23]. Hadron-hadron data from Fermilab and CERN are in good

[a] See references 8 to 21.

agreement with this expectation[16,18,19]. The HERA electroproduction data present an opportunity to study pion exchange in photoproduction and deep inelastic scattering, and to investigate the partonic structure of the pion at low x. The H1 collaboration has already published a preliminary study[2]. Here comments are made on the implications of ZEUS preliminary data[25].

6.1 *Effective flux method*

For the charge exchange reactions $p \to n$ (or $n \to p$) an effective splitting function (the effective flux of pions in the proton) can be determined by deconvolution ($\oslash$)

$$f_{eff}(x_L, t) \equiv \frac{d\sigma^{pn \to Xp}}{dx_L dt} \oslash \sigma^{\pi p \to X}$$

of the measured semi-inclusive differential cross section for $pn \to Xp$ and the measured πp total cross section. By construction, f_{eff} corrects for ρ and a_2 exchange, Δ production, absorptive rescattering and off-mass-shell effects.

Experiments[b] have shown that the effective flux can be presented by

$$f_{eff}(x_L, t) = \frac{1}{4\pi}\frac{g^2}{4\pi}\frac{-t}{(t - m_\pi^2)^2}(1 - x_L)^{1-2t}$$

where m_π is the mass of the pion, t is the invariant momentum transfer in $p \to n$, and $g^2/4\pi = 29$. f_{eff} is not expected to change significantly if the proton projectile is changed to another hadron (such as a pion), or to a real photon; however, for electroproduction, absorptive rescattering will decrease as Q^2 increases and the virtual photon becomes point-like[26,27]. Therefore, the effective flux will increase correspondingly: $f_{eff} \to f_{eff}^{\rm DIS} = 1.3 f_{eff}$ (see Fig. 2).

6.2 F_2^π *at low* x

ZEUS measurements show that about 2.5% of the DIS events, independent of x and Q^2, have a leading neutron with $0.64 < x_L < 0.82$ and $p_T < 0.66 x_L$ GeV. The constancy of this ratio with respect to Q^2 and x implies that $F_2^\pi(x, Q^2)$ is proportional to $F_2((1 - x_L)x, Q^2) = F_2(0.27x, Q^2)$ with a constant of proportionality given by

$$\frac{0.025}{\int\int f_{eff}^{\rm DIS}(x_L, t) dx_L dt} = 0.25$$

[b] See references 15 to 21.

where the integration is over the x_L and t range spanned by the measurement (see Eqn. 1). Because[28] $F_2(x) \propto (1/x)^\lambda$ with $\lambda \approx 0.2$

$$F_2^\pi(x, Q^2) \approx F_2(x, Q^2)/3$$

7 Conclusions

For leading protons the shape of the x_L spectra and angular distributions are independent of the kinematic variables Q^2, x and x_L. Neutron production is also approximately independent of Q^2 and x, but has a rapidly falling x_L distribution near the kinematic limit. The neutron data also shows evidence for absorption when $Q^2 \lesssim 4$ GeV^2. Both the proton and neutron data are in good conformance with hadroproduction data from Fermilab and CERN, and can be described by simple-particle exchange models. An effective flux analysis of the leading neutron data implies that F_2^π has a similar x and Q^2 behavior as F_2 at low x, but only about one third its magnitude.

Acknowledgments

I thank the Special Research Centre for the Subatomic Structure of Matter for the invitation to visit Adelaide and present this talk. I also thank A. Thomas and W. Melnitchouk for interesting and helpful discussions.

References

1. ZEUS Collaboration, M. Derrick et al., *Phys. Lett.* B **384**, 388 (1996).
2. H1 Collaboration, C. Adoff et al., *Eur. Phys. J.* C **6**, 587 (1999).
3. ZEUS Collaboration, J. Breitweg et al., *Nucl. Phys.* B **596**, 3 (2001).
4. H1 Collaboration, *Measurement of dijet cross sections with leading neutrons in photoproduction at HERA*, contributed paper to the XXXth International Conference on High Energy Physics ICHEP2000, Osaka, Japan,, July 27 - August 2, 2000.
5. ZEUS Collaboration, *Leading baryon production in ep scattering at HERA*, contributed paper to the International Europhysics Conference on High Energy Physics 99, Tampere, Finland, 15-21 July 1999.
6. ZEUS Collaboration, *Leading baryon production in ep scattering at HERA*, contributed paper to the XXIX International Conference on High Energy Physics ICHEP98, Vancouver, Canada, 23-29 July 1998.
7. J. Benecke, T. T. Chou, C. N. Yang, and E. Yen, *Phys. Rev.* **188**, 2159 (1969).

8. S. J. Barish et al., *Phys. Rev. Lett.* **31**, 1080 (1973).
9. J. Whitmore et al., *Phys. Rev.* D **11**, 3124 (1975).
10. F. C. Winkelmann et al., *Phys. Rev. Lett.* **32**, 121 (1974).
11. F. Sannes et al., *Phys. Rev. Lett.* **30**, 766 (1973).
12. J. W. Chapman et al., *Phys. Rev. Lett.* **32**, 257 (1974).
13. K. Abe et al., *Phys. Rev. Lett.* **31**, 1527 (1973).
14. F. T. Dao et al., *Phys. Lett.* B **45**, 399 (1973).
15. J. Engler et al., *Nucl. Phys.* B **84**, 70 (1975).
16. B. Robinson et al., *Phys. Rev. Lett.* **34**, 1475 (1975).
17. W. Flauger and F. Mönnig, *Nucl. Phys.* B **109**, 347 (1976).
18. G. Hartner, *A study of pn interactions at 205 GeV/c*, Doctoral Thesis (McGill University, 1977), unpublished.
19. Y. Eisenberg et al., *Nucl. Phys.* B **135**, 189 (1978).
20. V. Blobel et al., *Nucl. Phys.* B **135**, 379 (1978).
21. H. Abramowicz et al., *Nucl. Phys.* B **166**, 62 (1980).
22. J. D. Sullivan, *Phys. Rev.* D **5**, 1732 (1972).
23. M. Bishari, *Phys. Lett.* B **38**, 510 (1972).
24. A. Szczurek, N. N. Nikolaev, and J. Speth, *Phys. Lett.* B **428**, 383 (1998).
25. ZEUS Collaboration, *Energetic neutron production in e^+p collisions*, contributed paper to the XXXth International Conference on High Energy Physics ICHEP2000, Osaka, Japan,, July 27 - August 2, 2000.
26. N. N. Nikolaev, J. Speth and B. G. Zakharov, *Absorptive corrections to one pion exchange and measurability of the small-x pion structure function at HERA*, hep-ph/9708290 (1997).
27. U. D'Alesio and H. J. Pirner, *Eur. Phys. J.* A **7**, 109 (2000).
28. ZEUS Collaboration, J. Breitweg et al., *Eur. Phys. J.* C **7**, 1999 (609).

NUCLEAR MEDIUM EFFECTS AT HERMES

PASQUALE DI NEZZA (on behalf of the HERMES Collaboration)
Istituto Nazionale di Fisica Nucleare
Laboratori Nazionali di Frascati, via E.Fermi 40 I-00044 Frascati - Italy

Nuclear effects have been studied at HERMES in inclusive and semi-inclusive reactions in D, ^{3}He and ^{14}N unpolarized targets using a 27.5 GeV positron beam. In the inclusive case, cross section ratios for deep-inelastic scattering from ^{14}N and ^{3}He with respect to D have been measured. The data provide evidence for an A-dependence of the ratio $R = \sigma_L/\sigma_T$ of longitudinal to transverse DIS cross sections at low x and low Q^2. In the semi-inclusive case, the influence of the nuclear medium on the production of charged hadrons has been studied. The differential multiplicity of charged hadrons and pions from ^{14}N relative to that from D has been measured as a function of the virtual photon energy ν and the fraction z of this energy transferred to the hadron. A substantial reduction of the multiplicity ratio R_M at low ν and high z is observed. Moreover, a significant difference of R_M is found for positive and negative hadrons.

1 Introduction

HERMES is able to perform both inclusive (DIS) and semi-inclusive (SIDIS) deep inelastic scattering studies on nuclear targets with high accuracy in the optimal energy range. From inclusive studies the A-dependence of the structure function F_2 is already known, which in the Quark-Parton Model represents the quark momentum distribution inside the nucleon. This phenomenon is known as the EMC effect [1] at large values of the Bjorken scaling variable x and as shadowing at low values of x [2]. Little information is available on the A-dependence of the longitudinal (σ_L) and the transverse (σ_T) components of the structure function F_2.

In the semi-inclusive case, the nucleus, which has the size of a few fm, acts as an ensemble of targets with which the struck quark or the produced hadron may interact. These interactions can provide information on the space-time structure of the hadronization process.

The data presented in this analysis were collected using D, ^{3}He and ^{14}N internal gas targets in the 27.57 GeV positron storage ring at HERA. The target gases were injected into a tubular open-ended storage cell inside the positron ring and have areal densities of up to 6×10^{15} nucleus/cm^2 for ^{14}N. In the HERMES spectrometer [3] (Fig.1) both the scattered positron and the produced hadrons can be detected and identified within an angular acceptance ±170 mrad horizontally, and 40-140 mrad vertically.

The trigger for scattered positrons was formed from a coincidence between

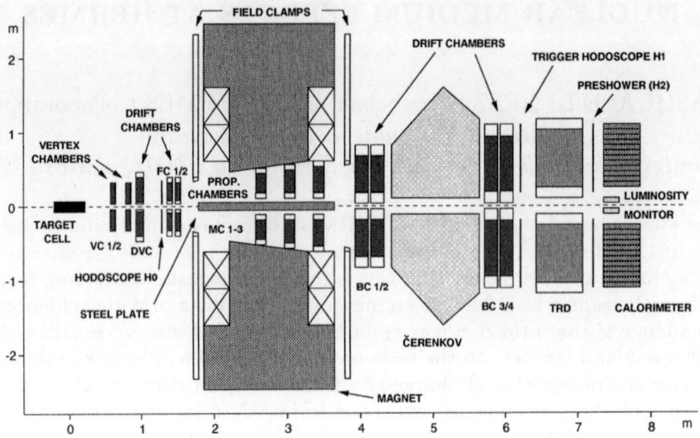

Figure 1: Longitudinal view of the HERMES spectrometer.

three scintillator hodoscope planes and a lead-glass calorimeter where a deposited energy of more than 3.5 GeV is required. Positron identification was accomplished using the calorimeter, the third hodoscope, which functioned as a preshower counter, a transition-radiation detector, and a threshold gas Čerenkov counter. This system provided positron identification with an average efficiency of 99 % and a hadron contamination of less than 1 %. Hadrons with an energy $E_h \geq 1.4$ GeV were distinguished from leptons. Furthermore, in the momentum range between 4 and 13.5 GeV, pions were identified with the help of the Čerenkov counter.

The energy transfer (ν) of the virtual photon to the target nucleon (of mass M) and the four-momentum transfer (Q^2) constrain the kinematics of an inclusive event completely. They are related to the energy of the incident (E) and scattered (E') positrons and to the scattering angle (θ) in the laboratory frame:

$$\nu = E - E', \qquad Q^2 = 4EE' sin^2(\theta/2).$$

Three dimensionless variables x, y and z are used to describe the kinematics of DIS and SIDIS processes:

$$x = Q^2/2M\nu, \quad y = \nu/E', \quad z = E_h/\nu;$$

where x is the nucleon momentum fraction of the struck quark, y is the energy fraction of the virtual photon and z is the fraction of the photon energy ν transferred to the hadron h.

2 Inclusive measurement

A possible small A-dependence of $R = \sigma_L/\sigma_T$ has been suggested by several authors. The Fermi motion of the nucleons is seen to enhance higher-twist effects[4] at low values of x and Q^2. It has been also argued[5] that an increase of the nuclear gluon distribution may lead to an enhancement of R. In addition, it has also been shown[6] that nuclear σ, ω and π mesons can contribute coherently to enhance the electroproduction cross section on nuclei for longitudinal virtual photons at low Q^2 while depleting the cross section for transverse photons. However, until the HERMES measurement [7], no experimental evidence for a nuclear medium effect on R was found.

The DIS double-differential cross section per nucleon of an unpolarised target can be written, in the one-photon exchange approximation, in terms of the cross sections for the absorption of transversely (σ_T) and longitudinally (σ_L) polarized virtual photons as:

$$\frac{d^2\sigma}{dx\,dQ^2} = \Gamma(\sigma_T + \epsilon\sigma_L),$$

where $\Gamma = \frac{\alpha(\nu - Q^2/2M)E'}{2\pi^2 Q^2 E(1-\epsilon)}$ is the virtual photon flux and $\epsilon = \frac{4(1-y)-Q^2/E^2}{4(1-y)+2y^2+Q^2/E^2}$ is the virtual photon polarization parameter.

This cross section can be also expressed in terms of the unpolarized structure functions $F_2 = \frac{(2M\nu - Q^2)\nu}{8\pi^2\alpha(M+\nu/2x)}(\sigma_L + \sigma_T)$ and R as well:

$$\frac{d^2\sigma}{dx\,dQ^2} = \frac{\sigma_{Mott}}{E'E}\frac{\pi F_2(x,Q^2)}{x\epsilon}\frac{1+\epsilon R(x,Q^2)}{1+R(x,Q^2)}$$

The study of the A-dependence involves measurements of the ratio of DIS cross sections for a nucleus (σ_A) and deuterium (σ_D):

$$\frac{\sigma_A}{\sigma_D} = \frac{F_2^A}{F_2^D}\frac{1+\epsilon R_A}{1+R_A}\frac{1+R_D}{1+\epsilon R_D}$$

where R_A and R_D represent the ratio σ_L/σ_T for a nucleus (A) and deuterium (D), respectively. For $\epsilon \to 1$ the cross section ratio equals the ratio of the structure functions F_2^A/F_2^D. For smaller values of ϵ the cross section ratio is equal to F_2^A/F_2^D only if $R_A = R_D$. A difference between R_A and R_D will thus introduce an ϵ-dependence of σ_A/σ_D. Hence, measurements of σ_A/σ_D as a function of ϵ can be used to extract experimental information on R_A/R_D.

The ratio of the inclusive DIS cross section off ^{14}N and ^{3}He with respect to D from the HERMES experiment is presented in Fig.2 (left) compared to similar ratios[8]. A significant difference between HERMES data and previous data is observed for $x < 0.06$. In this domain the HERMES data for both nuclei are smaller then the NMC and E665 data and the deviation increases towards smaller values of x. At high values of x the HERMES data are in agreement with the SLAC data. Apart from the data shown in fig.2 (left), other data show a strong reduction of σ_A/σ_D for $0.01< x <0.1$ and $0.05< Q^2 <1.5$ GeV^2

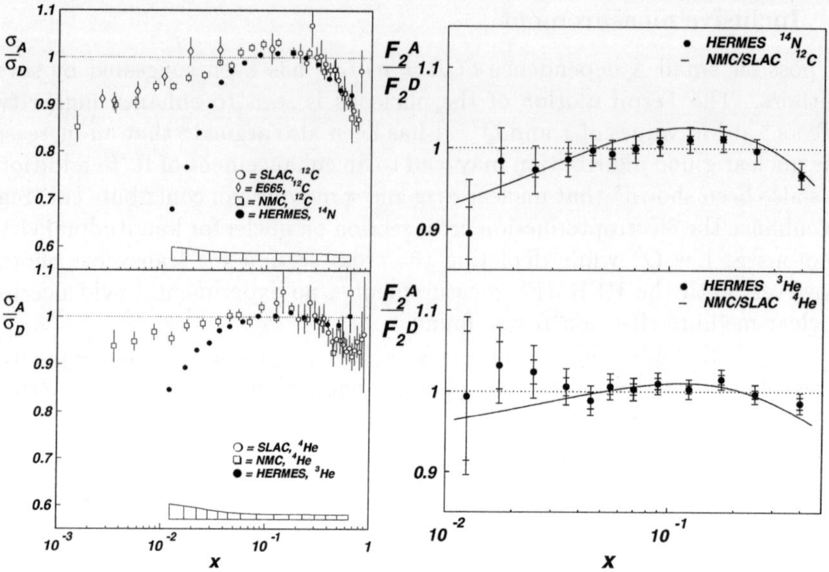

Figure 2: (Left) Ratio of inclusive DIS cross sections from nucleus (A) and deuterium (D) versus x. (Right) Ratio of the inclusive structure functions F_2^A/F_2^D.

that is similar to that of the HERMES data on ^{14}N [9]. However, these data were never used to study a possible A-dependence of R, because of insufficient statistics or because of their limited kinematic coverage.

The values of F_2^A/F_2^D derived from the HERMES σ_A/σ_D data are displayed in Fig.2 (right) for both ^{14}N and ^{3}He. The data are in agreement with the curves representing parameterizations of F_2^A/F_2^D ratios determined using the NMC and SLAC data. The experimental uncertainties on F_2^A/F_2^D are too large to observe any systematic deviation between neighbouring nuclei.

The resulting values of R_A/R_D are shown in Fig.3 (left). The data are plotted at the average values of Q^2 for a given x-bin. The values of R_A/R_D derived from the HERMES data are considerable larger than unity for x <0.06 and Q^2 <1.5 GeV^2. In this region the deviation from unity is bigger for ^{14}N than for ^{3}He, as expected for a medium dependent effect. All data above Q^2=1.5 GeV^2 are seen to be consistent with unity.

The mechanism that gives rise to an enhancement of R_A with respect to R_D at low Q^2 and x does not significantly change the ratio of the structure functions F_2^A/F_2^D. Since F_2^A and R_A depend differently on the longitudinal and transverse DIS cross sections, the different effects on σ_L and σ_T can be

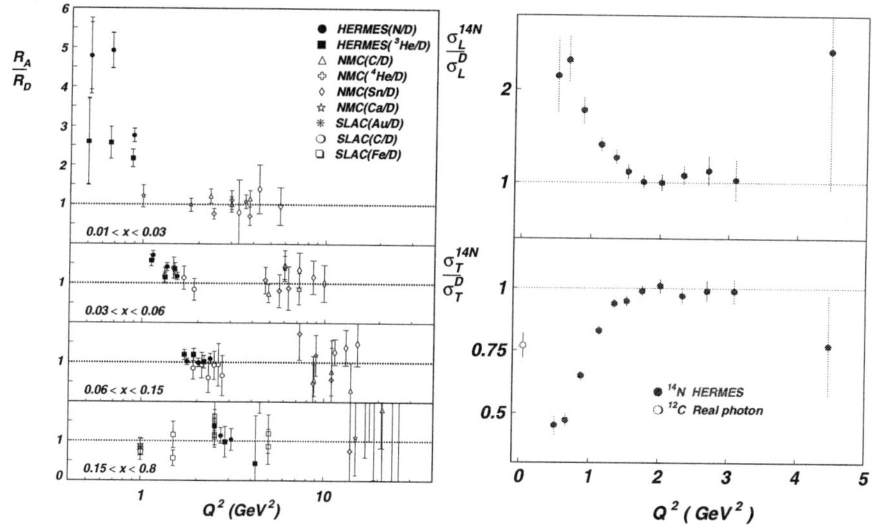

Figure 3: (Left) Ratio R_A/R_D for nucleus A and D as a functions of Q^2 for four different x bins. (Right) The extracted ratio of longitudinal and transverse virtual photon cross section from ^{14}N relative to D as a function of Q^2 with the real photon result for ^{12}C.

distinguished. In Fig.3 (right) it is shown that the longitudinal virtual photon cross section rises in conjunction with a reduction of the transverse virtual photon cross section in the low Q^2 range.

As explained above, possible interpretations of the observed A-dependence of σ_L and σ_T suggest that the effect is either due to coherent scattering from nuclear σ, ω and π mesons, or dynamical exchange of quarks and gluons in nuclei [6,10].

3 Semi-inclusive measurements

Besides the study of the internal structure of the nucleon, deep inelastic lepton nucleus scattering offers a direct way to study the space-time development of the hadronization process, because one can use the secondary reactions of the final state particles for measuring the so-called hadron formation time (τ_f) or the corresponding hadron formation length ($l_f = c\tau_f$).

In semi-inclusive DIS the struck quark will propagate over l_f before interacting with other quarks. When the quark propagation is embedded in the nuclear medium, the number of hadrons produced will depend on the interactions of the propagating quarks. Therefore the comparison of hadron distributions in DIS of different nuclei offers the best way to extract information on the formation time of the hadrons [11,12].

At present the formation time of hadrons is not calculable within QCD, as the hadronization process takes place at a scale where perturbative field theory techniques are not applicable anymore. Therefore, final-state distributions can be calculated only with the help of phenomenological models.

HERMES has been able to perform these kind of measurements [13] due to the fact that at this energies ($4< \nu < 24$ GeV) nuclear effects may be significant since l_f is of the same order of magnitude as the diameter of the target nucleus. In addition, previous experiments performed at SLAC with deep-inelastic electron scattering [14], at CERN with high-energy muons by SMC [15] and with neutrinos by WA21/WA59 [16], indicate that the optimal energy range of the lepton to study hadronization dynamics with nuclei equals a few tens of GeV.

Moreover, in order to discriminate among the different models for the formation time, it is important to study the hadronization process as a function of hadron species and this is actually possible at HERMES distinguishing pions from hadrons. About the pions, kaons and protons identification, provided by the RICH detector, a new analysis is actually underway at HERMES.

The directly observable quantity is the attenuation ratio for ^{14}N, defined as the ratio between the differential charged hadron multiplicity N_h normalized by the number of scattered electrons N_{DIS}, for ^{14}N and D targets:

$$R_M^h(z,\nu) = \frac{N_h^{^{14}N}(z,\nu)}{N_e^{^{14}N}(\nu)} / \frac{N_h^D(z,\nu)}{N_e^D(\nu)}.$$

The attenuation ratio as function of z for charged hadrons and pions is shown in Fig.4 (left). A significant attenuation of hadrons in the reaction on the ^{14}N nucleus is seen in the whole region $0.2 < z < 1$. Pion attenuation can be studied by using the Čerenkov detector in a restricted energy region. The z behavior of the pion attenuation is similar to that of the unidentified hadron ones. The data also cover the unexplored region $0.85 < z < 1$ showing a decrease of the attenuation ratio. This finding is in qualitative agreement with the predictions from different models [17,18] reported in the figure. The attenuation ratio as a function of ν is shown in Fig.4 (right) for charged hadrons and pions. The ratio is observed to increase with increasing ν. This can be understood considering that at high ν the hadron formation takes place mainly outside the nucleus and the attenuation process is suppressed. The data also show that the multiplicity ratio R_M^{π} is the same for positive and negative pions, Fig.4(right). This observation, together with the fact that the $\pi^{\pm}$–N interaction cross sections are about equal, is consistent with the likely

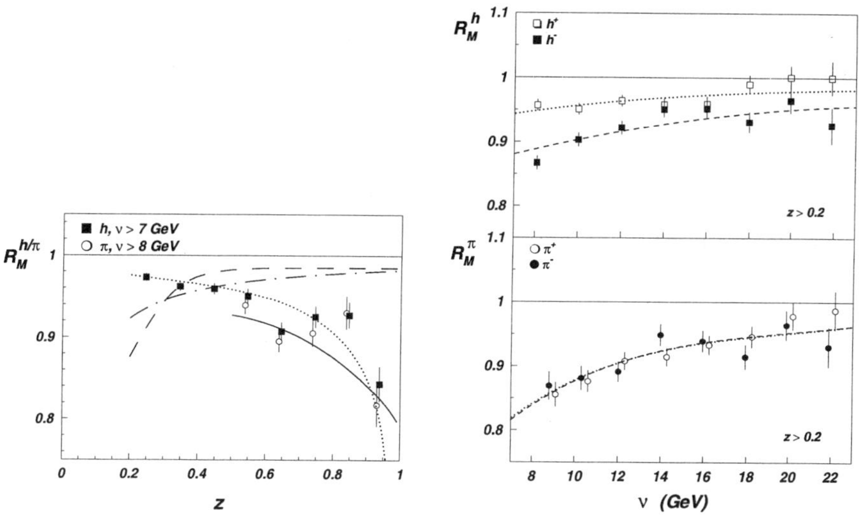

Figure 4: (Left) The multiplicity ratio as a function of z. The full curve represents a gluon-bremsstrahlung model calculation for pions. The other curves represent phenomenological formation-time calculations. (Right) The multiplicity ratio as a function of ν. The curves are parameterizations of the data using the one time-scale.

assumption that the formation times for the two charge states are the same. In contrast, a significant difference is observed between R_M^h of positive and negative hadrons. Since the multiplicity ratios of positive and negative pions are measured to be equal, this difference must be caused by other hadrons, such as protons, antiprotons and kaons.

In the absence of a theoretical model, the observed differences among R_M have been parameterized using the one time-scale model[19] and the assumption that the formation time is given by $t_f^h = c_h(1-z)\nu$. The following set of fit parameters has been determined: $c_{\pi^-} = 1.38 \pm 0.21$ fm/(GeVc), $c_{\pi^+} = 1.37 \pm 0.18$ fm/(GeVc), $c_{h^-} = 1.32 \pm 0.16$ fm/(GeVc), and $c_{h^+} = 3.49 \pm 0.51$ fm/(GeVc).

4 Conclusions

Inclusive and semi-inclusive deep-inelastic positron scattering data on D, ^{3}He and ^{14}N have been presented. In the inclusive case, at low x and Q^2, a large difference is observed between the presently measured cross section ratios $\sigma_{^3He}/\sigma_D$ and $\sigma_{^{14}N}/\sigma_D$, and those reported by previous experiments at higher values of Q^2. The data show also a strong Q^2-dependence of R_A/R_D at low x and Q^2 and represent the first observation of nuclear effect in the ratio of longitudinal to transverse photoabsorbtion cross sections.

In the semi-inclusive case the attenuation of charged hadrons and identified pions in ^{14}N relative to D has been determined with high accuracy. In contrast to what is observed for positive and negative pions, a difference is found in the ν-dependence of R_M^h for positive and negative hadrons. This difference suggests that a proton has a larger formation time than a pion.

References

1. J.J.Aubert et al. (EMC), *Phys. Lett.* B **123**, 275 (1983).
2. M.Arneodo et al. (EMC), *Phys. Rep.* **240**, 301 (1994).
3. K.Ackerstaff et al. (HERMES), *Nucl. Instrum. Methods* A **417**, 230 (1998).
4. J.Milana, *Phys. Rev.* C **49**, 2820 (1994).
5. T.Gousset and H.J.Pirner, *Phys. Lett.* B **375**, 349 (1996).
6. G.A. Miller, S.J. Brodsky and M. Karliner, *Phys. Lett.* B **481**, 245 (2000).
7. K. Ackerstaff et al. (HERMES), *Phys. Lett.* B **475**, 386 (2000).
8. P.Amaudruz et al.(NMC), *Nucl. Phys.* B **441**, 3 (1995); M.R. Adams et al. (E665), *Z. Phys.* C **67**, 403 (1995); J. Gomez et al., *Phys. Rev.* D **49**, 4348 (1994).
9. J. Franz et al., *Z. Phys.* C **10**, 105 (1981); J. Bailey et al., *Nucl. Phys.* B **151**, 367 (1979); J. Eickmeyer et al., *Phys. Rev. Lett.* **36**, 289 (1976); S. Stein et al., *Phys. Rev.* D **12**, 1884 (1975).
10. S.Barshay and K.Kreyerhoff, hep-ph/0002233.
11. G.V.Davidenkov and N.N.Nikolaev, *Nucl. Phys.* B **135**, 333 (1978).
12. A.Bialas and E.Bialas, *Phys. Rev.* D **21**, 675 (1980).
13. A. Airapetian et al. (HERMES), *Eur. Phys. J.* C , ()(in print), hep-ex/0012049.
14. L.S. Osborne et al., *Phys. Rev. Lett.* **40**, 25 (1978).
15. J.Ashman et al., *Z. Phys.* C **52**, 1 (1991).
16. W. Burkort et al., *Z. Phys.* C **70**, 47 (1996).
17. B.Z.Kopeliovich, J.Nemchik and E. Predazzi, 1st ELFE Summer School - Cambridge, UK; hep-ph/9511214.
18. J.Czyzewski and P.Sawicki, *Z. Phys.* C **56**, 243 (1992).
19. A.Bialas and T.Chmay, *Phys. Lett.* B **133**, 241 (1983).

NON-SINGLET STRUCTURE FUNCTION OF THE ^{3}HE-^{3}H SYSTEM AND DIVERGENCE OF THE GOTTFRIED INTEGRAL

V. GUZEY, A.W. THOMAS, K. TSUSHIMA

Department of Physics and Mathematical Physics, and Special Research Centre for the Subatomic Structure of Matter (CSSM), Adelaide University, 5005, Australia

K. SAITO

Tohoku College of Pharmacy, Sendai 981-8558, Japan

M. STRIKMAN

The Pennsylvania State University, University Park, PA 16802, USA

We study nuclear shadowing for the flavor non-singlet structure function $F_2^{^3\mathrm{He}} - F_2^{^3\mathrm{H}}$. The violation of isospin symmetry for the ^{3}He and ^{3}H wave functions induces, via nuclear shadowing terms, the charge symmetry breaking for $F_2^{^3\mathrm{He}}$ and $F_2^{^3\mathrm{H}}$ which significantly increases, at small x, $F_2^{^3\mathrm{He}} - F_2^{^3\mathrm{H}}$ over $F_2^p - F_2^n$ and leads to the divergence of the Gottfried integral.

1 Introduction

This talk is based on the original publication by V. Guzey *et al.* [1].

Initially, the prime motivation for this work was to extend the analysis of the structure functions of ^{3}He and ^{3}H nuclei, $F_2^{^3\mathrm{He}}$ and $F_2^{^3\mathrm{H}}$, by K. Saito *et al.* [2], to the domain of small values of Bjorken x. Having obtained $F_2^{^3\mathrm{He}}$ and $F_2^{^3\mathrm{H}}$ in the whole range of x, $0 \leq x \leq 3$, one could study the effect of the nuclear environment on different mechanisms leading to the sea quark flavor asymmetry. Any discrepancy between the theoretical predictions and the data would indicate that the mechanisms responsible for $\bar{u} \neq \bar{d}$ for the free proton, are modified by the nuclear medium.

However, it should be noted that studies of small x nuclear effects, nuclear shadowing and antishadowing, for $F_2^{^3\mathrm{He}}$ and $F_2^{^3\mathrm{H}}$ and their non-singlet combination, $F_2^{^3\mathrm{He}} - F_2^{^3\mathrm{H}}$, are interesting in their own right. Up to now, most of the theoretical and experimental attention was given to deep inelastic scattering (DIS) on flavor-singlet nuclei, i.e. nuclei with an equal number of protons and neutrons. An example of the flavor-singlet structure function is the combination $F_2^{^3\mathrm{He}} + F_2^{^3\mathrm{H}}$. In this case, nuclear shadowing is determined by Feynman diagrams including the Pomeron exchange; sea quark and gluon nuclear parton distributions play a major role. In contrast, the nuclear shadowing correction to the non-singlet structure functions ($F_2^{^3\mathrm{He}} - F_2^{^3\mathrm{H}}$, for example) is given by the diagrams describing interference of the Pomeron and Reggeon contributions; the valence nuclear parton distributions are probed. Different dynamics of nuclear shadowing for the singlet and non-singlet structure functions results in the prediction that the nuclear shadowing correction to the non-singlet structure functions is twice as large as that to the singlet structure functions.

One used to think that the nuclear shadowing corrections to $F_2^{^3\mathrm{He}}$, denoted $\delta F_2^{^3\mathrm{He}}$, and to $F_2^{^3\mathrm{H}}$, denoted $\delta F_2^{^3\mathrm{H}}$, are not large. For example, for the flavor singlet combination, $\delta(F_2^{^3\mathrm{He}} + F_2^{^3\mathrm{H}})/3(F_2^p + F_2^n) \approx 1.05$ at $x = 10^{-4}$ and $Q = 2$ GeV. The major new result of this work is that, if charge symmetry breaking in the ^{3}He-^{3}H system is included, the nuclear shadowing corrections to $F_2^{^3\mathrm{He}} - F_2^{^3\mathrm{H}}$ become very important: they significantly increase $F_2^{^3\mathrm{He}} - F_2^{^3\mathrm{H}}$ relative to $F_2^p - F_2^n$ and lead to the divergence of the Gottfried integral $I_G^{^3\mathrm{He},^3\mathrm{H}}(z)$ at $z \to 0$:

$$I_G^{^3\mathrm{He},^3\mathrm{H}}(z) = \int_z^3 \frac{dx}{x}\Big(F_2^{^3\mathrm{He}}(x,Q^2) - F_2^{^3\mathrm{H}}(x,Q^2)\Big) \to \infty\,. \tag{1}$$

In the future, our predictions can be confronted with experiments planned at RIKEN [3], Electron-Ion Collider [4] and TJNAF [5].

2 Nuclear shadowing for $F_2^{^3\mathrm{He}} - F_2^{^3\mathrm{H}}$

The nuclear shadowing correction to the structure function F_2, δF_2, is an effect associated with the virtual photon interaction with several nucleons of the target. It was understood by Gribov [6,7] that, at high energies, the elementary virtual photon-nucleon amplitude should be of diffractive nature. For ^{3}He and ^{3}H, the major shadowing effect is the interaction with 2 nucleons. The relevant Feynman diagrams for $\delta F_2^{^3\mathrm{He}}$ are presented in Fig. 1. The corresponding analytic expressions read

$$\begin{aligned}\delta F_2^{^3\mathrm{He}} &\propto \Big((P+R/2)^2 + 2(P+R/2)(P-R/2)\Big) \otimes \rho_\alpha\,,\\ \delta F_2^{^3\mathrm{H}} &\propto \Big((P-R/2)^2 + 2(P-R/2)(P+R/2)\Big) \otimes \rho_{\alpha'}\,.\end{aligned} \tag{2}$$

Note that $\delta F_2^{^3\mathrm{H}}$ is readily found from $\delta F_2^{^3\mathrm{He}}$ by replacing $R \to -R$ and $\rho_\alpha \to \rho_{\alpha'}$. In Eq. (2), P (R) stand for the amplitude of the elementary diffractive process $\gamma^* + p \to X + P$ driven by the Pomeron (Reggeon) exchange; ρ_α

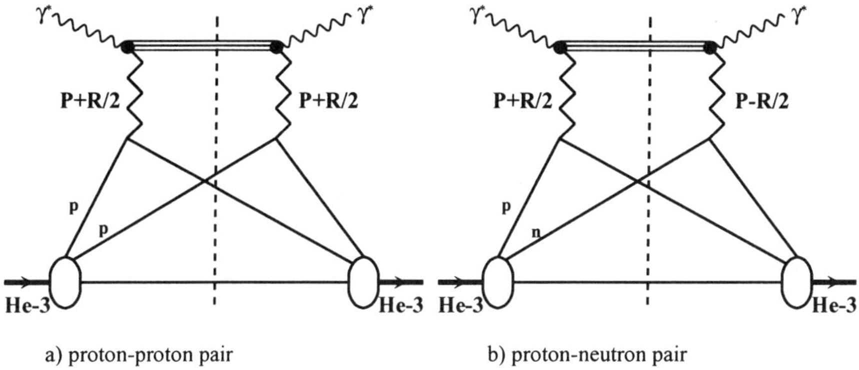

Figure 1. Feynman diagrams describing the principal contribution to nuclear shadowing for $F_2^{^3\mathrm{He}}$. The symbol P (R) stands for the Pomeron (Reggeon) exchange. Note the different signs of the Reggeon contribution in the proton and neutron diffractive amplitudes.

$(\rho_{\alpha'})$ is the two-body nucleon density in ^{3}He (^{3}H); $\otimes$ is a symbol for the convolution. The isospin breaking (charge asymmetry) means that $\rho_\alpha \neq \rho_{\alpha'}$. It is assumed that diffraction on the proton and neutron are different, and that diffraction on a flavor singlet target, proton + neutron, is driven by the Pomeron contribution only.

Equations (2) can be studied further by introducing the effective cross sections σ_{eff} and $\bar{\sigma}$

$$\sigma_{eff} \propto \frac{P^2}{F_2^N}\,, \qquad \bar{\sigma} \propto \frac{PR}{F_2^p - F_2^n}\,, \tag{3}$$

where F_2^p (F_2^n) is the proton (neutron) structure function; $F_2^N = (F_2^p + F_2^n)/2$. From Fig. 1 one can see that σ_{eff} $(\bar{\sigma})$ describes the cross section of the virtual photon interaction with two nucleons of the target mediated by 2 Pomerons (Pomeron-Reggeon interference). While σ_{eff} is constrained very well by the HERA diffractive data [8], $\bar{\sigma}$ is a new quantity. For simplicity, we use $\bar{\sigma} = \sigma_{eff}$ in this work. This means that the virtual photon interacts with valence and sea quarks in nuclei with the same strength.

Using the notation of Eq. (3), $\delta F_2^{^3\mathrm{He}}$ and $\delta F_2^{^3\mathrm{H}}$ can be presented in the following useful form (we ignore numerically unimportant triple scattering terms and do not write explicitly the x-dependence of f_α and $f_{\alpha'}$)

$$\delta F_2^{^3\text{He}} = \left(\tfrac{5}{2}F_2^p + \tfrac{1}{2}F_2^n\right) f_\alpha \ ,$$
$$\delta F_2^{^3\text{H}} = \left(\tfrac{5}{2}F_2^n + \tfrac{1}{2}F_2^p\right) f_{\alpha'} \ , \tag{4}$$

where the factor f_α ($f_{\alpha'}$) is proportional to σ_{eff} and ρ_α ($\rho_{\alpha'}$). It is worth emphasising that charge asymmetry in the ^{3}He-^{3}H system manifests itself as $f_\alpha \neq f_{\alpha'}$. Thus, the violation of the charge symmetry in the tri-nucleon wave functions induces the isospin symmetry breaking in the structure functions ($F_2^{^3\text{He}}$ is not related to $F_2^{^3\text{H}}$ by a simple exchange $p \leftrightarrow n$) and in the nuclear quark parton densities (e.g., $u^{^3\text{He}} \neq d^{^3\text{H}}$).

Let us ignore the charge asymmetry effects for the moment, i.e. use $f_\alpha = f_{\alpha'}$. Then one can easily compare the nuclear shadowing correction to the flavor singlet and non-singlet structure functions of the tri-nucleon system:

$$\frac{\delta F_2^{^3\text{He}} + \delta F_2^{^3\text{H}}}{3(F_2^p + F_2^n)} = f_\alpha \ ,$$
$$\frac{\delta F_2^{^3\text{He}} - \delta F_2^{^3\text{H}}}{F_2^p - F_2^n} = 2 f_\alpha \ . \tag{5}$$

This proves our statement that, due to different dynamics, the nuclear shadowing effect in the non-singlet structure function is twice as large as that in the singlet structure function. This observation was confirmed in the analysis of polarized DIS on ^{2}He, ^{3}He and ^{7}Li.

Nuclear shadowing for the non-singlet structure function $F_2^{^3\text{He}} - F_2^{^3\text{H}}$ becomes much more important when charge asymmetry is taken into account. Due to the Coulomb repulsion, the ^{3}He nucleus has a larger size than ^{3}H and, as a result, $f_\alpha < f_{\alpha'}$. At small values of Bjorken x, where $F_2^p + F_2^n \gg F_2^p - F_2^n$, one immediately obtains from Eq. (4)

$$F_2^{^3\text{He}} - F_2^{^3\text{H}} \approx \delta F_2^{^3\text{He}} - \delta F_2^{^3\text{H}} \approx 3F_2^N (f_{\alpha'} - f_\alpha) \ . \tag{6}$$

The result of Eq. (6) is presented in Fig. 2. One can see that, when charge asymmetry is included, nuclear shadowing significantly increases, at small x, $(F_2^{^3\text{He}} - F_2^{^3\text{H}})/x$ (solid line) relative to $(F_2^p - F_2^n)/x$ (dashed line). Nuclear shadowing for $(F_2^{^3\text{He}} - F_2^{^3\text{H}})/x$ (dotted line) is not very important when charge asymmetry is ignored, i.e when $f_{\alpha'} = f_\alpha$. The proton and neutron structure functions are evaluated at $Q = 2$ GeV using the CTEQ5L parametrization. The details of calculations with several sets of (f_α, $f_{\alpha'}$) can be found in our original publication [1].

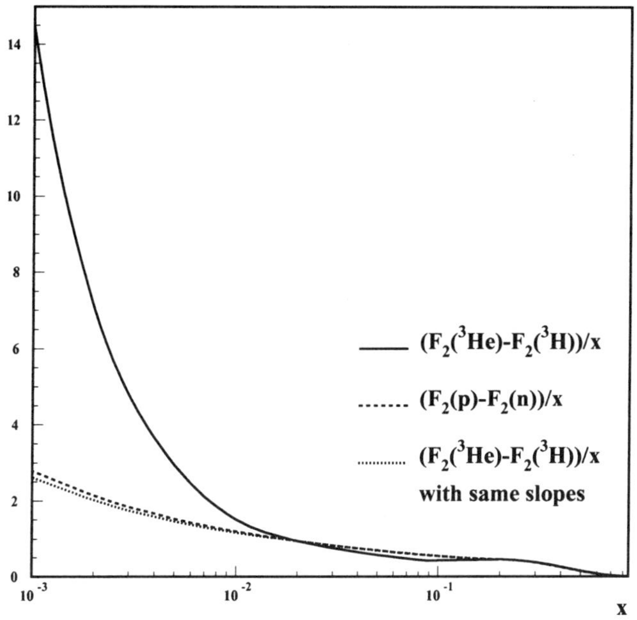

Figure 2. Non-singlet structure functions as a function of x at $Q = 2$ GeV.

One immediate consequence of Eq. (6) is that the Gottfried integral $I_G^{^3\mathrm{He},^3\mathrm{H}}(z)$ (see Eq. (1)) diverges at least logarithmically at $z \to 0$:

$$\int_z^3 \frac{dx}{x}\left(F_2^{^3\mathrm{He}}(x,Q^2) - F_2^{^3\mathrm{H}}(x,Q^2)\right) \approx 3F_2^N(z)(f_{\alpha'} - f_\alpha)\ln(z) \to \infty\ , \quad (7)$$

One should note that the divergence of $I_G^{^3\mathrm{He},^3\mathrm{H}}(z=0)$ is not paradoxical since the Gottfried integral is not constrained by current algebra - as, for example, the Bjorken sum rule. Thus, the value of $I_G^{^3\mathrm{He},^3\mathrm{H}}(z=0)$ is not related to any physical observable and can be infinite.

It is interesting to consider briefly another explanation of this divergence. At very small values of Bjorken x, the virtual photon-target cross section increases and eventually reaches its maximal [7] possible value $2\pi R_T^2$, where R_T is the target radius. This regime of strong interactions is called the black body limit [9] (BBL). The application of the BBL to DIS is better justified for very heavy nuclear targets. However, even for DIS on light nuclei such as ^{3}He and ^{3}H, some portion of the cross section (structure functions) can be calculated within the BBL approximation

$$F_2 \propto 2\pi R_T^2 Q^2 \ln(1/x) + \textit{subleading terms} \ . \quad (8)$$

Applying Eq. (8) to DIS on ^{3}He and ^{3}H, one readily obtains

$$F_2^{^3\mathrm{He}}(x,Q^2) - F_2^{^3\mathrm{H}}(x,Q^2) \propto 2\pi(R^2_{^3\mathrm{He}} - R^2_{^3\mathrm{H}})Q^2 \ln(1/x) + \textit{sublead. terms} \ . \quad (9)$$

The isospin symmetry breaking in the ^{3}He-^{3}H system manifests itself as the non-equality of ^{3}He and ^{3}H matter radii, $R^2_{^3\mathrm{He}} \neq R^2_{^3\mathrm{H}}$. The use of Eq. (9) in the Gottfried integral (1) yields an integral divergent as $\ln^2(1/z)$ at $z \to 0$.

The discussed phenomena should also be relevant for $F_2^p - F_2^n$. In this case, we also expect that, due to the isospin symmetry breaking between the proton and neutron, their sizes should be different. This should lead to the divergence of the Gottfried integral for the nucleon. If, indeed, this is the case, modern parametrizations of parton distributions need to be revised since they impose the condition $F_2^p - F_2^n \to 0$ as $x \to 0$ and, hence, give a finite value of the corresponding Gottfried integral.

3 Conclusions

First detailed estimate [1] of nuclear shadowing and antishadowing for the non-singlet structure function of the tri-nucleon system, $F_2^{^3\mathrm{He}} - F_2^{^3\mathrm{H}}$, is presented. Charge asymmetry in the nuclear wave functions of ^{3}He and ^{3}H induces charge symmetry breaking for $F_2^{^3\mathrm{He}}$ and $F_2^{^3\mathrm{H}}$. This results in a very significant increase of $F_2^{^3\mathrm{He}} - F_2^{^3\mathrm{H}}$ over $F_2^p - F_2^n$ and in the divergence of the Gottfried integral $I_G^{^3\mathrm{He},^3\mathrm{H}}(z=0)$. Our analysis within the framework of the black body limit supports these findings and suggests that the Gottfried integral for the free nucleon should be also divergent.

References

1. V. Guzey, K. Saito, M. Strikman, A. W. Thomas and K. Tsushima, preprint hep-ph/0102133 (February 2001).
2. K. Saito, C. Boros, K. Tsushima, F. Bissey, I.R. Afnan and A.W. Thomas, *Phys. Lett.* B **493**, 288 (2000).
3. MUSES (Multi-Use Experimental Storage-rings) project at RIKEN, see the RIKEN home page (http://www.rarf.riken.go.jp).
4. R. Venugopalan, preprint hep-ph/0102087 (February 2001).
5. "The Science Driving the 12 GeV Upgrade of CEBAF", edited by the 12 GeV upgrade white paper steering committee (TJLAB, 2000).

6. V.N. Gribov, *Sov. Phys. JETP* **29**, 483 (1969).
7. V.N. Gribov, *Sov. Phys. JETP* **30**, 709 (1970).
8. L. Frankfurt and M. Strikman, *Eur. Phys. J.* **A 5**, 293 (1999).
9. L. Frankfurt, V. Guzey, M. McDermott, M. Strikman, preprint hep-ph/0104252 (April 2001).

QUARK-DIQUARK STRUCTURE OF THE NUCLEON: STRUCTURE FUNCTIONS, STATIC PROPERTIES AND NUCLEAR MATTER EQUATION OF STATE

W. BENTZ

Department of Physics, School of Science, Tokai University
Hiratsuka-shi, Kanagawa 259-1292, Japan

H. MINEO

Department of Physics, Faculty of Science, University of Tokyo
Bunkyo-ku, Tokyo 113-0033, Japan

A.W. THOMAS

Special Research Center for the Subatomic Structure of Matter and
Department of Physics and Mathematical Physics, Adelaide University
Adelaide, SA 5005, Australia

K. YAZAKI

Department of Physics, Faculty of Humanities and Sciences, Tokyo Woman's Christian University
Suginami-ku, Tokyo 167-8585, Japan

In the first part of this work we investigate the structure functions, in particular their flavor dependencies, and static properties of the nucleon in a relativistic three-body approach to the NLJ model, using a simple quark-diquark approximation to the full Faddeev method. In the second part we consider the problem of nuclear matter stability in chiral models like the NJL model.

1 Introduction

The study of modifications of nucleon properties inside the nuclear medium is currently a very active field both experimentally and theoretically. The relativistic Faddeev approach to nucleon structure based on effective chiral quark theories [1] seems to be a powerful candidate to study these problems. For this purpose, first one should be able to account for the properties of a free nucleon, and second one should construct an equation of state of nuclear matter which gives the density dependence of the effective nucleon and quark masses. In this work we consider these two aspects in the Faddeev approach to the Nambu-Jona-Lasinio (NJL) model [2], having in mind future applications of the model to study the medium modifications. We will, however, for technical reasons limit ourselves to the "static approximation" to the full Faddeev equation [3] when describing the nucleon as a quark-diquark state.

Concerning the properties of a free nucleon, we concentrate on the spin independent structure functions [4] and static properties of the nucleon [5]. The valence (v) and sea quark distributions extracted from the experimental nucleon structure functions show an interesting flavor dependence [6]: $d_v(x)$ is 'softer' than $u_v(x)$, i.e; it is more concentrated at smaller x, while $\overline{d}(x)$ is larger than $\overline{u}(x)$ in the whole region of x. The first feature can naturally be explained [7] as an effect of quark-quark (diquark) correlations in the scalar ($J^P = 0^+$, $T = 0$) channel, while the second feature is usually attributed to the pion cloud around the valence quarks [8]. In the first part of this work we investigate how attractive quark-quark correlations in the axial vector (a.v.) channel ($J^P = 1^+$, $T = 1$) influence the flavor dependence of the valence quark distributions and the static properties of the nucleon. The aim is to extract information on the strength of the interaction in the a.v. channel [9].

In the second part we discuss a simple hybrid model for the nuclear matter equation of state in the NJL model, where the single nucleon is described as a quark-diquark state [10]. We will point out that a simple method based on the introduction of an infrared cut-off to mimic confinement effects leads to a saturating equation of state.

2 Structure functions and static properties of the nucleon

Our approach is based on the relativistic Faddeev approach to the Nambu-Jona-Lasinio (NJL) model. The relativistic Faddeev equation in the NJL model can be reduced to a Bethe-Salpeter (BS) equation for a quark and a structured 'diquark' interacting via quark exchange [1]. A very simple approximation to the full Faddeev equation arises if one neglects the momentum dependence of the quark exchange kernel. This 'static approximation' has been shown to be not unreasonable for the nucleon mass and some static properties of the nucleon [3,5]. In the actual calculation we include the scalar and a.v. diquark channels with the effective 4-Fermi coupling constants g_s and g_a, respectively. Since any 4-Fermi interaction of the NJL type can be decomposed into the various diquark channels by means of Fierz transformations, the ratios $r_s = g_s/g_\pi$ and $r_a = g_a/g_\pi$ of these effective coupling constants to the one in the pionic $q\overline{q}$ channel (g_π) reflect different possible forms of the original 4-Fermi interaction lagrangian [1]. (For example, one has $r_s = 0.5$, $r_a = 0.25$ for the 'color current interaction lagrangian' [3].) We will use r_a as a free parameter, and adjust r_s (for each r_a) such as to reproduce the nucleon mass. We use the constituent quark mass $M = 400$ MeV, which is related to the current quark mass m via the gap equation, and employ the invariant mass regularization scheme [11], which is well suited for the use of light cone momentum variables and equivalent to the familiar 3-momentum cut-off scheme when applied for the usual momentum variables [12]. We will consider the three parameter sets I, II, III for r_s and r_a shown in Table 1. The case I (III) involves relatively strong (weak) scalar diquark correlations, while the set II describes an intermediate situation. The value of r_a used in case III reproduces also a bound state for the delta isobar with a mass of $M_\Delta = 1140$ MeV.

Table 1: Three different parameter sets for r_a and r_s, the corresponding scalar diquark masses (M_s, from the quark-quark BS equation) and weights of the scalar diquark channel in the nucleon state (W_s, from the quark-diquark BS equation), and some static properties of the nucleon in comparison to their experimental values. The a.v. diquark is unbound (continuum state) for all 3 cases.

case	I	II	III	exp.
r_a	0	0.25	0.66	
r_s	0.73	0.63	0.50	
M_s[MeV]	600	684	766	
W_s[%]	100	90	66	
μ_p	2.32	2.84	2.79	2.79
μ_n	-1.39	-1.79	-1.83	-1.91
$g_A^{(3)}$	0.66	0.77	0.81	1.26
$g_{\pi NN}$	7.5	12.68	15.25	13.2
$g_A^{(0)}$	0.60	0.41	0.31	$0.2 \sim 0.3$

The twist-2 light cone momentum distributions of the quarks (flavor q) in the proton (momentum p) are defined by [13]

$$q(x) = \frac{1}{4\pi} \int dz^- e^{i(p_- x)z^-} \langle p|T\left(\overline{\psi}(0)\gamma^+ \psi(z^-)\right)|p\rangle, \tag{1}$$

where x is the Bjorken variable, and the light cone variables are defined as $a_\pm = (a_0 \pm a_3)/\sqrt{2}$. The antiquark distributions are obtained as $\overline{q}(x) = -q(-x)$ for $0 < x < 1$, and the valence and sea quark distributions are then given by $q_v(x) = q(x) - \overline{q}(x)$ and $q_s(x) = \overline{q}(x)$. Since Eq.(1) is just the quark propagator in the proton (traced with γ^+) for fixed quark light cone momentum component $k_- = p_- x$, the evaluation of (1) amounts to a straight forward Feynman diagram calculation [4] using the quark-diquark vertex functions including the scalar and a.v. diquark channels. The effects of the pion cloud around the valence quarks are taken into account by using the standard one-dimensional convolution formalism [8]. We then perform the Q^2 evolution in the next-to-leading order [14] from a low energy scale Q_0^2 to the value of Q^2 where experimental data, or parametrizations of quark distributions [15], are available.

If we choose $Q_0^2 = 0.16$ GeV^2, the overall features of the empirical valence quark distributions can be reproduced.

In Fig.1 we plot the ratio of structure functions F_2^n/F_2^p in comparison with the experimental data [16]. Strong scalar diquark correlations (case I) lead to a strong flavor asymmetry of the valence quark distributions, i.e., $d_v(x)$ becomes softer than $u_v(x)$ and the ratio F_2^n/F_2^p becomes very small as $x \to 1$. The axial vector diquark correlations work towards more flavor symmetric distributions. We see from Fig. 1 that reasonable values of r_a are in the range between 0.0 and 0.25, i.e., to reproduce the observed flavor dependence the weight of the a.v. diquark channel should not be larger than about 10% (see Table 1).

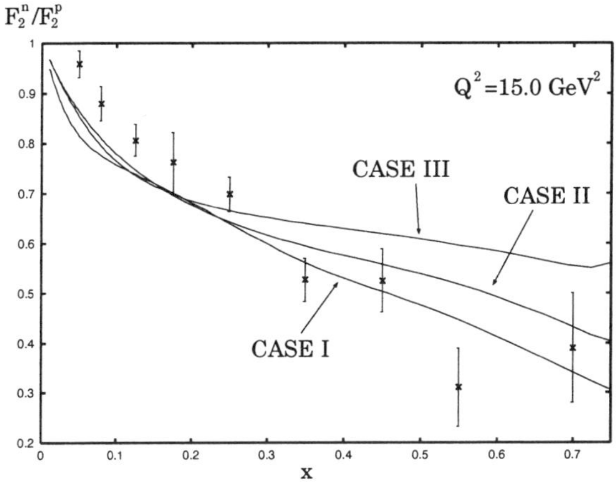

Figure 1: The ratio of structure functions F_2^n/F_2^p for $Q^2 = 15$ GeV^2 is shown for the three cases of Table 1 and compared to the experimental data of ref. [16].

The valence quark and antiquark distributions for the case II ($r_a = 0.25$) are compared to the empirical distributions of ref. [15] in Figs. 2 and 3, respectively. Due to the pion cloud effects we also obtain a flavor asymmetry of the antiquark distributions ($\overline{d} > \overline{u}$), but the simple convolution formalism used here underestimates this asymmetry for intermediate values of x, although the Gottfried sum is reproduced rather well ($S_G = 0.262$ compared to the experimental value of 0.235 ± 0.026).

In conclusion of this part we can say that the overall description of the structure functions in the present quark-diquark model is reasonable as long as the weight of the a.v. diquark component is less than about 10%.

In order to see whether this conclusion is consistent with other observables, we show some static properties of the nucleon, including corrections due to the

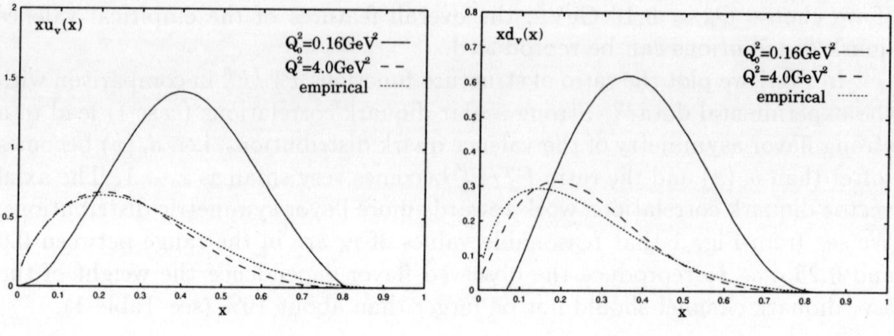

Figure 2: Valence quark distributions in the proton including pionic cloud effects for case II of Table 1. The calculated distributions (solid lines) are evolved from $Q_0^2 = 0.16$ GeV^2 to $Q^2 = 4$ GeV^2 (dashed lines) and compared to the empirical distributions of ref. [15] (dotted lines).

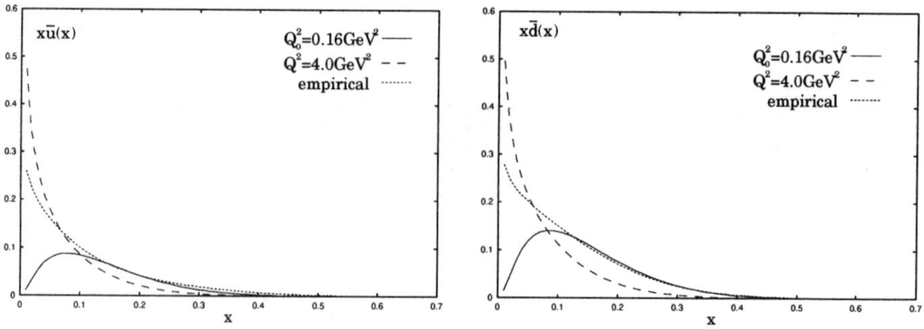

Figure 3: Same as Fig. 2 for the antiquark distributions in the proton.

pion cloud, in the lower part of Table 1. We see that the a.v. diquark channel has beneficial effects for all static properties considered here, but it is not necessary to introduce very strong (case III) correlations. (In particular, the fact that $g_A^{(3)}$ is too small cannot be improved significantly by introducing very strong a.v. correlations.) Therefore the static nucleon properties support the conclusion that $r_a \simeq 0.25$ ($W_a \simeq 10\%$) is reasonable.

3 Nuclear matter equation of state

It is well known that models based on the linear realization of chiral symmetry, like the linear σ model or the NJL model, do not lead to a saturating nuclear matter equation of state [17]. To see where the basic problem is, we note that these models are characterized by a vacuum effective potential (or energy density) $\mathcal{E}_V(\sigma)$ of the Mexican hat form. (Here σ is the nuclear matter expectation value of the scalar field.) Since the curvature of this potential decreases as one moves away from the vacuum value σ_0 towards smaller values of σ, this implies attractive contributions ('vacuum fluctuation terms') to the sigma meson mass $M_\sigma^2 \propto \frac{\partial^2 \mathcal{E}}{\partial \sigma^2}$ and therefore also to the effective NN interaction in the medium (Landau-Migdal interaction), leading to a collapse of the system for decreasing σ (increasing density).

A possible solution of this problem is provided by the internal quark structure of the nucleon: If the scalar field is coupled to the quarks inside the nucleon instead of an elementary nucleon, the nucleon mass in the scalar field $M_N(\sigma)$ can have a finite curvature $\frac{\partial^2 M_N}{\partial \sigma^2}$. If this 'scalar polarizability'[18] of the nucleon is positive and large enough, it can lead to a repulsive contribution to $\frac{\partial^2 \mathcal{E}}{\partial \sigma^2}$, i.e., the sigma meson mass, and can avoid the collapse discussed above. It is, however, very easy to see that this mechanism can work effectively only if the confinement is taken into account: As the scalar potential $\Phi \equiv \sigma_0 - \sigma$ becomes strong the quark mass tends to zero. On the other hand, a positive curvature of M_N persisting also for large scalar potentials means that $M_N(\sigma)$ should become flat, i.e., the nucleon mass should *not* tend to zero for large scalar potentials. Such a situation ($M \to 0$ but $M_N \to$ finite) is possible only if the confinement is taken into account, or, if at least there are no unphysical thresholds which force the nucleon mass to vanish as $M \to 0$. This aspect of the confinement, namely the absence of unphysical quark- diquark or 3-quark thresholds, can be achieved in the NJL model by introducing an infrared cut-off (μ) in addition to the standard ultraviolet one in the framework of the proper-time regularization scheme[19].

In order to show how the nuclear matter equation of state is stabilized by taking into account confinement effects via an infrared cut-off μ, we plot the nucleon mass in the scalar field in Fig. 4 for the cases $\mu = 0$ and $\mu = 0.1$ GeV. The positive curvature for large scalar potentials in the case $\mu = 0.1$ GeV is clearly seen. In Fig. 4 we also show the results for the binding energy per nucleon as a function of the density for $\mu = 0$ and $\mu = 0.1$ GeV. The details of this 'hybrid model' for the nuclear matter energy density, which is based on the quark-diquark description of the free nucleon discussed above and very similar in spirit to the successful model of Guichon and collaborators[20], are described in ref. [10]. The case $\mu = 0$ leads to the collapse of the system due to the reasons explained above, while the case $\mu = 0.1$ GeV gives a saturating equation of state. Some implications of this equation of state for the density dependence of the sigma mass and the πN and πNN sigma terms are discussed in ref. [10]. As an example, we show the πN sigma term $\Sigma_{\pi N} = m \frac{\mathrm{d}\epsilon_F}{\mathrm{d}m}$ and the πNN

sigma term $\Sigma_{\pi NN} = m\frac{\mathrm{d}f_0}{\mathrm{d}m}$, which was introduced in ref.[10] for the first time, as functions of the density in Fig. 5. (Here ϵ_F the Fermi energy, and f_0 the spin-isospin independent $L = 0$ Landau-Migdal parameter.) The decrease of $\Sigma_{\pi N}$ is caused by the decrease of the sigma-nucleon coupling ($\propto \frac{\partial M_N}{\partial \sigma}$) with increasing scalar potential, while the behaviour of $\Sigma_{\pi NN}$ reflects the dependence of the sigma coupling to a 2-nucleon state ($\propto \frac{\partial f_0}{\partial \sigma}$). Finally we would like to mention that the reduction of the sigma-nucleon coupling (reduction of the slope of $M_N(\sigma)$ due to the confinement effects) leads also to a suppression of the famous 'Z-graph contributions'.

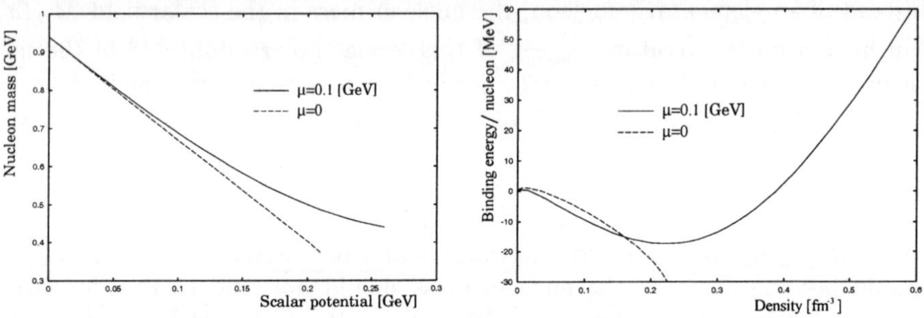

Figure 4: The nucleon mass in the quark-diquark model as a function of the scalar potential (left figure) and the binding energy per nucleon as a function of the density (right figure) for the cases $\mu = 0$ (dashed lines) and $\mu = 0.1$ GeV (solid lines).

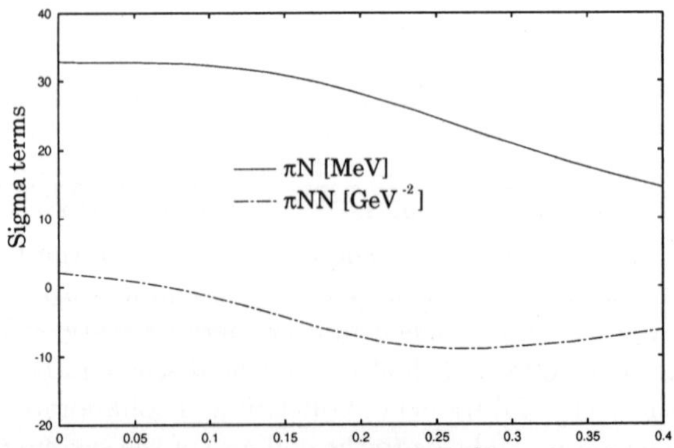

Figure 5: The πN sigma term (solid line, in [MeV]), and the πNN sigma term (dot-dashed line, in [GeV^{-2}]) for the case $\mu = 0.1$ GeV as functions of the density.

Acknowledgements

This work was supported in part by the Australian Research Council, Adelaide University, the Japanese Society for the Promotion of Science, and the Grant in Aid for Scientific Research of the Japanese Ministry of Education, Culture, Sports, Science and Technology, Project No. C2-13640298.

References

1. N. Ishii, W. Bentz and K. Yazaki, *Nucl. Phys.* A **587**, 617 (1995); U. Zückert, R. Alkofer, H. Weigel and H. Reinhardt, *Phys. Rev.* C **55**, 2030 (1997).
2. Y. Nambu and G. Jona-Lasinio, *Phys. Rev.* **122**, 345 (1960); **124**, 246 (1961).
3. A. Buck, R. Alkofer and H. Reinhardt, *Phys. Lett.* B **286**, 29 (1992).
4. H. Mineo, W. Bentz and K. Yazaki, *Phys. Rev.* C **60**, 065201 (1999).
5. H. Mineo, W. Bentz and K. Yazaki, to be published.
6. A. W. Thomas, *Phys. Lett.* B **126**, 97 (1983).
7. F.E. Close and A.W. Thomas, *Phys. Lett.* B **312**, 227 (1988).
8. A.W. Schreiber, P.J. Mulders, A.I. Signal and A.W. Thomas, *Phys. Rev.* D **45**, 3069 (1992).
9. M Oettel, R. Alkofer and L. von Smekal, *Eur. Phys. J.* A **8**, 553 (2000).
10. W. Bentz and A.W. Thomas, The Stability of Nuclear Matter in the Nambu-Jona-Lasinio Model, nucl-th/0105022, Nucl. Phys. **A**, in press.
11. G.P. Lepage and S.J. Brodsky, *Phys. Rev.* D **22**, 2157 (1980).
12. W. Bentz, T. Hama, T. Matsuki and K. Yazaki, *Nucl. Phys.* A **651**, 143 (1999).
13. R.L. Jaffe,*1985 Los Alamos School on Relativistic Dynamics and Quark Nuclear Physics* (eds. M.B. Johnson and A. Pickleseimer, Wiley, New York, 1985).
14. M. Miyama and S. Kumano, *Comp. Phys. Commun.* **94**, 185 (1996).
15. A. D. Martin, R. G. Roberts, W. J. Stirling and R.S. Thone, *Eur. Phys. J.* C **14**, 133 (2000).
16. NMC Collaboration, P. Amaudruz et al., *Nucl. Phys.* B **371**, 185 (1992).
17. W. Bentz, L.G. Liu and A. Arima, *Ann. Phys.* **188**, 61 (1988).
18. M.C. Birse, *Phys. Rev.* C **51**, R1083 (1995) K. Saito and A.W. Thomas, *Phys. Rev.* C **52**, 2789 (1995).
19. G. Hellstern, R. Alkofer and H. Reinhardt, *Nucl. Phys.* A **625**, 697 (1997).
20. P.A.M. Guichon, *Phys. Lett.* B **200**, 235 (1988); K. Saito, K. Tsushima and A.W. Thomas, *Phys. Rev.* C **55**, 2637 (1997).

PHYSICS MOTIVATION FOR A POLARIZED ELECTRON-ION COLLIDER

J.M. CAMERON, J.T. LONDERGAN
Indiana University Cyclotron Facility & Nuclear Theory Center,
2401 Sampson Road, Bloomington, IN 47408, USA
Email: cameron@iucf.indiana.edu, tlonderg@indiana.edu

1 A "Next-Generation" Facility for Partonic Physics

A few decades studying Quantum Chromodynamics (QCD), the theory of the strong interactions, has provided us with increasingly precise information about the partonic (quark and gluon) structure of hadronic matter. However, our knowledge is still far from complete. Some fundamental questions in this field remain open: *(1) What is the structure of hadrons in terms of their quark and gluon constituents? (2) How do quarks and gluons evolve into hadrons? (3) What is the role of quarks and gluons in the structure of atomic nuclei?* The answer to these questions will provide the key to our ultimate understanding of the microscopic structure of matter. Several experimental facilities have provided data which is essential to our current understanding of particle structure, beginning with the initial experiments at SLAC. The $e - p$ facility at HERA has extended our knowledge of nucleon structure to unprecedented small x and high Q^2; however, HERA does not have particularly high luminosity, polarized beams, or beams of atomic nuclei. HERMES and Jefferson Laboratory are capable of performing semi-inclusive measurements but over a small and limited Q^2 range. The experimental situation is summarized in Fig. 1 (L). The boxed area compares the luminosity vs. CM energy range for an "*E*lectron-*I*on *C*ollider", or *EIC* facility, with several existing and proposed electron facilities. Existing facilities tend to cluster into high luminosity but relatively low CM energy machines, and high energy but low luminosity facilities. A dedicated EIC facility, designed from the outset to probe the essential $x - Q^2$ region with high luminosity, with beams ranging from polarized protons (and light ions) to heavy ions, would open up a unique regime for electromagnetic and hadronic physics.

The physics potential of such a collider has been reviewed in a series of conferences and workshops. Initially, these began as two independent ideas: one ("eRHIC") focused on physics with nuclei, particularly inclusive measurements at low x. The second avenue ("EPIC") stressed physics with protons and light nuclei, and focused on semi-inclusive and exclusive measurements.

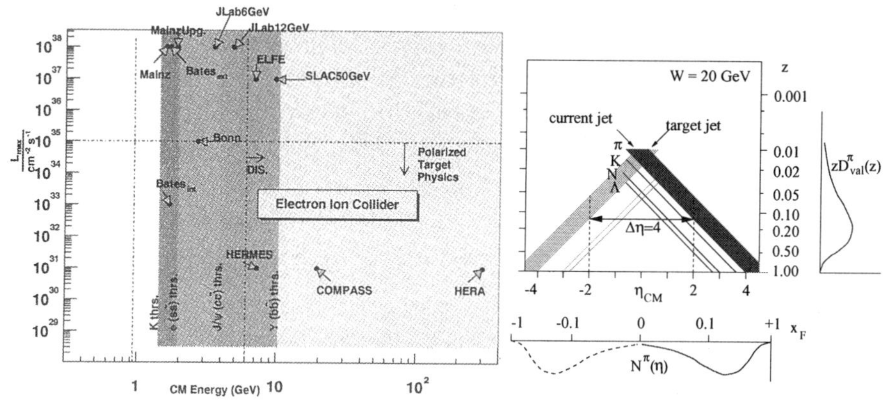

Figure 1: Left: Maximum luminosity vs. CM energy for existing and proposed lepton scattering facilities worldwide. Boxed area: range of proposed EIC parameters. Right: Relation between z-values in fragmentation and CM rapidity for $W = 20$ GeV.

These two streams have now merged in the current "EIC" collider initiative. This paper will focus on the physics possibilities available with light ion beams, and the prospects for semi-inclusive and exclusive processes; a detailed discussion of the novel physics accessible with heavy nuclei is given in a White Paper submitted to this year's nuclear physics long range plan [1]. This document contains an extensive list of all prior workshops on this topic.

A next-generation EIC facility must have sufficiently high energy to guarantee access to the scaling regime. It should also have optimal control of spin-flavor degrees of freedom. An EIC facility should allow full kinematic coverage, to observe both the fast current jets, and the slow target fragments, and it should provide full azimuthal coverage. EIC CM energies would be substantially greater than at fixed target facilities such as Jefferson Lab or SLAC, and well below the high energy range of HERA. Inclusive measurements on nuclear targets at low x will probe entirely new regimes of partonic physics. Such a facility would provide high luminosity (an $e-p$ luminosity of at least 10^{33} $\mathrm{cm}^{-2}\mathrm{s}^{-1}$), variable CM energy with hermetic detectors, to enable the study of semi-inclusive and exclusive reactions. For inclusive measurements and heavy-nucleus interactions higher energies, up to $\sqrt{s} \sim 100$ GeV, are desirable. The asymmetric collider geometry would allow both current fragment and target fragmentation events to be measured and kinematically separated. For partonic physics at moderate x, in semi-inclusive and exclusive measurements, a large solid-angle detector is required with excellent particle identification and tracking capabilities. The need for high luminosity, plus the ability to detect particles at very small scattering angle, require that the accelerator and detector design be closely correlated. An aggressive R&D effort over the next few years will enable us to evaluate competing accelerator designs, and to demonstrate that the luminosity and detector requirements can be simultaneously satisfied. The favorable response of the US nuclear physics long-range planning group to the EIC concept makes it possible that EIC R&D funding may soon become available.

2 Semi-Inclusive Measurements of Parton Structure

In semi-inclusive processes induced by leptons, one measures both the scattered lepton, and a fast outgoing final hadron. The resulting cross section is given by $\sigma \sim \sum_q e_q^2\, q(x)\, D_q^h(z)$, proportional to the "fragmentation function" $D_q^h(z)$, which describes the probability that quark q will fragment to final hadron h with fractional energy z.

The EIC facility could probe parton spin distributions at quite small values of x. Figure 2 shows the precision which could be obtained for spin probabilities for various quark flavors, with one month's running at the EIC projected luminosity. One can obtain impressive statistical measurements, in a kinematic region clearly complementary to that probed at Hermes, and at an order of magnitude smaller values of x.

Another possibility is model-independent studies of neutron structure using the technique of *spectator tagging.* With a deuteron beam, one can measure protons in coincidence with the scattered electron in the process $e + D \to e' + p + X$. The spectator proton will emerge with momentum $k_p \sim k_D/2$, at very small lab angles. The neutron structure function can thus be measured on-shell, over the kinematic range $1 < Q^2 < 200$ GeV^2. Such techniques could extend measurements of the neutron spin structure function $g_1^n(x)$ to very low x and large Q^2, reducing dramatically experimental errors in the determination of the Bjorken sum rule. With this technique, one can obtain accurate measurements of $d^p(x)/u^p(x)$ at large x. In addition, with either polarized deuteron or ^{3}He beams, one can obtain accurate measurements of nucleon spin asymmetries A_1^N at large x, as measurements of this quantity appear to be crucial tests of QCD models.

An electron collider would also allow precise measurements of pion and kaon structure functions. At present, pion valence quark distributions are poorly known for $x_\pi \leq 0.2$ [2]. A collider could obtain accurate values through the reaction $e + p \to e' + n + X$, by detecting forward neutrons in $e - p$ scattering. The photon scatters from a virtual π^+ very near its mass shell, and the resulting cross section is proportional to $F_\pi(x_\pi)$. Simulations by Holt and Reimer[3] show that F_π can be determined with excellent precision in the region

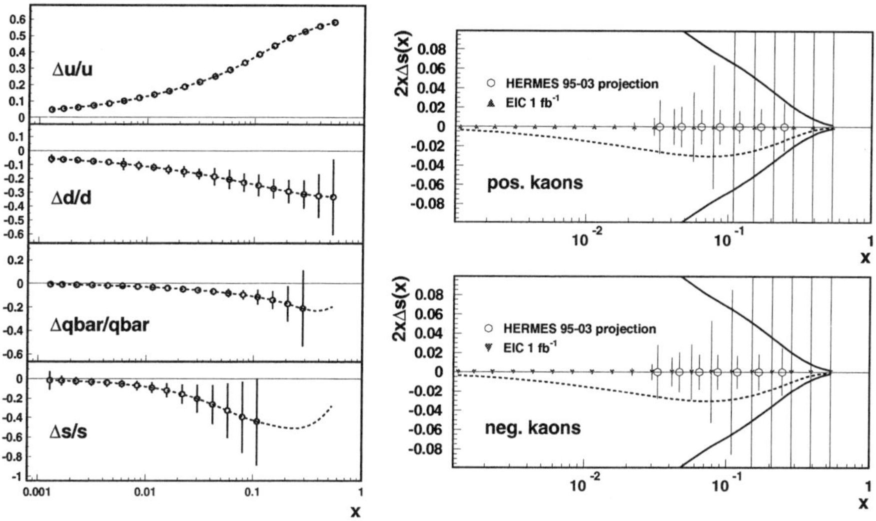

Figure 2: Left: statistical errors and range of x which could be covered by EIC collider for quark spin densities. Right: statistical accuracy which could be reached by EIC in one month of running at design luminosity, for strange quark spin distribution, if spin densities of non-strange quarks were accurately determined.

$0.01 \leq x_\pi \leq 0.9$. At low x, one can answer the question, *"Are sea quarks and gluons as important for pions at low x as for the nucleon?"*. One can also obtain accurate measurements of the poorly-known kaon structure functions using the reaction $e + p \rightarrow e' + \Lambda + X$.

The nucleon's polarized gluon distribution could be accessed at EIC via open charm production. This is an essential quantity in understanding the origin of nucleon spin. The COMPASS facility will provide measurements of this quantity, which will also be investigated in $p - p$ collisions at RHIC. EIC appears to be an excellent facility for such studies [7].

A polarized EIC facility could make definitive studies of *transversity*, a leading-twist structure function. The transversity $h_1(x)$ is proportional to $\delta q(x)$, the transverse spin difference for nucleons. One reason that transversity is interesting is that gluons do not contribute to h_1. Consequently, h_1 should evolve much more like the valence quark polarization than the longitudinal spin structure function g_1, which is believed to receive a large contribution from polarized glue. The first moment of h_1 is equal to the nucleon's tensor charge, a quantity which may be calculable with lattice QCD. As h_1 is a chiral-odd operator, it is inaccessible in inclusive DIS, but in semi-inclusive

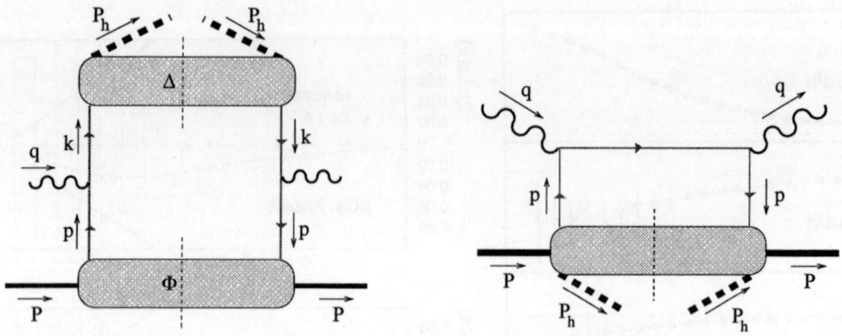

Figure 3: Leading contribution to fragmentation processes, in which quark q evolves to final hadron h. Left: current fragmentation; Right: target fragmentation, where the quark is struck by a virtual photon while baryon remnants fragment to final hadron.

processes (e.g., $e + \vec{p} \rightarrow e' + \pi + X$) this effect can be extracted from the azimuthal asymmetry of leading pions from transversely polarized nucleons. The asymmetry will be proportional to $A \sim h_1(x)\, H_1^{\perp}(z) \sin\phi$, a product of the transversity distribution times the chiral-odd Collins fragmentation function $H_1^{\perp}$. With sufficiently precise data, one could separately determine the x-dependence of h_1 and the z-dependence of $H_1^{\perp}$. Both Hermes and COMPASS will make exploratory measurements of this quantity; however the collider geometry is optimal for this type of study, since one can detect the full hadronic final state, both current and target fragmentation regimes are accessible, and one can measure complete angular distributions.

The EIC facility can provide unique insights into the *fragmentation* of quarks into hadrons. A hermetic detector will allow detailed studies of fragmentation. The most commonly studied process is *current fragmentation*, shown schematically on the left in Fig. 3, where the quark struck by the virtual photon decays to a final hadron h. There is also *target fragmentation*, shown schematically on the right-hand side of Fig. 3, where a quark is struck with a virtual photon, and the nucleon remnants subsequently decay. Target fragmentation is a largely unexplored regime of QCD. Observing such processes requires separating the decay fragments from the current jet by a large interval in rapidity. As a result, the collider geometry is probably essential for studies of the target fragmentation region. The right side of Fig. 1 shows a plot of rapidity vs. fragmentation energy fraction z, for $\gamma^* N$ invariant mass $W = 20$ GeV [4]. Existing experimental data shows that current and target fragmentation regions extend over a CM rapidity range $\Delta\eta \approx 2$, where $\eta \equiv 0.5 \ln(P_h^- / P_h^+)$. Fig. 1 (R) shows that for $W = 20$ GeV, current and target fragmentation regions for meson production can be separated for all $z \geq 0.02$. Increasing W

greatly lowers those z-values; this demonstrates the ability of the EIC collider to access and separate these two fragmentation regimes.

There are two types of target fragmentation. In the first case, the momentum fraction x carried by the struck parton is small, and the momentum fraction of the remnant, $1 - x$, is large. In this case the subsequent decay is not correlated with the initial parton. Target fragmentation processes are described by *fracture functions* [5], which represent the probability of finding a parton of a certain flavor i, together with a hadron h, in the target nucleon. QCD equations describe the subsequent evolution of the fracture function in Q^2. As the fracture functions are universal, they can be measured in other processes, for example diffraction. The first experimental data for these processes has recently been measured at HERA. EIC could investigate this regime in detail. As this field is largely unexplored, the discovery potential is quite high. An EIC facility could measure the extent to which the initial state polarization affects the quantum numbers of the final hadron h. The current fragmentation regime is "meson-rich," since the quark can become a meson by picking up a single antiquark. Naively, one expects the target fragmentation region to be relatively "baryon-rich", since after removal of a single quark the two remaining valence quarks can decay to a baryon by picking up one additional quark. An EIC facility could also obtain unique information regarding the regime of large x for the struck parton in target fragmentation. In this case the momentum fraction $1-x$ of the hadronic remnants is small. The subsequent remnant decay is correlated with the struck parton, and does not evolve with Q^2.

3 Hard Exclusive Processes at a Collider

An asymmetric polarized collider provides the opportunity to study partonic correlations in nucleons and nuclei. A particularly promising area is the study of hard exclusive processes; a thorough review is given in the talk by Mosse [6]. A virtual photon γ^* from the lepton produces either a real photon (Deeply Virtual Compton Scattering, or DVCS) or a meson (meson electroproduction). The amplitude depends on the quantity x, related to the parton momentum fraction, the "skewedness" ξ which measures the different momentum fractions carried by initial and final partons, and the four-momentum transfer $\Delta^2 = t$ at the photon vertex. Amplitudes for this process depend upon four *generalized parton distributions* or GPD's, which are functions of the aforementioned three kinematic variables. The GPD's *interpolate* between the quantities measured in inclusive deep inelastic scattering, and the particle form factors measured in nuclear physics processes . For example, in the forward direction, characterized by $\xi = \Delta^2 = 0$, two of the GPD's reduce to the helicity-averaged and helicity-dependent single-parton densities, the quantities measured in inclusive DIS reactions. The first moment of the GPD's with respect to x, at fixed four-momentum transfer, gives the nucleon form factors.

An asymmetric collider, capable of measuring all final hadrons in exclusive reactions, has the possibility of directly sampling effects of partonic correlations. One promising area would be to compare amplitudes for elastic processes, where the struck nucleon remains in its ground state, with inelastic excitations resulting in a final N^*, Δ or hyperon. This would enable us to determine, *what is the partonic structure of light baryons?* For example, valence-quark models of baryons would suggest that diquark correlations are more important in octet than in decuplet baryons. The reasoning is that one expects a very strong attraction between quarks in the spin-isospin zero channel. In contrast, in chiral models which employ an expansion about the large-N_C limit of QCD, the octet and decuplet baryons are different rotational excitations of the same soliton, and one would not expect significant differences from diquark correlations. The collider is uniquely suited to access partonic correlations, because of the control over spin variables, plus the ability to measure flavor effects in final state baryons and mesons. Λ hyperon production is useful because the Λ is self-analyzing. Comparison of hyperon, octet and decuplet baryon production can answer the question, *Is SU(3) flavor symmetry valid for baryon structure?* Despite the widespread use of $SU(3)$ arguments to relate light (u,d,s) processes, it is quite possible that polarized fragmentation functions will reveal large flavor-dependent effects.

The area of GPD's is quite new, as factorization proofs for hard exclusive processes were first derived in 1997. Such processes can be described as the product of a hard scattering term calculable from pQCD, times soft amplitudes (the GPD's) which are universal but not calculable from QCD. Extremely rapid progress is being made in this field. Preliminary experimental studies are being carried out at HERA and Hermes, and are proposed for Jefferson Lab. In addition, theoretical studies are needed to determine which experiments most directly reveal the important physical information accessible with hard exclusive processes, e.g.: *How does the transverse momentum of partons influence the GPD amplitudes?*, and *Which observables are most sensitive to quark-quark correlations?*

4 Hadronic Physics with Nuclei

With the addition of nuclear beams at a collider, there are a number of important questions which can be addressed. Measurement of the pion structure function (as described above) can also be considered for a pion in flight in a nucleus. This can answer the question, *"How is the pion field modified in the nucleus?"*. Studies of nuclear effects on pion structure functions can also check measurements of nuclear anti-quark distributions, which found no enhancement over anti-quark probabilities in a nucleon.

There are indications that gluon densities, at high partonic density and low x, undergo significant change. It has been suggested that experiments in this region will reveal a *"Colored Glass Condensate"*, a collective gluonic Bose condensate analogous to effects seen in spin glasses [8]. A number of important partonic phenomena can be explored with nuclei, e.g.: color transparency, color coherence, parton energy loss, hadronization in nuclei.

In conclusion, the scientific motivation for an Electron-Ion Collider is very strong. Previous experiments investigating hadron structure indicate the essential need for a machine with a large kinematic range, high luminosity and optimal control of spin and flavor. The collider geometry offers the crucial and unique capability of complete event reconstruction in hard scattering. The collider should be available in a timely fashion to build upon the insights gained from existing programs at BNL, CERN, DESY, and SLAC.

This work was supported in part by National Science Foundation grant NSF-PHY0070368.

References

1. This document is on the Web at URL *www.phenix.bnl.gov/WWW/publish/abhay/Home_of_EIC/*.
2. See the talk by C. Roberts, proceedings of this Workshop, p. 219.
3. R.J. Holt and P.E. Reimer, eprint nucl-ex/0010004.
4. P. Mulders, eprint hep-ph/0010199.
5. L. Trentadue and G. Veneziano, *Phys. Lett.* B323, 201 (1994).
6. See the talk by L. Mosse, proceedings of this Workshop, p. 61.
7. See the talk by I. Bojak, proceedings of this Workshop, p. 96.
8. See the talk by R. Venugopalan, proceedings of this Workshop, p. 281.

LIST OF PARTICIPANTS

R. Alkofer	U. Tübingen, Germany	reinhard.alkofer@uni-tuebingen.de
M. Anselmino	U. Torino, Italy	anselmino@to.infn.it
E. Ruiz Arriola	U. Granada, Spain	earriola@ugr.es
J. Ashley	CSSM	jashley@physics.adelaide.edu.au
W. Bentz	Tokai U., Japan	bentz@keyaki.cc.u-tokai.ac.jp
N. Bianchi	INFN-Frascati, Italy	bianchi@hermes.desy.de
S. Bilson-Thompson	CSSM	sbilson@physics.adelaide.edu.au
F. Bissey	CSSM	fbissey@physics.adelaide.edu.au
I. Bojak	CSSM	ibojak@physics.adelaide.edu.au
F. Bonnet	CSSM	fbonnet@physics.adelaide.edu.au
M. Burkardt	New Mexico State U.	burkardt@nmsu.edu
J. Cameron	Indiana U.	cameron@iucf.indiana.edu
F.-G. Cao	Massey U., New Zealand	f.g.cao@massey.ac.nz
S. Cotanch	North Carolina State U.	cotanch@ncsu.edu
B. Crouch	CSSM	bcrouch@physics.adelaide.edu.au
M. Derrick	ANL, Illinois	mxd@hep.anl.gov
W. Detmold	CSSM	wdetmold@physics.adelaide.edu.au
P. Di Nezza	INFN-Frascati, Italy	dinezza@hermes.desy.de
R. Ent	JLab, Virginia	ent@jlab.org
O. Gayou	William & Mary, Virginia	gayou@jlab.org
J. Gomez	JLab, Virginia	gomez@jlab.org

X.-H. Guo	CSSM	xhguo@physics.adelaide.edu.au
V. Guzey	CSSM	vguzey@physics.adelaide.edu.au
J. Hedditch	CSSM	jhedditc@physics.adelaide.edu.au
R. Hofmann	MPI Munich, Germany	ralfh@mppmu.mpg.de
N. Ishii	RIKEN, Japan	ishii@th.phys.titech.ac.jp
A. Kalloniatis	CSSM	akalloni@physics.adelaide.edu.au
W. Kamleh	CSSM	wkamleh@physics.adelaide.edu.au
M. Katramatou	Kent State U., Ohio	minadat@athens.kent.edu
A. Kızılersü	CSSM	akiziler@physics.adelaide.edu.au
J. Kodaira	Hiroshima U., Japan	jkodaira@hiroshima-u.ac.jp
D.-J. Kusterer	CSSM	djkuster@physics.adelaide.edu.au
S. Lawley	CSSM	slawley@physics.adelaide.edu.au
D. Leinweber	CSSM	dleinweb@physics.adelaide.edu.au
O. Leitner	CSSM	oleitner@physics.adelaide.edu.au
G. Levman	U. Toronto, Canada	levman@physics.utoronto.ca
S. Liuti	U. Virginia	simonetta@virginia.edu
M. Lohe	CSSM	mlohe@physics.adelaide.edu.au
T. Londergan	Indiana U.	tlonderg@indiana.edu
K. Maltman	CSSM & York U., Canada	kmaltman@physics.adelaide.edu.au
J.-F. Mathiot	U. Blaise-Pascal, France	mathiot@in2p3.fr
K. McCormick	JLab, Virginia	mccormic@jlab.org
W. Melnitchouk	CSSM & JLab, Virginia	wmelnitc@physics.adelaide.edu.au
H. Mineo	Tokyo U., Japan	mineo@nt.phys.s.u-tokyo.ac.jp
L. Mosse	CEA-Saclay, France	lmosse@cea.fr

J. Napolitano	RPI, New York	napolj@rpi.edu
M. Oettel	CSSM	moettel@physics.adelaide.edu.au
G. Petratos	Kent State U., Ohio	gpetrato@kent.edu
C. Roberts	ANL, Illinois	roberts@theory.phy.anl.gov
G. Schierholz	DESY, Germany	gerrit.schierholz@desy.de
A. Schreiber	CSSM	aschreib@physics.adelaide.edu.au
S. Sharpe	U. Washington	sharpe@phys.washington.edu
T. Signal	Massey U., New Zealand	A.I.Signal@massey.ac.nz
M. Stanford	CSSM	mstanfor@physics.adelaide.edu.au
A. Szczepaniak	Indiana U.	aszczepa@indiana.edu
M. Taiuti	U. Genoa, Italy	taiuti@ge.infn.it
P. Tandy	Kent State U., Ohio	tandy@faisun.kent.edu
A. Thomas	CSSM	athomas@physics.adelaide.edu.au
K. Tsushima	CSSM	ktsushim@physics.adelaide.edu.au
R. Venugopalan	BNL, New York	raju@quark.phy.bnl.gov
A. Williams	CSSM	awilliam@physics.adelaide.edu.au
S. Wright	CSSM	swright@physics.adelaide.edu.au
R. Young	CSSM	ryoung@physics.adelaide.edu.au
J. Zanotti	CSSM	jzanotti@physics.adelaide.edu.au
J. Zhang	CSSM	jzhang@physics.adelaide.edu.au